STEM CELL ENGINEERING
PRINCIPLES AND PRACTICES

STEM CELL ENGINEERING
PRINCIPLES AND PRACTICES

Edited by
David Schaffer
Joseph D. Bronzino
Donald R. Peterson

CRC Press
Taylor & Francis Group
Boca Raton London New York

CRC Press is an imprint of the
Taylor & Francis Group, an **informa** business

CRC Press
Taylor & Francis Group
6000 Broken Sound Parkway NW, Suite 300
Boca Raton, FL 33487-2742

First issued in paperback 2019

ISBN-13: 978-1-4398-7204-8 (hbk)
ISBN-13: 978-0-367-38064-9 (pbk)

Library of Congress Cataloging-in-Publication Data

Stem cell engineering : principles and practices / edited by David Schaffer, Joseph D. Bronzino, and Donald R. Peterson.
    p. ; cm.
Includes bibliographical references and index.
ISBN 978-1-4398-7204-8 (hardcover : alk. paper)
I. Schaffer, David V. II. Bronzino, Joseph D., 1937- III. Peterson, Donald R.
[DNLM: 1. Stem Cells. 2. Regenerative Medicine--methods. 3. Stem Cell Transplantation--methods. 4. Tissue Engineering--methods. QU 325]

616.02'774--dc23
    2012032221

Visit the Taylor & Francis Web site at
http://www.taylorandfrancis.com

and the CRC Press Web site at
http://www.crcpress.com

Contents

Introduction

Stem cells are in many ways like other cells. They proliferate, remain quiescent, apoptose, senesce, adhere, and migrate. However, they have two more properties that have been the crux of considerable biomedical attention and imagination: the capacity to self-renew or proliferate in an immature state and the ability to differentiate into one or more specialized lineages. If these cellular talents can be sufficiently understood and controlled, they can yield insights into mechanisms of organismal development and adult homeostasis and serve as the basis for cell replacement therapies to treat injured and diseased tissues and organs.

The field of stem cells is both mature and new. It originated with the discovery of hematopoietic stem cells by McCulloch and Till in the 1960s (Becker et al., 1963), which lie at the heart of the development of bone marrow transplant-based treatments for a broad range of blood and other disorders for the past few decades (Thomas et al., 1957). Multipotent stem cells with the capacity to differentiate into a subset of adult lineages have since been discovered in numerous additional adult tissues, including skeletal muscle (Muir et al., 1965), intestine (Troughton and Trier, 1969, Winton and Ponder, 1990), brain (Altman, 1962, Gage, 2000), and numerous others. In addition, pluripotent mouse (Evans and Kaufman, 1981) and subsequently human (Thomson et al., 1998) embryonic stem (ES) cells were derived with the capacity to differentiate into every cell of an adult organism. In addition to investigating mechanisms by which such immature cells undergo differentiation, the field has become increasingly interested in the phenomenon that mature cells can de-differentiate or be reprogrammed into an immature state. Initially, somatic cell nuclear transfer demonstrated that the genome from a differentiated mammalian cell could be reprogrammed to an embryonic state by factors present in oocyte cytoplasm (Wilmut et al., 1997). More recently, it was shown that a differentiated mammalian cell can be induced into pluripotency through the overexpression of just four transcription factors (Takahashi and Yamanaka, 2006). Finally, just as basic advances in hematopoietic stem cell biology have occurred in parallel to clinical translation of bone marrow transplant and cord blood-based therapies, the Food and Drug Administration has approved clinical trials to explore the therapeutic potential of human ES cell-derived cells to treat spinal cord injury and age-related macular degeneration.

Understanding the properties and processes of stemness, differentiation, and reprogramming at a molecular level will progressively advance our knowledge of human development. In addition, learning how to control these processes will increasingly aid the translation of stem cell biology toward regenerative medicine, further building upon the clinical success of blood stem cells (Appelbaum, 2003). Alternatively, stem cells are increasingly being utilized as the basis for high-throughput drug discovery and toxicology screens to enhance the therapeutic potential of traditional pharmaceuticals. However, in general clinical translation does face a number of challenges. While the potential for many stem cell classes to differentiate into one or more lineages is recognized, their proliferation and differentiation must be more precisely controlled to both maximize the production of therapeutically relevant cells, and minimize contamination with residual cells that can give rise to tumors or other side effects for cell replacement therapies. Furthermore, the continued development of robust processes to scale up the

production of desired cells in high yield and purity is required. Finally, engrafting the resulting differentiated stem cells *in vivo* typically results in extremely low viability, so better means to enhance their
functional integration into the target tissue must be developed.

How can engineers make contributions to address these challenges? This book provides a broad view
of engineering efforts in a number of these areas. For example, the analysis or reverse engineering of
complex systems can yield an understanding of how complex behaviors arise from the collective interactions of numerous interacting parts. In the field of stem cells, such complex systems occur both outside
and inside the cell. Specifically, in both developing embryos and adult tissues, stem cells reside within
highly complex microenvironments or niches in which they are continuously exposed to signals including small molecules, soluble proteins, extracellular matrix proteins and proteoglycans, proteins immobilized to the extracellular matrix, signals immobilized on the surface of adjacent cells, and mechanical
properties of the tissue. Controlling stem cell expansion and differentiation, either *in vitro* or *in vivo*,
requires a precise understanding of these complex, interacting cues.

In Chapter 1, Bauwens et al. describe the temporal evolution of cell differentiation state and overall tissue structure in the early developing embryo as well as some parallels with embryonic stem cell
differentiation within embryoid bodies. Furthermore, they present the biochemical and biophysical
signals that guide this process and that can be harnessed to differentiate an ES cell into a therapeutically
valuable cell type such as a cardiomyocyte. It is well recognized within the stem cell biology field that
biochemical signals such as growth factors, morphogens, and cytokines regulate stem cell fate and function, but it is becoming increasingly appreciated that biophysical cues can also play key roles (Keung
et al., 2010). In Chapter 2, Ananthanarayanan and Kumar discuss how mechanical cues regulate cellular processes such as proliferation and migration as well as highlight parallels for how what has been
learned about the mechanobiology of tumors can be applied to stem cells.

A stem cell residing within such complex repertoires of biochemical and biophysical signals must
sense this extracellular regulatory information, though the activation of receptors on the surface and
interior of cells, process these signals via the activation of complex signal transduction networks, and
thereby transform inputs into outputs or fate decisions. These signal transduction processes are dense
and highly interconnected, and individual molecular interactions within them are often nonlinear in
nature. Janes and colleagues have made considerable advances in developing data-driven modeling
approaches to elucidate key features of signal transduction processes that in general drive fate choices
(Janes and Lauffenburger, 2006, Janes et al., 2006), and in Chapter 3, Holmberg et al. describe how these
methods and principles can be applied to understand cellular signal processing events that drive stem
cell fate decisions.

In addition to the analysis of complex systems, engineers are versed in applying basic information
toward the synthesis or forward engineering of approaches to control the behavior of such systems. For
example, the broadest activities of engineers in this field have been focused on applying the advances in
biomaterials development over the past several decades to better control stem cell behavior. In Chapter 4,
Willerth and Schaffer describe recent efforts to develop culture systems that can support the indefinite
expansion of human embryonic stem cells, potentially by emulating the microenvironment that pluripotent cells experience during their brief window of existence in early embryogenesis. Advances by the
scientific community in general over the past decade have led to increasingly defined culture systems for
growing human ES cells, starting from co-culture with feeder cells in the presence of serum to growth
on synthetic substrates in defined medium. In general, such better defined systems enhance the reproducibility, safety, and scaleability of stem cell culture as the field moves toward the clinic.

In parallel with progress on expanding immature pluripotent stem cells, systems to support their
differentiation into therapeutically valuable lineages must be developed. Within Chapter 5, Sakiyama-
Elbert describes a number of important considerations in the application of stem cell therapies to the
central nervous system, including cell sources, disease and injury targets, and practical issues associated
with cell differentiation and implantation. In Chapter 6, Gupta et al. provide an in-depth analysis of the
design of stem cell-based cardiac therapies, including sources of resident and exogenous cells, as well as

natural and synthetic biomaterials for the differentiation and importantly for the implantation of cells to enhance their viability and engraftment.

Fridley and Roy (Chapter 7) provide a strong overview of the best characterized stem and progenitor cells, those of the hematopoietic system. This chapter highlights recent advances in the understanding of the cellular and molecular composition of the hematopoietic stem cell niche as well as approaches to build upon this basic information to direct stem cell differentiation into blood cell lineages. Moreover, in Chapter 8, Phadke and Varghese provide deep insights into the development of advanced materials that can interface with stem cells for the repair of connective tissues, including bone, cartilage, tendons, and ligaments.

Furthermore, numerous stem cell applications will require large numbers of cells, requiring the continuing development of scaleable technologies for cell expansion and differentiation. Palecek (Chapter 9) reviews considerable advances in the derivation and characterization of human pluripotent stem cells as well as the development of bioreactors and culture systems for their large-scale expansion. Furthermore, in Chapter 10, Rodrigues and colleagues provide a strong overview of numerous classes of bioreactor systems for cell expansion and differentiation as well as principles for their implementation to a number of multipotent and embryonic stem cell types.

In addition to highlighting many recent advances, these chapters describe the need for future work. The progressive identification of key biochemical and biophysical regulatory signals will benefit basic stem cell and developmental biology as well as regenerative medicine, and the application of quantitative approaches will deepen our understanding of intracellular mechanisms that govern cellular decisions. In parallel, the application of this basic information will aid the development of bioactive materials and ideally synthetic microenvironments to control and aid cell expansion, differentiation, and implantation. Furthermore, these engineered culture systems will increasingly become integrated into large-scale culture systems as stem cells progressively move toward the clinic. The growing recognition of stem cells as an important and exciting field will continue to draw investigators with diverse backgrounds—biology, engineering, and the physical sciences—and thereby drive further progress in these and other new directions.

References

Altman, J. 1962. Are new neurons formed in the brains of adult mammals? *Science,* 135, 1127–1128.

Appelbaum, F. R. 2003. The current status of hematopoietic cell transplantation. *Annual Review of Medicine,* 54, 491–512.

Becker, A. J., Mcculloch, E. A., and Till, J. E. 1963. Cytological demonstration of the clonal nature of spleen colonies derived from transplanted mouse marrow cells. *Nature,* 197, 452–454.

Evans, M. J. and Kaufman, M. H. 1981. Establishment in culture of pluripotential cells from mouse embryos. *Nature,* 292, 154–156.

Gage, F. H. 2000. Mammalian neural stem cells. *Science,* 287, 1433–1438.

Janes, K. A., Gaudet, S., Albeck, J. G., Nielsen, U. B., Lauffenburger, D. A., and Sorger, P. K. 2006. The response of human epithelial cells to Tnf involves an inducible autocrine cascade. *Cell,* 124, 1225–1239.

Janes, K. A. and Lauffenburger, D. A. 2006. A biological approach to computational models of proteomic networks. *Current Opinion in Chemical Biology,* 10, 73–80.

Keung, A. J., Kumar, S., and Schaffer, D. V. 2010. Presentation counts: Microenvironmental regulation of stem cells by biophysical and material cues. *Annual Reviews in Cell Developmental Biology,* 26, 533–556.

Muir, A. R., Kanji, A. H., and Allbrook, D. 1965. The structure of the satellite cells in skeletal muscle. *Journal of Anatomy,* 99, 435–444.

Takahashi, K. and Yamanaka, S. 2006. Induction of pluripotent stem cells from mouse embryonic and adult fibroblast cultures by defined factors. *Cell,* 126, 663–676.

Thomas, E. D., Lochte, H. L., Lu, W. C., and Ferrebee, J. W. 1957. Intravenous infusion of bone marrow in patients receiving radiation and chemotherapy. *New England Journal of Medicine,* 257, 491–496.

Thomson, J. A., Itskovitz-Eldor, J., Shapiro, S. S., Waknitz, M. A., Swiergiel, J. J., Marshall, V. S., and Jones, J. M. 1998. Embryonic stem cell lines derived from human blastocysts. *Science*, 282, 1145–1147.

Troughton, W. D. and Trier, J. S. 1969. Paneth and goblet cell renewal in mouse duodenal crypts. *Journal of Cell Biology*, 41, 251–268.

Wilmut, I., Schnieke, A. E., Mcwhir, J., Kind, A. J., and Campbell, K. H. 1997. Viable offspring derived from fetal and adult mammalian cells. *Nature*, 385, 810–813.

Winton, D. J. and Ponder, B. A. 1990. Stem-cell organization in mouse small intestine. *Proceedings of the Royal Society of London B: Biological Sciences*, 241, 13–18.

Editors

David V. Schaffer, PhD, is a professor of chemical and biomolecular engineering, bioengineering, and neuroscience at the University of California, Berkeley, where he also serves as the director of the Berkeley Stem Cell Center. He graduated from Stanford University with a BS in chemical engineering in 1993. Afterward, he attended the Massachusetts Institute of Technology (MIT) and earned his PhD, also in chemical engineering, in 1998 with Professor Doug Lauffenburger. While at MIT, Dr. Schaffer minored in molecular and cell biology. Finally, he did a postdoctoral fellowship in the laboratory of Fred Gage at the Salk Institute for Biological Studies in La Jolla, California, before moving to UC Berkeley in 1999. At Berkeley, Dr. Schaffer applies engineering principles to enhance stem cell and gene therapy approaches for neuroregeneration. This work includes mechanistic investigation of stem cell control as well as molecular evolution and engineering of viral gene delivery vehicles.

Dr. Schaffer has received an NSF Career Award, the Office of Naval Research Young Investigator Award, and the Whitaker Foundation Young Investigator Award, and he was named a Technology Review Top 100 Innovator. He was also awarded the American Chemical Society BIOT Division Young Investigator Award in 2006 and the Biomedical Engineering Society Rita Shaffer Young Investigator Award in 2000. He was elected to the College of Fellows of the American Institute of Medical and Biological Engineering in 2010.

Joseph D. Bronzino earned a BSEE from Worcester Polytechnic Institute, Worcester, Massachusetts, in 1959, an MSEE from the Naval Postgraduate School, Monterey, California, in 1961, and a PhD in electrical engineering from Worcester Polytechnic Institute in 1968. He is presently the Vernon Roosa Professor of Applied Science, an endowed chair at Trinity College, Hartford, Connecticut, and president of the Biomedical Engineering Alliance and Consortium (BEACON), a nonprofit organization consisting of academic and medical institutions as well as corporations dedicated to the development and commercialization of new medical technologies (www.beaconalliance.org).

Dr. Bronzino is the author of over 200 articles and 11 books, including *Technology for Patient Care* (C.V. Mosby, 1977), *Computer Applications for Patient Care* (Addison-Wesley, 1982), *Biomedical Engineering: Basic Concepts and Instrumentation* (PWS Publishing Co., 1986), *Expert Systems: Basic Concepts* (Research Foundation of State University of New York, 1989), *Medical Technology and Society: An Interdisciplinary Perspective* (MIT Press and McGraw-Hill, 1990), *Management of Medical Technology* (Butterworth/Heinemann, 1992), *The Biomedical Engineering Handbook* (CRC Press, 1st Ed., 1995; 2nd Ed., 2000; 3rd Ed., 2005), and *Introduction to Biomedical Engineering* (Academic Press 1st Ed., 1999; 2nd Ed., 2005),

Dr. Bronzino is a fellow of IEEE and the American Institute of Medical and Biological Engineering (AIMBE), an honorary member of the Italian Society of Experimental Biology, past chairman of the Biomedical Engineering Division of the American Society for Engineering Education (ASEE), a charter member and former vice president of the Connecticut Academy of Science and Engineering (CASE), and a charter member of the American College of Clinical Engineering (ACCE), the Association for the Advancement of Medical Instrumentation (AAMI), past president of the IEEE-Engineering in Medicine and Biology Society (EMBS), past chairman of the IEEE Health Care Engineering Policy Committee (HCEPC), past chairman of the IEEE Technical Policy Council in Washington, DC, and presently editor-in-chief of Elsevier's BME Book Series and CRC Press's *The Biomedical Engineering Handbook*.

Dr. Bronzino received the Millennium Award from IEEE/EMBS in 2000 and the Goddard Award from Worcester Polytechnic Institute for Professional Achievement in June 2004.

Donald R. Peterson is an associate professor of medicine and the director of the biodynamics laboratory in the School of Medicine at the University of Connecticut (UConn). He serves jointly as the director of the biomedical engineering undergraduate program in the School of Engineering and recently served as the director of the graduate program and as the BME Program chair. He earned a PhD in biomedical engineering and an MS in mechanical engineering at UConn and a BS in aerospace engineering and a BS in biomechanical engineering from Worcester Polytechnic Institute. Dr. Peterson has 16 years of experience in biomedical engineering education and offers graduate-level and undergraduate-level courses in BME in the areas of biomechanics, biodynamics, biofluid mechanics, and ergonomics, and he teaches in medicine in the subjects of gross anatomy, occupational biomechanics, and occupational exposure and response. Dr. Peterson's scholarly activities include over 50 published journal articles, 3 textbook chapters, and 12 textbooks, including his new appointment as co-editor-in-chief for *The Biomedical Engineering Handbook* by CRC Press.

Dr. Peterson has over 21 years of experience in biomedical engineering research and has been recently focused on measuring and modeling human, organ, and/or cell performance, including exposures to various physical stimuli and the subsequent biological responses. This work also involves the investigation of human–device interaction and has led to applications on the design and development of tools and various medical devices. Dr. Peterson is faculty within the occupational and environmental medicine group at the UConn Health Center, where his work has been directed toward the objective analysis of the anatomic and physiological processes involved in the onset of musculoskeletal and neuromuscular diseases, including strategies of disease mitigation. Recent applications of his research include human interactions with existing and developmental devices such as powered and non-powered tools, space-suits and space tools for NASA, surgical and dental instruments, musical instruments, sports equipment, and computer-input devices. Other overlapping research initiatives focus on cell mechanics and cellular responses to fluid shear stress, the acoustics of hearing protection and communication, human exposure and response to vibration, and the development of computational models of biomechanical performance.

Dr. Peterson is also the co-executive director of the Biomedical Engineering Alliance and Consortium (BEACON; www.beaconalliance.org), which is a nonprofit entity dedicated to the promotion of collaborative research, translation, and partnership among academic, medical, and industry people in the field of biomedical engineering to develop new medical technologies and devices.

Contributors

Badriprasad Ananthanarayanan
Department of Bioengineering
University of California, Berkeley
Berkeley, California

Céline L. Bauwens
Department of Chemical Engineering and
    Applied Chemistry
University of Toronto
Toronto, Canada

Anjun K. Bose
Department of Biomedical Engineering
University of Virginia
Charlottesville, Virginia

Joaquim M. S. Cabral
Institute for Biotechnology and Bioengineering
and
Centre for Biological and Chemical Engineering
Instituto Superior Técnico
Lisbon, Portugal

Cláudia Lobato da Silva
Institute for Biotechnology and Bioengineering
and
Centre for Biological and Chemical Engineering
Instituto Superior Técnico
Lisbon, Portugal

Maria Margarida Diogo
Institute for Biotechnology and Bioengineering
and
Centre for Biological and Chemical Engineering
Instituto Superior Técnico
Lisbon, Portugal

Tiago G. Fernandes
Institute for Biotechnology and Bioengineering
and
Centre for Biological and Chemical Engineering
Instituto Superior Técnico
Lisbon, Portugal

Krista M. Fridley
Department of Biomedical Engineering
University of Texas, Austin
Austin, Texas

Rohini Gupta
Department of Bioengineering
University of California, Berkeley
Berkeley, California

Kevin E. Healy
Department of Bioengineering
and
Department of Materials Science and Engineering
University of California, Berkeley
Berkeley, California

Kevin A. Janes
Department of Biomedical Engineering
University of Virginia
Charlottesville, Virginia

Karin J. Jensen
Department of Biomedical Engineering
University of Virginia
Charlottesville, Virginia

Kimberly R. Kam
Department of Materials Science and
    Engineering
University of California, Berkeley
Berkeley, California

Sanjay Kumar
Department of Bioengineering
University of California, Berkeley
Berkeley, California

Kunal Mehtani
Division of Internal Medicine
Kaiser Permanente
San Francisco, California

Sean P. Palecek
Department of Chemical and Biological
    Engineering
University of Wisconsin, Madison
and
WiCell Research Institute
Madison, Wisconsin

Ameya Phadke
Department of Bioengineering
University of California, San Diego
San Diego, California

Kelly A. Purpura
Department of Chemical Engineering
    and Applied Chemistry
University of Toronto
Toronto, Canada

Carlos A. V. Rodrigues
Institute for Biotechnology and Bioengineering
and
Centre for Biological and Chemical Engineering
Instituto Superior Técnico
Lisbon, Portugal

Krishnendu Roy
Department of Biomedical
    Engineering
University of Texas, Austin
Austin, Texas

Shelly Sakiyama-Elbert
Department of Biomedical Engineering
Washington University in St. Louis
St. Louis, Missouri

David V. Schaffer
Department of Chemical Engineering
and
Department of Bioengineering
and
Helen Wills Neuroscience Institute
University of California, Berkeley
Berkeley, California

Shyni Varghese
Department of Bioengineering
University of California, San Diego
San Diego, California

Stephanie Willerth
Department of Mechanical Engineering
and
Division of Medical Sciences
University of Victoria
Victoria, British Columbia, Canada

and

International Collaboration On Repair
    Discoveries (ICORD)
University of British Columbia
Vancouver, British Columbia, Canada

Peter W. Zandstra
Department of Chemical Engineering and
    Applied Chemistry
University of Toronto

and

McEwen Centre for Regenerative Medicine
University Health Network

and

Heart and Stroke Richard Lewar Centre of
    Excellence
Toronto, Canada

1

Engineering the Pluripotent Stem Cell Niche for Directed Mesoderm Differentiation

Céline L. Bauwens
University of Toronto

Kelly A. Purpura
University of Toronto

Peter W. Zandstra
University of Toronto
University Health Network
Heart and Stroke Richard
Lewar Centre of Excellence

1.1 Introduction

The excitement surrounding research on pluripotent stem cells (PSCs) (Martin et al. 1977; Evans and Kaufman 1981; Martin 1981; Thomson et al. 1995, 1998) is largely based on the expectation that these cells may one day provide a renewable source of human tissue for cell-based therapies and for drug screening in disease models. Equally important, however, is the opportunity these cells present for studying human embryonic development *in vitro*, which for ethical reasons would not otherwise be possible. Insight into stem cell differentiation along the mesoderm lineage is of particular interest in developing treatments for heart disease and blood disorders, as current treatments are limited by the necessity of donated patient-matched whole organs, blood, or bone marrow (BM).

For example, heart failure is associated with massive irreversible loss of cardiomyocytes, and cell trans-plantation is emerging as a potential alternative to organ transplantation as a number of studies have dem-onstrated improved heart function upon engraftment of different cell types—including cardiomyocytes, skeletal myoblasts, and BM-derived cells—into the heart (Menasche 2003, 2004). As another example, a variety of malignant and genetic blood diseases are treated by transplanting autologous or allogeneic hematopoietic stem cells (HSCs) from BM, peripheral blood, or umbilical cord blood (UCB). However, cell sources are limited by donor availability and low frequencies in the UCB (Barker and Wagner 2002).

Production of cardiomyocytes or HSCs from PSCs is an attractive option. By definition, PSCs are able to self-renew while maintaining the capacity to develop into all somatic cell types. PSCs also have unlim-ited expansion capabilities. Traditionally, the term "pluripotent stem cell" referred to embryonic stem cells (ESCs), but in 2006 it was demonstrated that adult cells could be genetically manipulated to take on a pluripotent state (Takahashi and Yamanaka 2006). Human PSCs have been widely demonstrated to differentiate to cardiomyocytes as well as to other noncardiac cell types present in the heart (Kehat et al. 2001; Xu et al. 2002; Mummery et al. 2003). Similarly, $CD34^+$, $CD45^+$, and hematopoietic colony-forming cells (CFCs) have been identified from human embryonic stem cells (hESCs) Cocultured on irradiated mouse BM stroma, yolk sac endothelial cells, or on OP9 or S17 stromal cells (Kaufman et al. 2001; Tian et al. 2004; Vodyanik et al. 2005). Despite these successes, however, significant improvements are required to efficiently and robustly produce target cell types and to appropriately mature them to the stage where they are functional in the adult.

1.2 Embryoid Body Differentiation: Capturing Aspects of Embryonic Development

In vitro ESC differentiation is routinely carried out by forming embryoid bodies (EBs), which are 3D aggregates of ESCs in suspension. Embryonic development is recapitulated within the EB, wherein cells of the three embryonic germ layers—endoderm, ectoderm, and mesoderm—develop (Doetschman et al. 1985; Itskovitz-Eldor et al. 2000; Xu et al. 2001) and subsequently differentiate into committed cell types, including neurons, glia, skeletal and cardiac muscle cells, hematopoietic cells, hepatic cells, and insulin-secreting (pancreatic) cells (Itskovitz-Eldor et al. 2000; Xu et al. 2001).

The early events that occur during embryogenesis are described in Figure 1.1. In the eight-cell morula-stage embryo, the inner blastomeres form a tight ball, through a process called compaction, producing the inner cell mass (ICM). Meanwhile, blastomeres located on the exterior of the morula flatten and form the trophoblast. Following compaction, a fluid-filled cavity called the blastocoele develops inside the embryo (blastocyst stage), and ICM cells in contact with the blastocoele differentiate to primitive endoderm (PE) and then visceral endoderm; ICM cells in contact with the trophoblast differentiate to parietal endoderm. At this stage, the remaining cells within the ICM begin to differentiate into an epithelial layer that is referred to as the epiblast or primitive ectoderm (Johnson and Ziomek 1981; O'Shea 2004). Development of the three primary germ layers occurs in the epiblast by a process called gastrulation. Gastrulation is initiated in the posterior epiblast by movement of cells through the primitive streak (PS) where cells undergo an epithelial to mesenchymal transition, exiting the PS as mesoderm in the proximal–anterior region of the epiblast and as definitive endoderm in the distal–anterior region (Gadue et al. 2005; Rust et al. 2006; Murry and Keller 2008).

EBs are thought to mimic the environment of the peri-implantation embryo where interactions between various cell types facilitate inductive events. As in the embryo, one of the earliest events during EB devel-opment is the organization of the cells into an outer epithelial layer of PE surrounding an inner core of epiblast-like pluripotent cells (Coucouvanis and Martin 1995, 1999; Abe et al. 1996), followed by the expression of gene and protein markers that are associated with the PS such as brachyury (Kispert and Herrmann 1994) and Mixl1 (Hart et al. 2002). Many proteins are involved in commitment to both the endoderm and the mesoderm lineage in the PS, including transforming growth factor (TGF)-β, activin/nodal, bone morphogenetic protein (BMP), and Wingless Int (Wnt) (Conlon et al. 1994; Hogan 1996;

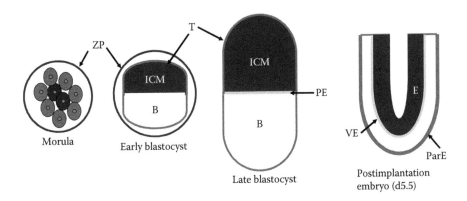

FIGURE 1.1   (**See color insert.**) Development of the early mouse embryo. In the morula, the inner (blue) cells will form ICM and the outer (pink) cells will form trophoblast. In the early blastocyst, a cavity (the blastocoele, B) forms between the ICM and the trophoblast; the embryo is still enclosed in the zona pellucida (ZP). By the late blastocyst stage, the ICM cells in contact with the blastocoele differentiate into the PE, which later forms visceral endoderm on the epiblast side and parietal endoderm on the trophoblast side. At implantation the proamniotic cavity begins to form within the ICM. Cells of the ICM differentiate into an epithelial layer, the epiblast (E). (Adapted from O'Shea, K. S. 2004. *Biol Reprod* **71**(6): 1755–65.)

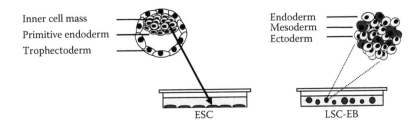

FIGURE 1.2   The *in vitro* EB model. Cells from the ICM can be expanded in tissue culture and differentiated in liquid suspension culture (LSC) into EBs that contain cells from the three germ layers, but in a disorganized fashion.

Yamaguchi 2001), and it is these different levels of pathway activation and inhibition of these proteins that regulate germ layer induction (Gadue et al. 2005). It is believed that in the embryo, distinct signaling environments exist that are defined by location in relation to extraembryonic and embryonic tissues which secrete signals that direct lineage commitment (Rust et al. 2006; Murry and Keller 2008). Temporally, the expression of genes associated with the PS and germ layer commitment in the EB recapitulates gastrulation in the embryo (St-Jacques and McMahon 1996; Dvash and Benvenisty 2004; Keller 2005; Murry and Keller 2008). However, while gastrulation occurs in a precise, spatially organized manner during embryogenesis, differentiation of PS-like cells in the EB is spatially chaotic. EBs lack polarity, and a deficiency in position-specific cues may underlie the resulting spatially disorganized germ layer induction (Figure 1.2).

1.3 Mesoderm Development: Similarities between the Embryo and the EB

1.3.1 Common Mesodermal Inductive Factors

Mesoderm develops from brachyury-expressing PS cells, and early mesoderm induction in this population is characterized by the upregulation of fetal liver kinase (Flk)-1 (kinase insert domain receptor (KDR) in human) and platelet-derived growth factor receptor (Kataoka et al. 1997; Ema et al. 2006;

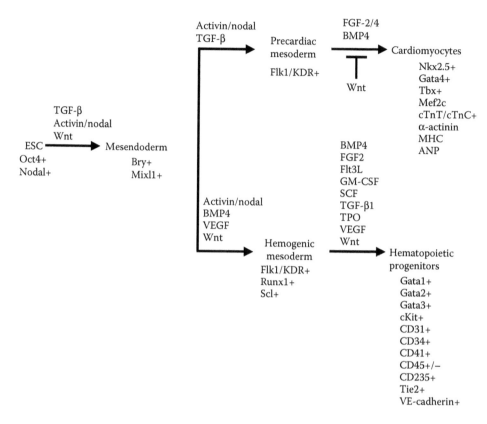

FIGURE 1.3 Signals known to be involved in directing cardiogenesis and hematopoiesis and the markers expressed at each stage of development.

Murry and Keller 2008). Studies using brachyury-GFP mouse (m)ESCs have shown that a combination of Wnt, activin/nodal, and BMP signaling (Nostro et al. 2008) correlates with an upregulation of Flk-1 and specifies commitment to a mesoderm fate giving rise to a population that can no longer develop to endoderm upon exposure to activin A. Cardiac and hematopoietic lineage commitment involve several similar signaling factors, and most mesoderm subpopulations can be derived from ESCs by manipulating the level of signaling and timing of BMP4 or a combination of BMP4 and activin A (Era et al. 2008; Yang et al. 2008b; Kattman et al. 2011). One of the main differences involves the β-catenin/Wnt pathway, which is required for mesoderm induction, but must subsequently be inhibited for cardiac specification (Naito et al. 2006; Ueno et al. 2007; Kattman et al. 2011) (Figure 1.3).

1.3.1.1 Cardiac Commitment

Cardiogenic morphogens from the TGF-β superfamily (TGF-β, nodal/activin A, and BMP) and the fibroblast growth factor (FGF-2, FGF-4) family activate cardiac transcription factors (Menard et al. 2004). Wnt-related signals and members of the Wnt family are also important for cardiac induction (Sachinidis et al. 2003), by playing both a repressive role, via the canonical Wnt/β-catenin pathway, and an inductive role, via the noncanonical Wnt/Ca²⁺ and c-Jun N-terminal kinase pathways (Povelones and Nusse 2002). Once optimal concentrations of activin and BMP4 are determined for a specific cell line, these signaling pathways need to be inhibited in a stage-specific fashion to promote cardiomyocytes (Kattman et al. 2011).

Binding of TGF-β family proteins to their receptors leads to activation of intracellular mediators of the Smad family. Smad2 and Smad3 transduce signals for TGF-β-like ligands, such as TGF-β and activin/nodal while Smad1, 5, and 8 transduce signals for BMP-like ligands (Lagna et al. 1996; Candia et al. 1997; Shi et al. 1997). Upon phosphorylation, these receptor-regulated (R)-Smads form complexes with Smad4, which are subsequently translocated to the nucleus to regulate activation of transcription factors, such as cardiac transcription factors Nkx-2.5, GATA-4, and Tbx factors (Massague and Chen 2000; Schlange et al. 2000; Moustakas et al. 2001; Attisano and Wrana 2002; Harvey 2002; Wakefield and Roberts 2002; Attisano and Labbe 2004; Menard et al. 2004). FGF family proteins (FGF-2 and FGF-4) have been shown to support cardiomyocyte induction during embryonic development mainly by stimulating proliferation of mesodermal cells *in vitro* (Mima et al. 1995; Lough et al. 1996; Schultheiss and Lassar 1997; Ladd et al. 1998; Barron et al. 2000; Kawai et al. 2004). Wnt signaling proteins play both a repressive and a supportive role in heart morphogenesis. Canonical Wnt signaling suppresses cardiac differentiation by degradation of β-catenin. Wnts' repressive activity is inhibited by antagonists Crescent and dickkopf homolog 1 (DKK1) (expressed in anterior endoderm during gastrulation), which subsequently results in the induction of beating muscle. Inhibition of Wnt signaling promotes heart formation in the anterior lateral mesoderm, whereas active Wnt signaling in the posterior lateral mesoderm promotes blood development (Marvin et al. 2001). In the noncanonical pathway, Wnt11 prevents signaling of other Wnts, amplifying the cardiogenic signal (Menard et al. 2004). During embryogenesis, these growth factors synergize in a precise spatial and temporal program to support and induce cardiogenesis in neighboring precardiac mesoderm (Figure 1.3).

1.3.1.2 Hematopoietic Commitment

The hematopoietic system first develops to provide differentiated cells for embryonic growth and later establishes the multipotent HSC for long-term survival. The hematopoietic needs of the embryo are distinct from those of the adult and initially relate to differences in oxygen transport and adaptive immunity (Kyba and Daley 2003). The first specialized cells produced are primitive erythrocytes expressing hemoglobin isoforms of higher oxygen affinity than the latter definitive (or adult) erythrocytes (Bauer et al. 1975; Brotherton et al. 1979), as the placenta provides immunological activity. These transient progenitors are followed by a second wave of definitive erythrocytes, macrophages, and granulocytes that can be detected in the yolk sac (Palis et al. 2001). A third wave of hematopoiesis generates HSCs that can self-renew and differentiate into all blood cell types (lymphoid and myeloid) and is associated with the intra- and extraembryonic arteries (Inman and Downs 2007; Samokhvalov et al. 2007; Van Handel et al. 2010). Hematopoiesis then shifts to the fetal liver (FL) and definitive red blood cells and other lineages supplant primitive erythrocytes in the circulation (Palis et al. 1999). The placenta also contains a large pool of pluripotent HSCs during midgestation (Alvarez-Silva et al. 2003; Ottersbach and Dzierzak 2005; Robin et al. 2009). Following birth, definitive multilineage hematopoiesis is primarily confined to the BM. Thus, embryonic blood development is complex as the process occurs in multiple sites that are spatially and temporally separated.

One point of debate is whether adult hematopoietic cells arise from the epiblast or from extraembryonic cells that migrate from the yolk sac to the aorta–gonad–mesonephros (AGM). Repopulation studies in the mouse and explant cultures with human yolk sac cells found that only the para-aortic splanchnopleura (PSp)-AGM tissues isolated prior to the onset of circulation contained cells with lymphoid–myeloid potential (Cumano et al. 1996; Cumano et al. 2001; Tavian et al. 2001), although, definitive HSC precursors derived from the yolk sac that require hemodynamic stresses to complete differentiation would be missed. Instead of using transplantation/explant assays to address the question of whether definitive HSCs arise locally from the embryo body or from migrant yolk sac cells, the cell-lineage relationship has been studied with *in vivo* cell tracing of pulse-labeled cells based on Cre/loxP recombination. This study showed that E7.5 Runx1+ cells develop into fetal lymphoid progenitors and adult HSCs (Samokhvalov et al. 2007). One caveat, however, is that *Runx1* is also expressed within 0.5–1 day of detection in the yolk sac at the base of the allantois and in the PSp (Zeigler et al. 2006; Nottingham et al. 2007).

As the origin of HSCs remains ambiguous, the direct precursors to definitive hematopoietic cells also remain elusive. Evidence suggests, however, that FL and adult BM HSCs originate from a subset of endothelial cells that line the blood vessels in the mouse, known as the hemogenic endothelium (Yoshimoto and Yoder 2009). Time-lapse photography has shown that 48 h after plating a transient cell population that expresses endothelial markers displays the potential to form both primitive and definitive hematopoietic colonies (Lancrin et al. 2009). Another time-lapse study tracked individually plated mESC-derived mesoderm cells and found that 1.2% displayed properties of adherent endothelial cells, giving rise to nonadherent HSCs (Eilken et al. 2009). Runx1 expression was shown to be essential (E8.25-11.5) for the formation of HSCs from hemogenic endothelium (Chen et al. 2009), and a genetic tracing study of the AGM endothelium demonstrated that this cell population and not the underlying mesenchyme was capable of generating HSCs that would then migrate to the FL and BM (Zovein et al. 2008). Together, these studies suggest that definitive HSCs can arise from hemangioblasts through a hemogenic intermediate.

BMPs were originally noted for their capacity to induce ectopic bone formation; however, multiple relations to the HSC system have been described and they play a critical role in the formation and patterning of mesoderm in the embryo, the commitment of embryonic mesodermal cells to a hematopoietic fate, and in blood island formation in the yolk sac (Marshall et al. 2000; Snyder et al. 2004). Additionally, BMP4 is secreted by pulmonary microvascular endothelial cells in response to hypoxia (Frank et al. 2005), and in differentiating ESC cultures hypoxia upregulates the mesodermal markers brachyury, BMP4, and Flk1 (Ramirez-Bergeron et al. 2004).

Vascular endothelial growth factor (VEGF) is critical for endothelial and hematopoietic development, as demonstrated by knock-out mice lacking specific components of the VEGF system. Mice homozygous for mutations that inactivate either tyrosine kinase receptor VEGFR1 (Flt1) or VEGFR2 (Flk1/KDR) die between day (d) 8.5 and 9.5 *in utero* (Fong et al. 1995; Shalaby et al. 1995). As ligands other than VEGF may activate these receptors, it was shown that embryos with functional inactivation of one VEGF allele (VEGF$^{+/-}$) die at d11–12 due to malformations in the vascular system (Carmeliet et al. 1996; Ferrara et al. 1996). VEGF expression is upregulated by hypoxia, activated oncogenes, and a variety of cytokines (for reviews, see Neufeld et al. 1999; Robinson and Stringer 2001). Once expressed, VEGF can be freely diffusible or sequestered within the extracellular matrix (ECM) by avidly binding heparin or heparin-like moieties. We demonstrated a novel role for soluble VEGFR-1 (sFlt-1) in modulating hemogenic mesoderm fate by measuring VEGF and VEGFRs (Purpura et al. 2008). Early transient Flk-1 signaling occurred in hypoxia due to low levels of sFlt-1 and high levels of VEGF, enhancing CFC generation, while sustained (or delayed) Flk-1 activation preferentially yielded hemogenic mesoderm-derived endothelial cells. In addition to the VEGF system, thrombopoietin (TPO) and its receptor c-Mpl play an important role in the maintenance and expansion of HSCs (Kobayashi et al. 1996; Ku et al. 1996; Sitnicka et al. 1996; Yagi et al. 1999; Ema et al. 2000; Huang et al. 2009). Generation of a true HSC *ex vivo* remains a significant challenge as the dynamic microenvironment and signals that stimulate primitive and definitive hematopoiesis during embryogenesis are difficult to deconvolute and apply.

1.3.2 Aspects of Embryogenesis That Influence the Expression of Mesodermal Inductive Factors

As discussed above, soluble signals direct mesoderm induction and development in the embryo, and consequently PSC differentiation strategies have typically involved the exogenous addition of these cytokines to direct mesoderm differentiation. Certain aspects of embryogenesis influence the temporal and spatial expression of these signals, including inductive tissue interactions (Sugi and Lough 1994; Climent et al. 1995; Schultheiss et al. 1995, 1997; Schultheiss and Lassar 1997; Raffin et al. 2000), oxygen tension (Rich and Kubanek 1982; Gassmann et al. 1996; Bichet et al. 1999; Ramirez-Bergeron and Simon 2001; Ramirez-Bergeron et al. 2004), and fluid shear stress (Hove et al. 2003; Lucitti et al. 2007; Adamo et al. 2009; North et al. 2009). Therefore, it may be that recapitulating these signals *in vitro* by engineering a niche for differentiating PSCs to mesodermal cells can lead to improved cardiac and blood differentiation efficiencies over the simple addition of exogenous factors alone.

1.3.2.1 Endogenous Tissue Induction

During embryogenesis, the heart is the first organ to fully form after gastrulation (Menard et al. 2004), when oxygen delivery by diffusion is no longer sufficient in the growing embryo. Forming the anatomical structure of the heart involves the precise spatiotemporal coordination of signals from neighboring tissues that promote or inhibit cardiac specification, proliferation, and migration of uncommitted precardiac mesoderm. Inductive cues originate from the anterior PE (Sugi and Lough 1994; Schultheiss et al. 1995, 1997; Schultheiss and Lassar 1997) and lateral regions of the embryo (Schultheiss et al. 1995, 1997; Schultheiss and Lassar 1997), while cardiogenesis is suppressed in the adjacent mesoderm by factors secreted by the neuronal tube (Climent et al. 1995; Schultheiss et al. 1997; Raffin et al. 2000). During embryonic development, precardiac mesoderm is in close contact with endoderm. A number of studies across various species have demonstrated that interactions between endoderm and overlying mesoderm are involved in cardiac differentiation (Orts Llorca 1963; Jacobson and Duncan 1968; Sugi and Lough 1994; Nascone and Mercola 1995; Schultheiss et al. 1995) and, more specifically, play an inductive role as evidenced by the generation of beating cardiac tissue in cocultures of noncardiogenic embryonic tissue explants and endodermal tissue (Schultheiss et al. 1995). The inductive characteristic of the endoderm can be attributed to TGF-β superfamily and FGF family growth factors, expressed by anterior lateral endoderm, that have been reported to be involved in cardiac differentiation.

1.3.2.2 Oxygen Tension

Oxygen tension is one aspect of the environment that may impact embryogenesis. However, it is difficult to measure the oxygen tension *in vivo*, such that specific differences pre- and postimplantation are largely unknown. The developmental competence of mouse oocytes cultured *in vitro* is significantly improved with low-oxygen conditions (5% oxygen tension) compared to atmospheric conditions (20% O_2), suggesting that the cells initially reside in a low-oxygen environment (Eppig and Wigglesworth 1995). This environmental preference was previously illustrated as pronuclear mouse embryos developed into blastocysts in 5% O_2 prior to transplantation into 3-day pseudopregnant females with a similar implantation rate and embryo viability upon comparison to *in vivo* developed blastocysts (Umaoka et al. 1991). Also, the relative abundance of a set of developmentally important gene transcripts in bovine morulae and blastocysts were similar to *in vivo* derived counterparts, showing that development was recapitulated under chemically defined conditions in 7% O_2 but not atmospheric conditions (Wrenzycki et al. 2001). Human studies indicate that there is low oxygen tension within the feto-placental unit until the start of the second trimester with the establishment of maternal circulation to the placenta (Burton and Jaunaiux 2001). This finding can be extended to the mouse, in which the labyrinth of the chorioallantoic placenta does not begin development until d8–9 and reliance on anaerobic glycolysis to meet metabolic demands has been previously demonstrated (Clough and Whittingham 1983).

1.3.2.3 Fluid Shear Stress

Interesting advances in our understanding of the dynamic aspects of the cellular environment have also been emerging. Blood flow may fundamentally impact the development or maturation potential of HSCs. Establishment of circulation (E8.5, in the mouse) delivers oxygen and nutrients more widely throughout the embryonic tissues and the resulting fluid shear stress or biomechanical forces are important in the formation of the heart and vessels (Hove et al. 2003; Lucitti et al. 2007). All vertebrate species have demonstrated functional HSCs arising from the dorsal aorta (Cumano and Godin 2007), and both mouse (Adamo et al. 2009) and zebrafish (North et al. 2009) models have shown that fluid shear stress enhances HSC number, suggesting that it is an evolutionarily conserved phenomenon. Pulsatile flow may induce nitric oxide (North et al. 2009) to trigger HSC maturation and this may lead to bioprocess improvements that enhance hematopoietic development *in vitro*. Alluding to this possibility, an earlier study showed higher engraftment and multilineage reconstitution with CD34+ UCB grown in stirred culture as opposed to static culture (Yang et al. 2008a). Additionally, megakaryocyte platelet production is increased in the presence of shear (Dunois-Larde et al. 2009).

1.4 ESC Differentiation: Strategies to Promote Mesoderm Subpopulations

As a differentiation system, the EB has often been viewed as a simple technique to demonstrate the capacity of PSCs to differentiate to a number of cell types, but a poor technique to control differentiation into one specific cell fate. A defining characteristic of the EB is that it gives rise to a heterogeneous population representing all the somatic tissue types. However, given that during embryogenesis spatial and temporal cues from neighboring tissues guide development, it is likely that the heterogeneity within the EB produces an environment with the necessary complexity of signals, regulated by the timing and proportion of emerging inductive tissue-associated cells, to produce all the cell types in the developing embryo. For this reason, the aggregate-based differentiation system is a valuable tool for studying the effect of endogenous signaling on induction efficiency of specific cell fates during PSC differentiation.

The term "embryoid body" is widely used to refer to uncontrolled differentiation induced by culturing aggregates of PSCs in suspension, and should be clearly distinguished from aggregate-based differentiation cultures wherein culture parameters are strictly controlled to promote differentiation of a specific cell type. Although the EB cannot give rise to a pure population of any specific cell fate, it is known that the proportion of cells belonging to the mesoderm lineage can be modulated in PSC aggregates by manipulating the culture system in a variety of ways, including the addition of exogenous factors (Yang et al. 2008b), controlling the macroenvironment (Bauwens et al. 2005; Purpura et al. 2008; Niebruegge et al. 2009), manipulating ECM signals or immobilized and/or time released growth factors (Ferreira et al. 2008; Chen et al. 2009), modulating PSC aggregate size (Burridge et al. 2007; Mohr et al. 2009; Niebruegge et al. 2009), or coculturing with an inductive cell type (Mummery et al. 2003; Passier et al. 2005; Xu et al. 2006; Xu et al. 2008).

1.4.1 Exogenous Growth Factors

Analysis of factors that may direct differentiation of hESCs as aggregates or in adherent culture indicates that multiple human cell types may be enriched *in vitro* by specific factors. One screen of eight potential factors showed that although none of the tested factors directs differentiation exclusively to one cell type, differentiation and/or cell selection, could be divided into three categories: growth factors (activin A and TGFβ1) that mainly induce mesodermal cells; factors (retinoic acid, EGF, BMP4, and bFGF) that activate ectodermal and mesodermal markers; and factors (NGF and HGF) that allow differentiation into the three embryonic germ layers, including endoderm (Schuldiner et al. 2000). Moreover, the method used to promote differentiation impacts hESC fate decisions and lineage survival.

Recently, it has been demonstrated that it is possible to efficiently generate hESC aggregate-derived cardiomyocytes without coculture and in the absence of serum or mouse cell conditioned medium (Yang et al. 2008b). The protocol, based on signaling that occurs during embryonic heart formation, consists of adding combinations of growth factors to hESC aggregates in stages temporally associated with the appropriate period of development. In the first stage (day 1–4), to induce a PS-like cell population, activin A, BMP4, and bFGF are added to the culture upregulating the PS markers brachyury and Wnt3A (Kispert and Herrmann 1994; Liu et al. 1999). In the next stage (day 4–8), to promote cardiac mesoderm commitment, DKK1 and VEGF are added. DKK1 is a Wnt inhibitor and VEGF promotes expansion and maturation of mesoderm cells (population expressing KDR/FLK1 in the mouse system). Although Wnt signaling is required at the onset of differentiation for PS induction (Lindsley et al. 2006; Naito et al. 2006; Ueno et al. 2007) after day 4, endogenous Wnt signaling promotes definitive endoderm and hematopoietic mesoderm at the expense of cardiac mesoderm (Gadue et al. 2006), thus DKK1 is added. In the final stage (starting on day 12), DKK1 and VEGF treatment are continued and FGF2 is readded to support expansion of the cardiac population. Human ESCs differentiated under these conditions yield a population of cardiac progenitors on day 6, identified by low expression of KDR and the absence of CKIT expression, and ultimately yield a population composed of 30% cardiomyocytes (cardiac Troponin T (cTnT)+ cells, express cardiac Troponin T) by day 14.

CD34+, CD45+, and hematopoietic CFCs are routinely derived in serum-free stromal cell coculture as stem cell factor (SCF), TPO, and Flt-3L are sufficient for progenitor support, whereas the PSC aggregate culture requires additional cytokines, VEGF and BMP4 (Tian et al. 2004). Multilineage hematopoietic CFU potential from hESCs has been shown with two cell lines (H1 and H9) following 15 days of aggregate-based differentiation with a mixture of cytokines, including SCF, Flt-3L, interleukin (IL)-3, IL-6, granulocyte-colony stimulating factor, and BMP4 (Chadwick et al. 2003). Recently, the addition of VEGF to this cytokine mix has been shown to promote erythropoietic differentiation, which can be further augmented by the addition of erythropoietin. The increased frequency of CD34+Flk1+ cells within VEGF-treated aggregates and their correlation to erythroid colonies provides evidence that factors are capable of regulating hematopoietic lineage development in hESCs, similar to mESCs (Cerdan et al. 2004). In addition to cell-type specificity, interest lies in microenvironmental controls of symmetric or asymmetric cell divisions during maintenance or differentiation culture.

1.5 Engineering the PSC Niche to Guide Mesoderm Development

Typically, attempts to control differentiation of PSCs involve the addition of exogenous factors as described in the previous section. Large screens are performed to determine the appropriate factors to add, in addition to the specific concentrations and timing of factor addition to maximize the efficiency of generating the cell type of interest (Schuldiner et al. 2000; Flaim et al. 2008). This form of "cytokine bingo" does not take advantage of the complex endogenous interactions that occur during embryonic development that are mirrored in the heterogeneous population of differentiating PSCs. Our focus in this review is to explore the role of the niche in directing PSC commitment and differentiation by manipulating the macroenvironment, PSC aggregate size, cell–cell interactions, and local microparticle delivery (Figure 1.4).

1.5.1 Macroenvironmental Controls: Oxygen Tension, Shear

Manipulation of oxygen concentration and fluid shear stress within PSC differentiation cultures is possible by employing controlled, stirred suspension bioreactors. Stirred suspension cultures (SSC) are well suited to control many aspects of the cellular environment. Stirring prevents formation of spatial concentration gradients within the bulk media, thus a point measurement reflects the conditions that all cells are exposed to. The ability to accurately measure culture conditions, such as oxygen tension or pH, allows control processes to maintain constant conditions or to change conditions as desired over time. SSC are also a practical means for scale-up, as vessel volume can be increased provided shear forces or sparging do not deleteriously affect the cells (van der Pol and Tramper 1998).

SSCs for the generation of mESC-derived cardiomyocytes and HSCs have been well developed (Zandstra et al. 2003; Dang et al. 2004; Bauwens et al. 2005; Schroeder et al. 2005; Purpura et al. 2008). The first major challenge to implementing stirred suspension bioreactors for PSC aggregate-based differentiation was that aggregates tended to agglomerate during the first 4 days of differentiation due to the expression of surface markers on PSCs that promote cell–cell aggregation (Dang et al. 2002). Initial attempts to prevent aggregate agglomeration focused on hydrogel encapsulation (Magyar et al. 2001; Dang et al. 2004; Bauwens et al. 2005; Dang and Zandstra 2005) of ESC aggregates to provide a barrier between aggregates. It was subsequently demonstrated that aggregate formation could be controlled by optimizing stirring conditions, specifically examining impeller type and stirring speed (Schroeder et al. 2005; Niebruegge et al. 2008). Purifying the heterogeneous differentiating cultures for a specific

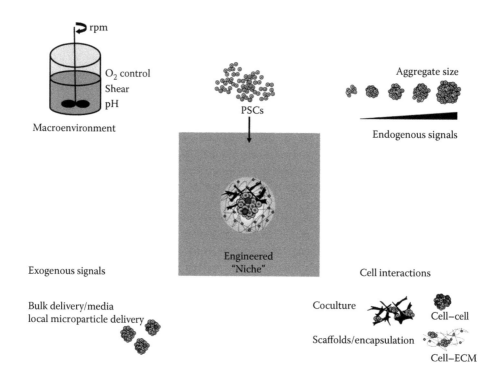

FIGURE 1.4 (See color insert.) The PSC differentiation niche can be manipulated in a variety of ways to direct mesoderm development. The macroenvironment can be controlled to manipulate medium oxygen tension and pH, and fluid shear stress. The level of endogenous signaling can be modulated by manipulating PSC aggregate size, thereby varying local cell density. Coculture systems, or ECM strategies, can be used to manipulate cell interactions that promote mesoderm induction. The most commonly employed method for directed mesoderm differentiation is the addition of exogenous signals that are known to promote mesoderm development. This is usually carried out by direct addition of cytokines to the bulk medium, but can be more precisely controlled by using local microparticle delivery.

cell type could be carried out using genetic selection, as has been demonstrated in the production of a purified mESC-derived cardiomyocyte (Klug et al. 1996; Li et al. 1998; Marchetti et al. 2002; Zandstra et al. 2003). ESCs were genetically engineered to be neomycin resistant upon expression of myosin heavy chain (MHC). This technique can efficiently enrich mESC-derived cardiomyoctes to greater than 70% in the stirred suspension system (Zandstra et al. 2003). Improved culture homogeneity has not only been achieved via stirring, but also by incorporating a settling tube to separate aggregates from the culture medium which permitted continuous medium perfusion thereby preventing wide variations in medium component concentrations, including glucose and lactate (Bauwens et al. 2005; Niebruegge et al. 2008).

Attempts to differentiate hESC aggregates in SSCs has generally consisted of an initial 24 h static aggregate formation step to prevent hESC aggregates from breaking apart under dynamic conditions (Cameron et al. 2006). Interestingly, when hESC aggregates are cultured in dynamic systems, such as SSCs, proliferation does occur, however, cell-fold expansion is still far lower (<20-fold) (Gerecht-Nir et al. 2004; Cameron et al. 2006) than what has been observed during mESC differentiation (>60-fold) (Zandstra et al. 2003; Dang et al. 2004; Bauwens et al. 2005). While the improved cell expansions may be attributed to the obvious benefits of stirred suspension such as medium homogeneity and reduced variations in metabolic by-products, the observation that under dynamic conditions hESC aggregate concentrations are maintained, while under static conditions aggregate concentrations sharply decrease in the first 4 days of culture indicate that cell expansion is largely due to the prevention of aggregate agglomeration in stirred suspension (Cameron et al. 2006). Further confirmation of reduced agglomeration under dynamic conditions is the visual observation that aggregates grown in spinner flasks are more homogenous in size and shape than those cultured statically (Cameron et al. 2006). Importantly, it was demonstrated that representative tissues from the three germ layers are produced in hESC aggregates cultured in stirred suspension and that differentiation efficiency to the hematopoietic (Cameron et al. 2006) and cardiac lineages (Niebruegge et al. 2008) are at least comparable to what is achieved under static conditions. Non-SSC dynamic cell culture systems have also been explored for hESC differentiation (Gerecht-Nir et al. 2004). Significant cell death and aggregate agglomeration was observed in high aspect rotating vessels. Aggregate agglomeration was prevented in slow-turning lateral vessels, and differentiation to cells representing the three germ layers was observed. These observations suggest that while mixing is crucial for cell expansion during differentiation of hESC aggregates, it is also essential to ensure that stirring is mild enough to prevent aggregate agglomeration and cell death.

Oxygen concentration has been shown to influence mouse PSC differentiation toward the hematopoietic and cardiac lineages (Gassmann et al. 1996; Sauer et al. 2000; Dang et al. 2004; Bauwens et al. 2005). Under hypoxic conditions, cardiac induction is enhanced, paralleling embryogenesis, wherein the development of the cardiovascular system takes place as diffusion of oxygen becomes limited by the growth of the embryo (Ramirez-Bergeron and Simon 2001). It is believed that the mechanism for the effect that hypoxia exerts on cardiomyocyte differentiation involves the activation of hypoxia inducible factor 1 (HIF-1) which activates a number of growth factors that are associated with cardiogenesis, including VEGF and FGF-2 (Gassmann et al. 1996; Ramirez-Bergeron and Simon 2001; Dang et al. 2004). The effect of hypoxia on hESC differentiation to the mesoderm lineages is still unclear. In one study examining cardiac differentiation of hESCs in an oxygen-controlled SSC, higher cell expansions and frequencies of beating aggregates were observed under 4% oxygen tension compared to 20% oxygen tension, but the difference was not statistically significant (Niebruegge et al. 2009). Another study comparing hematopoietic development in aggregates cultured in SSCs controlled at 5% and 21% oxygen tension revealed that while hypoxia upregulated the expression of HIF1-α protein and its downstream targets VEGF and GLUT-1, these hypoxia responsive molecules did not modulate differentiation of hESCs toward the hematopoietic lineage (Cameron et al. 2008).

In the future, realizing the therapeutic potential of stem cells may be facilitated by both the scaled-up production of undifferentiated and differentiated pluripotent cells. mESCs were first used in SSCs to demonstrate that pluripotency can be maintained in shear-controlled aggregates and in microcarrier suspension (Fok and Zandstra 2005; Cormier et al. 2006; Abranches et al. 2007; zur Nieden et al. 2007); the cells retain the ability to differentiate to cell types from all germ lineages upon shifting to differentiation conditions (Fok and Zandstra 2005). The development of an array of micro-bioreactors may also help define operating parameters and factors that impact scale-up (Figallo et al. 2007; Cimetta et al. 2009). Although processing challenges remain and mESC culture conditions often do not directly translate to human (reviewed in Kehoe et al. (2009)) most recently, hESCs have been maintained in SSC as aggregates (Cameron et al. 2006; Krawetz et al. 2010; Steiner et al. 2010), or in microcarrier suspension (Phillips et al. 2008; Fernandes et al. 2009; Nie et al. 2009; Oh et al. 2009).

1.5.2 Controlling Endogenous Signaling by Modulating the PSC Aggregate or Niche Size

The importance of initial aggregate size in regulating mesoderm specification and differentiation to mature subpopulations is likely related to the balance of endogenous promoters and inhibitors with exogenous factors, and neighboring cell interactions. Controlling aggregate size has been one of the major challenges in aggregate-mediated hESC differentiation because hESCs require cell–cell contact and paracrine and autocrine signaling for survival (Pyle et al. 2006), and as a result exhibit poor viability upon dissociation to single cells. Consequently, single-cell dissociation has typically been avoided during hESC aggregate formation (Kehat et al. 2001; Xu et al. 2002). Achieving reproducible, consistent, and efficient hESC differentiation has been difficult because cell aggregates are most often generated by partial enzymatic digestion of hESC colonies, resulting in variable aggregate sizes within and between cultures (Itskovitz-Eldor et al. 2000; Weitzer 2006). Attempts to control hESC aggregate size have involved either forced aggregation of defined cell numbers (Ng et al. 2005; Burridge et al. 2007; Ungrin et al. 2008) or the use of microwells to form 3D hESC aggregates of specified dimensions which can then be transferred to suspension to form monodisperse aggregates (Khademhosseini et al. 2006; Mohr et al. 2006, 2009).

In the first published report on forced aggregation of defined numbers of hESCs, it was observed that aggregate size influenced hematopoietic differentiation, with a minimum of 500 cells required for efficient blood formation and 1000 cells for optimum erythropoiesis (Ng et al. 2005). Building on this aggregate formation strategy, subsequent studies examined the effect of input hESC status and medium components on the efficiency of forced cell aggregation (Ungrin et al. 2008). It was observed that aggregate formation was inefficient when initiated with hESC input populations highly expressing Oct4 protein, a marker of pluripotency. Incorporating a "predifferentiation" step, in which maintenance medium is removed from hESC colonies and replaced with serum-containing medium 72 h prior to aggregation led to a significant improvement in aggregation efficiency, approaching 100%. This observation highlighted a separate issue that affects all differentiation studies. Variable input populations, with respect to expression levels of pluripotency markers as well as differentiation-associated markers result from variable hESC colony size, which is an inherent factor of passaging hESC colonies as cell clumps (Peerani et al. 2007). It was then established that the "predifferentiation" step could be eliminated and efficient aggregation of hESCs achieved in the presence of p160-Rho associated coiled-coil kinase (ROCK) inhibitor Y-27632. Human ESC aggregates formed by forced aggregation are not only size-specified but also display consistent shape, allowing for the reproducible observation of tissue-specific spatial organization within the aggregate. Uniform aggregates develop two distinct regions, an inner core that expresses Oct4, and a disordered outer layer that expresses a number of markers associated with PE (Ungrin et al. 2008).

In one study using forced aggregation to form hESC aggregates starting with 1000, 3000, and 10,000 cells per aggregate, the highest frequency of spontaneously contracting aggregates was observed in the cultures initiated with the highest aggregate size (Burridge et al. 2007). In another study, aggregates were formed by micropatterning round hESC colonies at 400 and 800 μm diameters and then transferring intact colonies to suspension to form consistent, size-specified hESC aggregates to culture in a stirred suspension bioreactor system (Niebruegge et al. 2009). Higher beating frequencies were observed in aggregates generated from 400 rather than 800 μm diameter colonies, and interestingly, micropatterned aggregates generated at both diameters achieved higher frequencies of beating aggregates than non-size-controlled aggregates. An alternate means to control aggregate size uses microwell-patterned surfaces. This method involves passaging hESC colonies as small clumps into size-specified microwells that have been either coated with Matrigel™ (Mohr et al. 2006, 2009) or mouse embryonic fibroblasts (Khademhosseini et al. 2006) and maintaining undifferentiated hESCs as 3D colonies. The colonies reach a maximum size defined by the volume of the microwell used, and can be transferred to suspension in differentiation medium to develop as size-specified aggregates. Aggregates cultured in this system have been proven to contain cells expressing proteins associated with each of the embryonic germ layers (Mohr et al. 2006), and have also been used to examine the effect of aggregate size on cardiac lineage induction from hESCs (Mohr et al. 2009).

The mechanism behind the influence of PSC aggregate size on differentiation trajectory is still unclear. A number of parameters are affected by varying aggregate size. Proteins may be differentially expressed as aggregate size is varied. For example, WNT5a was highly expressed in small aggregates while WNT11 was preferentially expressed in larger aggregates, impacting the canonical or noncanonical WNT signaling pathway, respectively, and promoting endothelial or cardiogenic differentiation (Hwang et al. 2009). Diffusion of oxygen and medium components becomes more limited as aggregate size increases. Observations made in studies examining the effect of hESC colony size on pluripotency, in which larger colonies maintained high levels of Oct4 expression while differentiation to the extraembryonic endoderm (ExE) lineage was promoted in small colonies (Peerani et al. 2007), are likely relevant in a 3D

system. The expectation being that larger aggregates would differentiate more slowly and ExE induction would occur at a higher frequency in smaller aggregates. Geometric relationships also vary with aggregate size, as the ratio of the surface area to volume of a sphere decreases with increasing sphere size (Bauwens et al. 2011).

1.5.3 Guiding PSC Differentiation to the Mesoderm Lineage by Mimicking Inductive Cellular Interactions That Occur during Embryonic Development

1.5.3.1 Coculture Systems

An alternate approach to aggregate-based differentiation has been to promote differentiation along a given lineage by coculturing ESCs with an inductive cell type. Cardiac induction has been demonstrated in hESCs cocultured with endoderm cells, perhaps mimicking the observation in the embryo where endodermal tissues appear to promote and/or support mesoderm and cardiac induction during embryogenesis. Human ESC aggregates cultured on a visceral endoderm feeder layer (END-2 cell line) are capable of cardiac induction, as evidenced by the appearance of beating areas in 35% of aggregates, and the emergence of cells expressing α-actinin, tropomyosin, and ryadonine receptors (Mummery et al. 2003; Passier et al. 2005). Building upon this system to enhance cardiac induction efficiency, aggregates were cultured in END-2-conditioned serum-free medium (Xu et al. 2008). In this system, beating was observed in 60–70% of aggregates after 12 days of differentiation. The positive effect that END-2-conditioned medium (END2-CM) exerted on cardiac induction of hESCs appeared to be due to the absence of insulin in the serum-free medium. An enzyme-linked immunosorbent assay (ELISA) analysis on END2-CM revealed a significant drop in insulin concentration to negligible levels after 3–4 days of exposure to END-2 cells, and beating activity and the expression of cardiac genes decreased with increasing concentrations of exogenously added insulin to END2-CM.

There is a large body of evidence that a variety of coculture systems can sustain and enhance HSC growth *in vitro* (Dexter et al. 1984; Nakano et al. 1994; Ohneda and Bautch 1997; Jung et al. 2005). Within the BM microenvironment, nonhematopoietic cells, including endothelial, fibroblastic stromal cells, and osteoblasts, compose an interactive niche that through cell–cell contact and/or localized delivery of factors, support or promote HSC expansion (Calvi et al. 2003; Zhang et al. 2003). There is some evidence that HSCs cocultured with endothelial cells increases HSC transplantation efficiency (Chute et al. 2004a,b). AGM stromal lines have also been isolated and shown to support mouse and human hematopoietic progenitor cells, as the AGM microenvironment contains all the necessary supportive cells for hematopoiesis (Ohneda et al. 1998; Xu et al. 1998; Matsuoka et al. 2001; Oostendorp et al. 2002; Takeuchi et al. 2002). The stromal cell line OP9, derived from BM of a macrophage colony-stimulating factor (M-CSF)-deficient (op/op) mouse supports lymphoid as well as myeloid differentiation from mESCs or hESCs *in vitro* (Kodama et al. 1994; Nakano et al. 1994; Cho et al. 1999; Vodyanik et al. 2005). Additionally, it has been shown that such influences are not dependent on cell contact (Oostendorp et al. 2005). The combination of low-dose hematopoietic cytokines (SCF, Flt-3L, VEGF) and human BM stromal cells during hESC development as aggregates promoted cell clusters with hematopoietic potential (8.81% Flk1+, 9.94% CD34+, 25.7% CD45+) and is an emerging method for the generation of hematopoietic cells (Wang et al. 2005).

Recently, using mESCs, aggregates were formed by mixing different ratios of ESCs that were transfected with a vector for Doxycycline (Dox)-inducible overexpression of GATA-4 and the parental untransfected cell line (Holtzinger et al. 2010). In the case where GATA-4 is induced in 2-day-old aggregates, enhanced cardiac differentiation was observed in aggregates generated with 50% GATA-4 inducible ESCs (0%, 50%, and 100% GATA-4-containing aggregates were examined). The cells that developed into cardiac cells were the non-GATA-4-induced cells, as GATA-4-overexpressing cells go on to express the endoderm-associated transcription factor Sry-related HMG box (SOX)17 and terminally differentiate to liver cells. It was concluded that the presence of these SOX17+ cells (that secrete DKK and BMPs) promoted cardiac induction of the non-GATA-4 induced cells.

1.5.3.2 Cell Scaffolds or Cell Encapsulation

The ECM can be thought of as a structural framework of secreted macromolecules that provides mechanical support, integrin-mediated signaling or adhesive interactions and that may sequester growth factors for proteolytic release to influence cell behavior or cell fates (Czyz and Wobus 2001; Engler et al. 2006). Both integrin and growth factors initiate intracellular signaling cascades upon binding with cellular receptors that influence gene expression and cell phenotype (Juliano and Haskill 1993).

The ECM consists primarily of collagens, other proteins, polysaccharides, and water, with a structure highly dependent on the location and tissue function. A range of materials and properties have been used to study stem cell interactions, fabricated as porous, fibrous, or hydrogel scaffolds (Nair and Laurencin 2006; Dawson et al. 2008). The scale of the cellular interactions and intent behind the scaffold types vary. For example, porous scaffolds provide macroscopic voids for the migration and infiltration of cells, whereas fibrous scaffolds may be fabricated on a size-scale to control cell alignment by mimicking native ECM, while water-swollen hydrogels may be fabricated from natural or synthetic materials to allow cell growth through the material (Burdick and Vunjak-Novakovic 2009).

One benefit of employing natural materials is their ability to provide signaling to encapsulated cells. 3D collagen gels have been used to provide integrin binding for mesenchymal stem cells (Battista et al. 2005; Chang et al. 2007) and hyaluronic acid, a polysaccharide, has been used for cell encapsulation and interacts with receptor CD44 (Burdick et al. 2005; Gerecht et al. 2007; Ifkovits and Burdick 2007), promoting ESC maintenance. Bioinert hydrogels such as alginate, which forms through ionic cross-linking, and thermoresponsive agarose contain no adhesive or native signaling and are resistant to nonspecific protein adsorption, preventing cell agglomeration and allowing the encapsulated cells to be differentiated for a variety of applications (Dang et al. 2004; Gerecht-Nir et al. 2004; Dean et al. 2006). Synthetic materials have also been investigated due to the versatility and adaptability of their physical properties although cytotoxicity may limit some applications. Two nontoxic synthetic materials that have been investigated and modified with tethered groups to alter cellular interactions such as adhesion peptides or phosphates are poly(ethylene glycol) hydrogels (Burdick and Anseth 2002; Nuttelman et al. 2004; Yang et al. 2005) and poly(hydroxyethyl)-methacrylate.

With the desire either to provide cells with pertinent physical cues or to guide appropriate cellular functions, scaffolds have been employed with many cell systems. For example, agarose covalently coupled to laminin 1 was embedded with nerve growth factor lipid microtubules before being used to bridge peripheral nerve gaps and enhance regeneration (Yu and Bellamkonda 2003); a poly(lactide-*co*-glycolide) (PLGA) scaffold was cross-linked to a BMP2-derived peptide to act as an inductive factor for osteogenesis (Duan et al. 2007); electrospun fibrous recombinant human (rh)BMP-2 loaded PLGA with hydroxyapatite was tested for cell attachment and cytotoxicity (Nie et al. 2008); and alginate covalently modified with the tripeptide arginine–glycine–aspartic acid (RGD) was used to investigate 3D mesenchymal stem cell properties (Duggal et al. 2009). The effect of various biomaterials on ESC differentiation have also been examined with either singly seeded cells or preformed aggregates/EBs (Battista et al. 2005; Flaim et al. 2005; Liu et al. 2006; Gerecht et al. 2007). Aggregates differentiated in collagen demonstrated a greater cardiomyocyte phenotype when mixed with high laminin, while high fibronectin enhanced an epithelial/vascular cell fate (Battista et al. 2005), as did incorporating the signaling peptide RGD and VEGF in modified dextran (Ferreira et al. 2007).

Not only the biochemical components affect stem cell differentiation, differences in the physical properties such as elasticity may also impact cell fate decisions. The mechanical stiffness of the hydrogels may influence cell viability, proliferation, and function as it was shown that soft, medium, and rigid substrates are neurogenic, myogenic, and osteogenic, respectively (Engler et al. 2006). A more in-depth survey of microencapsulation techniques and criteria for engineering the stem cell microenvironment can be found in recent reviews (Metallo et al. 2007; Schmidt et al. 2008; Burdick and Vunjak-Novakovic 2009).

1.5.4 Local Microparticle Delivery Systems

Overall, the application of cytokines and microenvironmental controls to regulate transcription factors and gene regulation point to a multifaceted approach to facilitate the development of a robust system capable of generating large numbers of cardiomyocytes or HSCs. The mode of cytokine presentation can impact downstream signaling, and thus stem and progenitor cell fates. Localized cytokine presentation may mimic *in vivo* mechanisms more closely, and provide a greater range of cell fate options than soluble delivery (Peerani and Zandstra 2010). Looking to develop greater control over aspects of differentiation from within the PSC aggregate itself, growth factor release from embedded particles is being explored. Access to the interior intercellular environment and molecular composition becomes progressively restricted, as the cell aggregates coalesce and proliferate (Sachlos and Auguste 2008). One approach to circumvent the developing barrier to differentiation signals is to use microparticles that mimic the natural environment by releasing growth factors; controlled release has been studied for applications in many areas. For delivery, molecules can be physically adsorbed or immobilized on a particle surface that would be incorporated into the aggregate, although large molecule delivery may be limited by the structure of the aggregate itself and steric diffusional barriers. Localized delivery would also make scale-up more economical by facilitating expansion processes and yields.

One model system for controlled release is to use biocompatible and biodegradable PLGA particles loaded with growth factors. The PLGA growth factor loaded microparticles show an initial burst followed by a slow or negligible release when agitated in phosphate buffered saline at 37°C (Ferreira et al. 2008). Upon testing within hESC aggregates, growth factors (including VEGF, bFGF, PlGF) were largely released over the first 2 days, but small concentrations of the factors (~55–285-fold less than soluble delivery) resulted in a 2–4-fold upregulation of PECAM-1, a definitive endothelial cell marker (Ferreira et al. 2008). Retinoic acid has also been delivered from PLGA particles in mESC aggregates, resulting in cystic spheroids of an epiblast nature (Carpenedo et al. 2009).

Another system for controlled release is to use gelatin, a degraded animal collagen that can be positively or negatively charged and that is also biocompatible and biodegradable. The production methods of gelatin microparticles have been evolving since first reported, and include spray-drying, precipitation, and emulsification (Bruschi et al. 2003; Sivakumar and Rao 2003; Vandervoort and Ludwig 2004). The disadvantage of these methods is that nonuniform particles with a broad size distribution result. Methodologies have also been described to generate size-controlled spheres with a narrow distribution, to increase the reproducibility of drug release from a more uniform carrier (Oner and Groves 1993; Huang et al. 2009). In contrast to the burst release observed with PLGA microparticles, BMP-2 gelatin microparticles exhibited minimal burst release with linear release kinetics *in vitro* for over 3 weeks (Patel et al. 2008b). Specific growth factor release depends on the effects of growth factor size, charge, and conformation (Patel et al. 2008b; Chen et al. 2009). VEGF release kinetics were also dependent on the extent of gelatin cross-linking (Patel et al. 2008a). These studies demonstrate the utility of gelatin microparticles as delivery vehicles for the controlled release of various growth factors for developing tissue engineering applications.

1.6 Conclusion

The basis of most previous studies examining lineage-specific induction of PSCs has focused on addition of exogenous growth factors and coculturing with an inductive cell type to direct differentiation and commitment. Continuing work must now focus on strategies and technologies that specifically optimize lineage-specific yields by more precisely controlling the differentiation environment through delivery of growth factors in controlled spatiotemporal manner, and by organizing cell–cell and cell–ECM interactions (Figure 1.4). These strategies should lead to enhanced robustness and reproducibility in cell generation studies, and to mechanistic insight into the processes which control PSC specification into mesoderm and its derivative tissues.

Acknowledgments

The authors wish to thank members of the Zandstra lab for helpful discussions. Support for this work was provided by NSERC, the CIHR, the HSFO, and the Canadian Stem Cell Network. Céline L. Bauwens is supported by an Ontario Graduate Scholarship in Science and Technology and Kelly A. Purpura is supported by an Ontario Graduate Scholarship. Peter W. Zandstra is the Canadian Research Chair in Stem Cell Bioengineering.

References

Abe, K., H. Niwa, K. Iwase et al. 1996. Endoderm-specific gene expression in embryonic stem cells differentiated to embryoid bodies. *Exp Cell Res* **229**(1): 27–34.

Abranches, E., E. Bekman, D. Henrique et al. 2007. Expansion of mouse embryonic stem cells on microcarriers. *Biotechnol Bioeng* **96**(6): 1211–21.

Adamo, L., O. Naveiras, P. L. Wenzel et al. 2009. Biomechanical forces promote embryonic haematopoiesis. *Nature* **459**(7250): 1131–5.

Alvarez-Silva, M., P. Belo-Diabangouaya, J. Salaun et al. 2003. Mouse placenta is a major hematopoietic organ. *Development* **130**(22): 5437–44.

Attisano, L. and E. Labbe. 2004. TGFbeta and Wnt pathway cross-talk. *Cancer Metastasis Rev* **23**(1–2): 53–61.

Attisano, L. and J. L. Wrana. 2002. Signal transduction by the TGF-beta superfamily. *Science* **296**(5573): 1646–7.

Barker, J. N. and J. E. Wagner. 2002. Umbilical cord blood transplantation: Current state of the art. *Curr Opin Oncol* **14**(2): 160–4.

Barron, M., M. Gao, and J. Lough. 2000. Requirement for BMP and FGF signaling during cardiogenic induction in non-precardiac mesoderm is specific, transient, and cooperative. *Dev Dyn* **218**(2): 383–93.

Battista, S., D. Guarnieri, C. Borselli et al. 2005. The effect of matrix composition of 3D constructs on embryonic stem cell differentiation. *Biomaterials* **26**(31): 6194–207.

Bauer, C., R. Tamm, D. Petschow et al. 1975. Oxygen affinity and allosteric effects of embryonic mouse haemolglobins. *Nature* **257**(5524): 333–4.

Bauwens, C., T. Yin, S. Dang et al. 2005. Development of a perfusion fed bioreactor for embryonic stem cell-derived cardiomyocyte generation: Oxygen-mediated enhancement of cardiomyocyte output. *Biotechnol Bioeng* **90**(4): 452–61.

Bauwens, C. L., H. Song, N. Thavandiran et al. 2011. Geometric control of cardiomyogenic induction in human pluripotent stem cells. *Tissue Eng Part A* **17**(15–16): 1901–9.

Bichet, S., R. H. Wenger, G. Camenisch et al. 1999. Oxygen tension modulates beta-globin switching in embryoid bodies. *FASEB J* **13**(2): 285–95.

Brotherton, T. W., D. H. Chui, J. Gauldie et al. 1979. Hemoglobin ontogeny during normal mouse fetal development. *Proc Natl Acad Sci USA* **76**(6): 2853–7.

Bruschi, M. L., M. L. Cardoso, M. B. Lucchesi et al. 2003. Gelatin microparticles containing propolis obtained by spray-drying technique: Preparation and characterization. *Int J Pharm* **264**(1–2): 45–55.

Burdick, J. A. and K. S. Anseth. 2002. Photoencapsulation of osteoblasts in injectable RGD-modified PEG hydrogels for bone tissue engineering. *Biomaterials* **23**(22): 4315–23.

Burdick, J. A., C. Chung, X. Jia et al. 2005. Controlled degradation and mechanical behavior of photopolymerized hyaluronic acid networks. *Biomacromolecules* **6**(1): 386–91.

Burdick, J. A. and G. Vunjak-Novakovic. 2009. Engineered microenvironments for controlled stem cell differentiation. *Tissue Eng Part A* **15**(2): 205–19.

Burridge, P. W., D. Anderson, H. Priddle et al. 2007. Improved human embryonic stem cell embryoid body homogeneity and cardiomyocyte differentiation from a novel V-96 plate aggregation system highlights interline variability. *Stem Cells* **25**(4): 929–38.

Burton, G. J. and E. Jaunaiux. 2001. Maternal vascularisation of the human placenta: Does the embryo develop in a hypoxic environment? *Gynecol Obstet Fertil* **29**(7–8): 503–8.

Calvi, L. M., G. B. Adams, K. W. Weibrecht et al. 2003. Osteoblastic cells regulate the haematopoietic stem cell niche. *Nature* **425**(6960): 841–6.

Cameron, C. M., F. Harding, W. S. Hu et al. 2008. Activation of hypoxic response in human embryonic stem cell-derived embryoid bodies. *Exp Biol Med (Maywood)* **233**(8): 1044–57.

Cameron, C. M., W. S. Hu, and D. S. Kaufman. 2006. Improved development of human embryonic stem cell-derived embryoid bodies by stirred vessel cultivation. *Biotechnol Bioeng* **94**(5): 938–48.

Candia, A. F., T. Watabe, S. H. Hawley et al. 1997. Cellular interpretation of multiple TGF-beta signals: Intracellular antagonism between activin/BVg1 and BMP-2/4 signaling mediated by Smads. *Development* **124**(22): 4467–80.

Carmeliet, P., V. Ferreira, G. Breier et al. 1996. Abnormal blood vessel development and lethality in embryos lacking a single VEGF allele. *Nature* **380**(6573): 435–9.

Carpenedo, R. L., A. M. Bratt-Leal, R. A. Marklein et al. 2009. Homogeneous and organized differentiation within embryoid bodies induced by microsphere-mediated delivery of small molecules. *Biomaterials* **30**(13): 2507–15.

Cerdan, C., A. Rouleau, and M. Bhatia. 2004. VEGF-A165 augments erythropoietic development from human embryonic stem cells. *Blood* **103**(7): 2504–12.

Chadwick, K., L. Wang, L. Li et al. 2003. Cytokines and BMP-4 promote hematopoietic differentiation of human embryonic stem cells. *Blood* **102**(3): 906–15.

Chang, C. F., M. W. Lee, P. Y. Kuo et al. 2007. Three-dimensional collagen fiber remodeling by mesenchymal stem cells requires the integrin-matrix interaction. *J Biomed Mater Res A* **80**(2): 466–74.

Chen, F. M., R. Chen, X. J. Wang et al. 2009. *In vitro* cellular responses to scaffolds containing two microencapsulated growth factors. *Biomaterials* **30**(28): 5215–24.

Chen, M. J., T. Yokomizo, B. M. Zeigler et al. 2009. Runx1 is required for the endothelial to haematopoietic cell transition but not thereafter. *Nature* **457**(7231): 887–91.

Cho, S. K., T. D. Webber, J. R. Carlyle et al. 1999. Functional characterization of B lymphocytes generated *in vitro* from embryonic stem cells. *Proc Natl Acad Sci USA* **96**(17): 9797–802.

Chute, J. P., J. Fung, G. Muramoto et al. 2004a. *Ex vivo* culture rescues hematopoietic stem cells with long-term repopulating capacity following harvest from lethally irradiated mice. *Exp Hematol* **32**(3): 308–17.

Chute, J. P., G. Muramoto, J. Fung et al. 2004b. Quantitative analysis demonstrates expansion of SCID-repopulating cells and increased engraftment capacity in human cord blood following *ex vivo* culture with human brain endothelial cells. *Stem Cells* **22**(2): 202–15.

Cimetta, E., E. Figallo, C. Cannizzaro et al. 2009. Micro-bioreactor arrays for controlling cellular environments: Design principles for human embryonic stem cell applications. *Methods* **47**(2): 81–9.

Climent, S., M. Sarasa, J. M. Villar et al. 1995. Neurogenic cells inhibit the differentiation of cardiogenic cells. *Dev Biol* **171**(1): 130–48.

Clough, J. R. and D. G. Whittingham. 1983. Metabolism of [14C]glucose by postimplantation mouse embryos *in vitro*. *J Embryol Exp Morphol* **74**: 133–42. (Abstract only).

Conlon, F. L., K. M. Lyons, N. Takaesu et al. 1994. A primary requirement for nodal in the formation and maintenance of the primitive streak in the mouse. *Development* **120**(7): 1919–28.

Cormier, J. T., N. I. zur Nieden, D. E. Rancourt et al. 2006. Expansion of undifferentiated murine embryonic stem cells as aggregates in suspension culture bioreactors. *Tissue Eng* **12**(11): 3233–45.

Coucouvanis, E. and G. R. Martin. 1995. Signals for death and survival: A two-step mechanism for cavitation in the vertebrate embryo. *Cell* **83**(2): 279–87.

Coucouvanis, E. and G. R. Martin. 1999. BMP signaling plays a role in visceral endoderm differentiation and cavitation in the early mouse embryo. *Development* **126**(3): 535–46.

Cumano, A., F. Dieterlen-Lievre, and I. Godin. 1996. Lymphoid potential, probed before circulation in mouse, is restricted to caudal intraembryonic splanchnopleura. *Cell* **86**(6): 907–16.

Cumano, A., J. C. Ferraz, M. Klaine et al. 2001. Intraembryonic, but not yolk sac hematopoietic precursors, isolated before circulation, provide long-term multilineage reconstitution. *Immunity* **15**(3): 477–85.

Cumano, A. and I. Godin. 2007. Ontogeny of the hematopoietic system. *Annu Rev Immunol* **25**: 745–85.

Czyz, J. and A. Wobus. 2001. Embryonic stem cell differentiation: The role of extracellular factors. *Differentiation* **68**(4–5): 167–74.

Dang, S. M., S. Gerecht-Nir, J. Chen et al. 2004. Controlled, scalable embryonic stem cell differentiation culture. *Stem Cells* **22**(3): 275–82.

Dang, S. M., M. Kyba, R. Perlingeiro et al. 2002. Efficiency of embryoid body formation and hematopoietic development from embryonic stem cells in different culture systems. *Biotechnol Bioeng* **78**(4): 442–53.

Dang, S. M. and P. W. Zandstra. 2005. Scalable production of embryonic stem cell-derived cells. *Methods Mol Biol* **290**: 353–64.

Dawson, E., G. Mapili, K. Erickson et al. 2008. Biomaterials for stem cell differentiation. *Adv Drug Deliv Rev* **60**(2): 215–28.

Dean, S. K., Y. Yulyana, G. Williams et al. 2006. Differentiation of encapsulated embryonic stem cells after transplantation. *Transplantation* **82**(9): 1175–84.

Dexter, T. M., P. Simmons, R. A. Purnell et al. 1984. The regulation of hemopoietic cell development by the stromal cell environment and diffusible regulatory molecules. *Prog Clin Biol Res* **148**: 13–33.

Doetschman, T. C., H. Eistetter, M. Katz et al. 1985. The *in vitro* development of blastocyst-derived embryonic stem cell lines: Formation of visceral yolk sac, blood islands and myocardium. *J Embryol Exp Morphol* **87**: 27–45.

Duan, Z., Q. Zheng, X. Guo et al. 2007. Experimental research on ectopic osteogenesis of BMP2-derived peptide P24 combined with PLGA copolymers. *J Huazhong Univ Sci Technolog Med Sci* **27**(2): 179–82.

Duggal, S., K. B. Fronsdal, K. Szoke et al. 2009. Phenotype and gene expression of human mesenchymal stem cells in alginate scaffolds. *Tissue Eng Part A* **15**(7): 1763–73.

Dunois-Larde, C., C. Capron, S. Fichelson et al. 2009. Exposure of human megakaryocytes to high shear rates accelerates platelet production. *Blood* **114**(9): 1875–83.

Dvash, T. and N. Benvenisty. 2004. Human embryonic stem cells as a model for early human development. *Best Pract Res Clin Obstet Gynaecol* **18**(6): 929–40.

Eilken, H. M., S. Nishikawa, and T. Schroeder. 2009. Continuous single-cell imaging of blood generation from haemogenic endothelium. *Nature* **457**(7231): 896–900.

Ema, H., H. Takano, K. Sudo et al. 2000. *In vitro* self-renewal division of hematopoietic stem cells. *J Exp Med* **192**(9): 1281–8.

Ema, M., S. Takahashi, and J. Rossant. 2006. Deletion of the selection cassette, but not cis-acting elements, in targeted Flk1-lacZ allele reveals Flk1 expression in multipotent mesodermal progenitors. *Blood* **107**(1): 111–7.

Engler, A. J., S. Sen, H. L. Sweeney et al. 2006. Matrix elasticity directs stem cell lineage specification. *Cell* **126**(4): 677–89.

Eppig, J. J. and K. Wigglesworth. 1995. Factors affecting the developmental competence of mouse oocytes grown *in vitro*: Oxygen concentration. *Mol Reprod Dev* **42**(4): 447–56.

Era, T., N. Izumi, M. Hayashi et al. 2008. Multiple mesoderm subsets give rise to endothelial cells, whereas hematopoietic cells are differentiated only from a restricted subset in embryonic stem cell differentiation culture. *Stem Cells* **26**(2): 401–11.

Evans, M. J. and M. H. Kaufman. 1981. Establishment in culture of pluripotential cells from mouse embryos. *Nature* **292**(5819): 154–6.

Fernandes, A. M., P. A. Marinho, R. C. Sartore et al. 2009. Successful scale-up of human embryonic stem cell production in a stirred microcarrier culture system. *Braz J Med Biol Res* **42**(6): 515–22.

Ferrara, N., K. Carver-Moore, H. Chen et al. 1996. Heterozygous embryonic lethality induced by targeted inactivation of the VEGF gene. *Nature* **380**(6573): 439–42.

Ferreira, L. S., S. Gerecht, J. Fuller et al. 2007. Bioactive hydrogel scaffolds for controllable vascular differentiation of human embryonic stem cells. *Biomaterials* **28**(17): 2706–17.

Ferreira, L., T. Squier, H. Park et al. 2008. Human embryoid bodies containing nano- and microparticlulate delivery vehicles. *Adv Mater* **20**: 2285–2291.

Figallo, E., C. Cannizzaro, S. Gerecht et al. 2007. Micro-bioreactor array for controlling cellular microenvironments. *Lab Chip* **7**(6): 710–9.

Flaim, C. J., S. Chien, and S. N. Bhatia. 2005. An extracellular matrix microarray for probing cellular differentiation. *Nat Methods* **2**(2): 119–25.

Flaim, C. J., D. Teng, S. Chien et al. 2008. Combinatorial signaling microenvironments for studying stem cell fate. *Stem Cells Dev* **17**(1): 29–39.

Fok, E. Y. and P. W. Zandstra. 2005. Shear-controlled single-step mouse embryonic stem cell expansion and embryoid body-based differentiation. *Stem Cells* **23**(9): 1333–42.

Fong, G. H., J. Rossant, M. Gertsenstein et al. 1995. Role of the Flt-1 receptor tyrosine kinase in regulating the assembly of vascular endothelium. *Nature* **376**(6535): 66–70.

Frank, D. B., A. Abtahi, D. J. Yamaguchi et al. 2005. Bone morphogenetic protein 4 promotes pulmonary vascular remodeling in hypoxic pulmonary hypertension. *Circ Res* **97**(5): 496–504.

Gadue, P., T. L. Huber, M. C. Nostro et al. 2005. Germ layer induction from embryonic stem cells. *Exp Hematol* **33**(9): 955–64.

Gadue, P., T. L. Huber, P. J. Paddison et al. 2006. Wnt and TGF-beta signaling are required for the induction of an *in vitro* model of primitive streak formation using embryonic stem cells. *Proc Natl Acad Sci USA* **103**(45): 16806–11.

Gassmann, M., J. Fandrey, S. Bichet et al. 1996. Oxygen supply and oxygen-dependent gene expression in differentiating embryonic stem cells. *Proc Natl Acad Sci USA* **93**(7): 2867–72.

Gerecht-Nir, S., S. Cohen, and J. Itskovitz-Eldor. 2004. Bioreactor cultivation enhances the efficiency of human embryoid body (hEB) formation and differentiation. *Biotechnol Bioeng* **86**(5): 493–502.

Gerecht-Nir, S., S. Cohen, A. Ziskind et al. 2004. Three-dimensional porous alginate scaffolds provide a conducive environment for generation of well-vascularized embryoid bodies from human embryonic stem cells. *Biotechnol Bioeng* **88**(3): 313–20.

Gerecht, S., J. A. Burdick, L. S. Ferreira et al. 2007. Hyaluronic acid hydrogel for controlled self-renewal and differentiation of human embryonic stem cells. *Proc Natl Acad Sci USA* **104**(27): 11298–303.

Hart, A. H., L. Hartley, K. Sourris et al. 2002. Mixl1 is required for axial mesendoderm morphogenesis and patterning in the murine embryo. *Development* **129**(15): 3597–608.

Harvey, R. P. 2002. Patterning the vertebrate heart. *Nat Rev Genet* **3**(7): 544–56.

Hogan, B. L. 1996. Bone morphogenetic proteins in development. *Curr Opin Genet Dev* **6**(4): 432–8.

Holtzinger, A., G. E. Rosenfeld and T. Evans. 2010. GATA4 directs development of cardiac-inducing endoderm from ES cells. *Dev Biol* **337**(1): 63–73.

Hove, J. R., R. W. Koster, A. S. Forouhar et al. 2003. Intracardiac fluid forces are an essential epigenetic factor for embryonic cardiogenesis. *Nature* **421**(6919): 172–7.

Huang, K. S., K. Lu, C. S. Yeh et al. 2009. Microfluidic controlling monodisperse microdroplet for 5-fluorouracil loaded genipin-gelatin microcapsules. *J Control Release* **137**(1): 15–9.

Hwang, Y. S., B. G. Chung, D. Ortmann et al. 2009. Microwell-mediated control of embryoid body size regulates embryonic stem cell fate via differential expression of WNT5a and WNT11. *Proc Natl Acad Sci USA* **106**(40): 16978–83.

Ifkovits, J. L. and J. A. Burdick. 2007. Review: Photopolymerizable and degradable biomaterials for tissue engineering applications. *Tissue Eng* **13**(10): 2369–85.

Inman, K. E. and K. M. Downs. 2007. The murine allantois: Emerging paradigms in development of the mammalian umbilical cord and its relation to the fetus. *Genesis* **45**(5): 237–58.

Itskovitz-Eldor, J., M. Schuldiner, D. Karsenti et al. 2000. Differentiation of human embryonic stem cells into embryoid bodies compromising the three embryonic germ layers. *Mol Med* **6**(2): 88–95.

Jacobson, A. G. and J. T. Duncan. 1968. Heart induction in salamanders. *J Exp Zool* **167**(1): 79–103.

Johnson, M. H. and C. A. Ziomek. 1981. Induction of polarity in mouse 8-cell blastomeres: Specificity, geometry, and stability. *J Cell Biol* **91**(1): 303–8.

Juliano, R. L. and S. Haskill. 1993. Signal transduction from the extracellular matrix. *J Cell Biol* **120**(3): 577–85.

Jung, Y., J. Wang, A. Havens et al. 2005. Cell-to-cell contact is critical for the survival of hematopoietic progenitor cells on osteoblasts. *Cytokine* **32**(3–4): 155–62.

Kataoka, H., N. Takakura, S. Nishikawa et al. 1997. Expressions of PDGF receptor alpha, c-Kit and Flk1 genes clustering in mouse chromosome 5 define distinct subsets of nascent mesodermal cells. *Dev Growth Differ* **39**(6): 729–40.

Kattman, S. J., A. D. Witty, M. Gagliardi et al. 2011. Stage-specific optimization of activin/nodal and BMP signaling promotes cardiac differentiation of mouse and human pluripotent stem cell lines. *Cell Stem Cell* **8**(2): 228–40.

Kaufman, D. S., E. T. Hanson, R. L. Lewis et al. 2001. Hematopoietic colony-forming cells derived from human embryonic stem cells. *Proc Natl Acad Sci USA* **98**(19): 10716–21.

Kawai, T., T. Takahashi, M. Esaki et al. 2004. Efficient cardiomyogenic differentiation of embryonic stem cell by fibroblast growth factor 2 and bone morphogenetic protein 2. *Circ J* **68**(7): 691–702.

Kehat, I., D. Kenyagin-Karsenti, M. Snir et al. 2001. Human embryonic stem cells can differentiate into myocytes with structural and functional properties of cardiomyocytes. *J Clin Invest* **108**(3): 407–14.

Kehoe, D. E., D. Jing, L. T. Lock et al. 2009. Scalable stirred-suspension bioreactor culture of human pluripotent stem cells. *Tissue Eng Part A* **16**(2): 405–21.

Keller, G. 2005. Embryonic stem cell differentiation: Emergence of a new era in biology and medicine. *Genes Dev* **19**(10): 1129–55.

Khademhosseini, A., L. Ferreira, J. Blumling, 3rd, et al. 2006. Co-culture of human embryonic stem cells with murine embryonic fibroblasts on microwell-patterned substrates. *Biomaterials* **27**(36): 5968–77.

Kispert, A. and B. G. Herrmann. 1994. Immunohistochemical analysis of the Brachyury protein in wild-type and mutant mouse embryos. *Dev Biol* **161**(1): 179–93.

Klug, M. G., M. H. Soonpaa, G. Y. Koh et al. 1996. Genetically selected cardiomyocytes from differentiating embronic stem cells form stable intracardiac grafts. *J Clin Invest* **98**(1): 216–24.

Kobayashi, M., J. H. Laver, T. Kato et al. 1996. Thrombopoietin supports proliferation of human primitive hematopoietic cells in synergy with steel factor and/or interleukin-3. *Blood* **88**(2): 429–36.

Kodama, H., M. Nose, S. Niida et al. 1994. Involvement of the c-kit receptor in the adhesion of hematopoietic stem cells to stromal cells. *Exp Hematol* **22**(10): 979–84.

Krawetz, R., J. T. Taiani, S. Liu et al. 2010. Large-scale expansion of pluripotent human embryonic stem cells in stirred suspension bioreactors. *Tissue Eng Part C Methods* **16**(4): 573–82.

Ku, H., Y. Yonemura, K. Kaushansky et al. 1996. Thrombopoietin, the ligand for the Mpl receptor, synergizes with steel factor and other early acting cytokines in supporting proliferation of primitive hematopoietic progenitors of mice. *Blood* **87**(11): 4544–51.

Kyba, M. and G. Q. Daley. 2003. Hematopoiesis from embryonic stem cells: Lessons from and for ontogeny. *Exp Hematol* **31**(11): 994–1006.

Ladd, A. N., T. A. Yatskievych, and P. B. Antin. 1998. Regulation of avian cardiac myogenesis by activin/TGFbeta and bone morphogenetic proteins. *Dev Biol* **204**(2): 407–19.

Lagna, G., A. Hata, A. Hemmati-Brivanlou et al. 1996. Partnership between DPC4 and SMAD proteins in TGF-beta signalling pathways. *Nature* **383**(6603): 832–6.

Lancrin, C., P. Sroczynska, C. Stephenson et al. 2009. The haemangioblast generates haematopoietic cells through a haemogenic endothelium stage. *Nature* **457**(7231): 892–5.

Li, M., L. Pevny, R. Lovell-Badge et al. 1998. Generation of purified neural precursors from embryonic stem cells by lineage selection. *Curr Biol* **8**(17): 971–4.

Lindsley, R. C., J. G. Gill, M. Kyba et al. 2006. Canonical Wnt signaling is required for development of embryonic stem cell-derived mesoderm. *Development* **133**(19): 3787–96.

Liu, H., S. F. Collins, and L. J. Suggs. 2006. Three-dimensional culture for expansion and differentiation of mouse embryonic stem cells. *Biomaterials* **27**(36): 6004–14.

Liu, P., M. Wakamiya, M. J. Shea et al. 1999. Requirement for Wnt3 in vertebrate axis formation. *Nat Genet* **22**(4): 361–5.

Lough, J., M. Barron, M. Brogley et al. 1996. Combined BMP-2 and FGF-4, but neither factor alone, induces cardiogenesis in non-precardiac embryonic mesoderm. *Dev Biol* **178**(1): 198–202.

Lucitti, J. L., E. A. Jones, C. Huang et al. 2007. Vascular remodeling of the mouse yolk sac requires hemodynamic force. *Development* **134**(18): 3317–26.

Magyar, J. P., M. Nemir, E. Ehler et al. 2001. Mass production of embryoid bodies in microbeads. *Ann N Y Acad Sci* **944**: 135–43.

Marchetti, S., C. Gimond, K. Iljin et al. 2002. Endothelial cells genetically selected from differentiating mouse embryonic stem cells incorporate at sites of neovascularization *in vivo. J Cell Sci* **115**(Pt 10): 2075–85.

Marshall, C. J., C. Kinnon, and A. J. Thrasher. 2000. Polarized expression of bone morphogenetic protein-4 in the human aorta-gonad-mesonephros region. *Blood* **96**(4): 1591–3.

Martin, G. R. 1981. Isolation of a pluripotent cell line from early mouse embryos cultured in medium conditioned by teratocarcinoma stem cells. *Proc Natl Acad Sci USA* **78**(12): 7634–8.

Martin, G. R., L. M. Wiley and I. Damjanov. 1977. The development of cystic embryoid bodies *in vitro* from clonal teratocarcinoma stem cells. *Dev Biol* **61**(2): 230–44.

Marvin, M. J., G. Di Rocco, A. Gardiner et al. 2001. Inhibition of Wnt activity induces heart formation from posterior mesoderm. *Genes Dev* **15**(3): 316–27.

Massague, J. and Y. G. Chen. 2000. Controlling TGF-beta signaling. *Genes Dev* **14**(6): 627–44.

Matsuoka, S., K. Tsuji, H. Hisakawa et al. 2001. Generation of definitive hematopoietic stem cells from murine early yolk sac and paraaortic splanchnopleures by aorta-gonad-mesonephros region-derived stromal cells. *Blood* **98**(1): 6–12.

Menard, C., C. Grey, A. Mery et al. 2004. Cardiac specification of embryonic stem cells. *J Cell Biochem* **93**(4): 681–7.

Menasche, P. 2003. Myoblast-based cell transplantation. *Heart Fail Rev* **8**(3): 221–7.

Menasche, P. 2004. Skeletal myoblast transplantation for cardiac repair. *Expert Rev Cardiovasc Ther* **2**(1): 21–8.

Metallo, C. M., J. C. Mohr, C. J. Detzel et al. 2007. Engineering the stem cell microenvironment. *Biotechnol Prog* **23**(1): 18–23.

Mima, T., H. Ueno, D. A. Fischman et al. 1995. Fibroblast growth factor receptor is required for *in vivo* cardiac myocyte proliferation at early embryonic stages of heart development. *Proc Natl Acad Sci USA* **92**(2): 467–71.

Mohr, J. C., J. J. de Pablo, and S. P. Palecek. 2006. 3-D microwell culture of human embryonic stem cells. *Biomaterials* **27**(36): 6032–42.

Mohr, J. C., J. Zhang, S. M. Azarin et al. 2009. The microwell control of embryoid body size in order to regulate cardiac differentiation of human embryonic stem cells. *Biomaterials* **31**(7): 1885–93.

Moustakas, A., S. Souchelnytskyi, and C. H. Heldin. 2001. Smad regulation in TGF-beta signal transduction. *J Cell Sci* **114**(Pt 24): 4359–69.

Mummery, C., D. Ward-van Oostwaard, P. Doevendans et al. 2003. Differentiation of human embryonic stem cells to cardiomyocytes: Role of coculture with visceral endoderm-like cells. *Circulation* **107**(21): 2733–40.

Murry, C. E. and G. Keller. 2008. Differentiation of embryonic stem cells to clinically relevant populations: Lessons from embryonic development. *Cell* **132**(4): 661–80.

Nair, L. S. and C. T. Laurencin. 2006. Polymers as biomaterials for tissue engineering and controlled drug delivery. *Adv Biochem Eng Biotechnol* **102**: 47–90.

Naito, A. T., I. Shiojima, H. Akazawa et al. 2006. Developmental stage-specific biphasic roles of Wnt/beta-catenin signaling in cardiomyogenesis and hematopoiesis. *Proc Natl Acad Sci USA* **103**(52): 19812–7.

Nakano, T., H. Kodama, and T. Honjo. 1994. Generation of lymphohematopoietic cells from embryonic stem cells in culture. *Science* **265**(5175): 1098–101.

Nascone, N. and M. Mercola. 1995. An inductive role for the endoderm in Xenopus cardiogenesis. *Development* **121**(2): 515–23.

Neufeld, G., T. Cohen, S. Gengrinovitch et al. 1999. Vascular endothelial growth factor (VEGF) and its receptors. *FASEB J* **13**(1): 9–22.

Ng, E. S., R. P. Davis, L. Azzola et al. 2005. Forced aggregation of defined numbers of human embryonic stem cells into embryoid bodies fosters robust, reproducible hematopoietic differentiation. *Blood* **106**(5): 1601–3.

Nie, H., B. W. Soh, Y. C. Fu et al. 2008. Three-dimensional fibrous PLGA/HAp composite scaffold for BMP-2 delivery. *Biotechnol Bioeng* **99**(1): 223–34.

Nie, Y., V. Bergendahl, D. J. Hei et al. 2009. Scalable culture and cryopreservation of human embryonic stem cells on microcarriers. *Biotechnol Prog* **25**(1): 20–31.

Niebruegge, S., C. L. Bauwens, R. Peerani et al. 2009. Generation of human embryonic stem cell-derived mesoderm and cardiac cells using size-specified aggregates in an oxygen-controlled bioreactor. *Biotechnol Bioeng* **102**(2): 493–507.

Niebruegge, S., A. Nehring, H. Bar et al. 2008. Cardiomyocyte production in mass suspension culture: Embryonic stem cells as a source for great amounts of functional cardiomyocytes. *Tissue Eng Part A* **14**(10): 1591–601.

North, T. E., W. Goessling, M. Peeters et al. 2009. Hematopoietic stem cell development is dependent on blood flow. *Cell* **137**(4): 736–48.

Nostro, M. C., X. Cheng, G. M. Keller et al. 2008. Wnt, activin, and BMP signaling regulate distinct stages in the developmental pathway from embryonic stem cells to blood. *Cell Stem Cell* **2**(1): 60–71.

Nottingham, W. T., A. Jarratt, M. Burgess et al. 2007. Runx1-mediated hematopoietic stem-cell emergence is controlled by a GATA/Ets/SCL-regulated enhancer. *Blood* **110**(13): 4188–97.

Nuttelman, C. R., M. C. Tripodi and K. S. Anseth. 2004. *In vitro* osteogenic differentiation of human mesenchymal stem cells photoencapsulated in PEG hydrogels. *J Biomed Mater Res A* **68**(4): 773–82.

O'Shea, K. S. 2004. Self-renewal vs. differentiation of mouse embryonic stem cells. *Biol Reprod* **71**(6): 1755–65.

Oh, S. K., A. K. Chen, Y. Mok et al. 2009. Long-term microcarrier suspension cultures of human embryonic stem cells. *Stem Cell Res* **2**(3): 219–30.

Ohneda, O. and V. L. Bautch. 1997. Murine endothelial cells support fetal liver erythropoiesis and myelopoiesis via distinct interactions. *Br J Haematol* **98**(4): 798–808.

Ohneda, O., C. Fennie, Z. Zheng et al. 1998. Hematopoietic stem cell maintenance and differentiation are supported by embryonic aorta-gonad-mesonephros region-derived endothelium. *Blood* **92**(3): 908–19.

Oner, L. and M. J. Groves. 1993. Optimization of conditions for preparing 2- to 5-micron-range gelatin microparticles by using chilled dehydration agents. *Pharm Res* **10**(4): 621–6.

Oostendorp, R. A., K. N. Harvey, N. Kusadasi et al. 2002. Stromal cell lines from mouse aorta-gonads-mesonephros subregions are potent supporters of hematopoietic stem cell activity. *Blood* **99**(4): 1183–9.

Oostendorp, R. A., C. Robin, C. Steinhoff et al. 2005. Long-term maintenance of hematopoietic stem cells does not require contact with embryo-derived stromal cells in cocultures. *Stem Cells* **23**(6): 842–51.

Orts Llorca, F. 1963. Influence of the endoblast in the morphogenesis and late differentiation of the chick heart. *Acta Anat (Basel)* **52**: 202–14.

Ottersbach, K. and E. Dzierzak. 2005. The murine placenta contains hematopoietic stem cells within the vascular labyrinth region. *Dev Cell* **8**(3): 377–87.

Palis, J., R. J. Chan, A. Koniski et al. 2001. Spatial and temporal emergence of high proliferative potential hematopoietic precursors during murine embryogenesis. *Proc Natl Acad Sci USA* **98**(8): 4528–33.

Palis, J., S. Robertson, M. Kennedy et al. 1999. Development of erythroid and myeloid progenitors in the yolk sac and embryo proper of the mouse. *Development* **126**(22): 5073–84.

Passier, R., D. W. Oostwaard, J. Snapper et al. 2005. Increased cardiomyocyte differentiation from human embryonic stem cells in serum-free cultures. *Stem Cells* **23**(6): 772–80.

Patel, Z. S., H. Ueda, M. Yamamoto et al. 2008a. *In vitro* and *in vivo* release of vascular endothelial growth factor from gelatin microparticles and biodegradable composite scaffolds. *Pharm Res* **25**(10): 2370–8.

Patel, Z. S., M. Yamamoto, H. Ueda et al. 2008b. Biodegradable gelatin microparticles as delivery systems for the controlled release of bone morphogenetic protein-2. *Acta Biomater* **4**(5): 1126–38.

Peerani, R., B. M. Rao, C. Bauwens et al. 2007. Niche-mediated control of human embryonic stem cell self-renewal and differentiation. *Embo J* **26**(22): 4744–55.

Peerani, R. and P. W. Zandstra. 2010. Enabling stem cell therapies through synthetic stem cell-niche engineering. *J Clin Invest* **120**(1): 60–70.

Phillips, B. W., R. Horne, T. S. Lay et al. 2008. Attachment and growth of human embryonic stem cells on microcarriers. *J Biotechnol* **138**(1–2): 24–32.

Povelones, M. and R. Nusse. 2002. Wnt signalling sees spots. *Nat Cell Biol* **4**(11): E249–50.

Purpura, K. A., S. H. George, S. M. Dang et al. 2008. Soluble Flt-1 regulates Flk-1 activation to control hematopoietic and endothelial development in an oxygen-responsive manner. *Stem Cells* **26**(11): 2832–42.

Pyle, A. D., L. F. Lock and P. J. Donovan. 2006. Neurotrophins mediate human embryonic stem cell survival. *Nat Biotechnol* **24**(3): 344–50.

Raffin, M., L. M. Leong, M. S. Rones et al. 2000. Subdivision of the cardiac Nkx2.5 expression domain into myogenic and nonmyogenic compartments. *Dev Biol* **218**(2): 326–40.

Ramirez-Bergeron, D. L., A. Runge, K. D. Dahl et al. 2004. Hypoxia affects mesoderm and enhances hemangioblast specification during early development. *Development* **131**(18): 4623–34.

Ramirez-Bergeron, D. L. and M. C. Simon. 2001. Hypoxia-inducible factor and the development of stem cells of the cardiovascular system. *Stem Cells* **19**(4): 279–86.

Rich, I. N. and B. Kubanek. 1982. The effect of reduced oxygen tension on colony formation of erythropoietic cells in vitro. *Br J Haematol* **52**(4): 579–88.

Robin, C., K. Bollerot, S. Mendes et al. 2009. Human placenta is a potent hematopoietic niche containing hematopoietic stem and progenitor cells throughout development. *Cell Stem Cell* **5**(4): 385–95.

Robinson, C. J. and S. E. Stringer. 2001. The splice variants of vascular endothelial growth factor (VEGF) and their receptors. *J Cell Sci* **114**(Pt 5): 853–65.

Rust, W. L., A. Sadasivam, and N. R. Dunn. 2006. Three-dimensional extracellular matrix stimulates gastrulation-like events in human embryoid bodies. *Stem Cells Dev* **15**(6): 889–904.

Sachinidis, A., B. K. Fleischmann, E. Kolossov et al. 2003. Cardiac specific differentiation of mouse embryonic stem cells. *Cardiovasc Res* **58**(2): 278–91.

Sachlos, E. and D. T. Auguste. 2008. Embryoid body morphology influences diffusive transport of inductive biochemicals: A strategy for stem cell differentiation. *Biomaterials* **29**(34): 4471–80.

Samokhvalov, I. M., N. I. Samokhvalova, and S. Nishikawa. 2007. Cell tracing shows the contribution of the yolk sac to adult haematopoiesis. *Nature* **446**(7139): 1056–61.

Sauer, H., G. Rahimi, J. Hescheler et al. 2000. Role of reactive oxygen species and phosphatidylinositol 3-kinase in cardiomyocyte differentiation of embryonic stem cells. *FEBS Lett* **476**(3): 218–23.

Schlange, T., B. Andree, H. H. Arnold et al. 2000. BMP2 is required for early heart development during a distinct time period. *Mech Dev* **91**(1–2): 259–70.

Schmidt, J. J., J. Rowley and H. J. Kong. 2008. Hydrogels used for cell-based drug delivery. *J Biomed Mater Res A* **87**(4): 1113–22.

Schroeder, M., S. Niebruegge, A. Werner et al. 2005. Differentiation and lineage selection of mouse embryonic stem cells in a stirred bench scale bioreactor with automated process control. *Biotechnol Bioeng* **92**(7): 920–33.

Schuldiner, M., O. Yanuka, J. Itskovitz-Eldor et al. 2000. Effects of eight growth factors on the differentiation of cells derived from human embryonic stem cells. *Proc Natl Acad Sci USA* **97**(21): 11307–12.

Schultheiss, T. M., J. B. Burch, and A. B. Lassar. 1997. A role for bone morphogenetic proteins in the induction of cardiac myogenesis. *Genes Dev* **11**(4): 451–62.

Schultheiss, T. M. and A. B. Lassar. 1997. Induction of chick cardiac myogenesis by bone morphogenetic proteins. *Cold Spring Harb Symp Quant Biol* **62**: 413–9.

Schultheiss, T. M., S. Xydas, and A. B. Lassar. 1995. Induction of avian cardiac myogenesis by anterior endoderm. *Development* **121**(12): 4203–14.

Shalaby, F., J. Rossant, T. P. Yamaguchi et al. 1995. Failure of blood-island formation and vasculogenesis in Flk-1-deficient mice. *Nature* **376**(6535): 62–6.

Shi, Y., A. Hata, R. S. Lo et al. 1997. A structural basis for mutational inactivation of the tumour suppressor Smad4. *Nature* **388**(6637): 87–93.

Sitnicka, E., N. Lin, G. V. Priestley et al. 1996. The effect of thrombopoietin on the proliferation and differentiation of murine hematopoietic stem cells. *Blood* **87**(12): 4998–5005.

Sivakumar, M. and K. P. Rao. 2003. Preparation, characterization, and *in vitro* release of gentamicin from coralline hydroxyapatite-alginate composite microspheres. *J Biomed Mater Res A* **65**(2): 222–8.

Snyder, A., S. T. Fraser and M. H. Baron. 2004. Bone morphogenetic proteins in vertebrate hematopoietic development. *J Cell Biochem* **93**(2): 224–32.

St-Jacques, B. and A. P. McMahon. 1996. Early mouse development: Lessons from gene targeting. *Curr Opin Genet Dev* **6**(4): 439–44.

Steiner, D., H. Khaner, M. Cohen et al. 2010. Derivation, propagation and controlled differentiation of human embryonic stem cells in suspension. *Nat Biotechnol* **28**(4): 361–4.

Sugi, Y. and J. Lough. 1994. Anterior endoderm is a specific effector of terminal cardiac myocyte differentiation of cells from the embryonic heart forming region. *Dev Dyn* **200**(2): 155–62.

Takahashi, K. and S. Yamanaka. 2006. Induction of pluripotent stem cells from mouse embryonic and adult fibroblast cultures by defined factors. *Cell* **126**(4): 663–76.

Takeuchi, M., T. Sekiguchi, T. Hara et al. 2002. Cultivation of aorta-gonad-mesonephros-derived hematopoietic stem cells in the fetal liver microenvironment amplifies long-term repopulating activity and enhances engraftment to the bone marrow. *Blood* **99**(4): 1190–6.

Tavian, M., C. Robin, L. Coulombel et al. 2001. The human embryo, but not its yolk sac, generates lympho-myeloid stem cells: Mapping multipotent hematopoietic cell fate in intraembryonic mesoderm. *Immunity* **15**(3): 487–95.

Thomson, J. A., J. Itskovitz-Eldor, S. S. Shapiro et al. 1998. Embryonic stem cell lines derived from human blastocysts. *Science* **282**(5391): 1145–7.

Thomson, J. A., J. Kalishman, T. G. Golos et al. 1995. Isolation of a primate embryonic stem cell line. *Proc Natl Acad Sci USA* **92**(17): 7844–8.

Tian, X., J. K. Morris, J. L. Linehan et al. 2004. Cytokine requirements differ for stroma and embryoid body-mediated hematopoiesis from human embryonic stem cells. *Exp Hematol* **32**(10): 1000–9.

Ueno, S., G. Weidinger, T. Osugi et al. 2007. Biphasic role for Wnt/beta-catenin signaling in cardiac specification in zebrafish and embryonic stem cells. *Proc Natl Acad Sci USA* **104**(23): 9685–90.

Umaoka, Y., Y. Noda, K. Narimoto et al. 1991. Developmental potentiality of embryos cultured under low oxygen tension with superoxide dismutase. *J In Vitro Fert Embryo Transf* **8**(5): 245–9.

Ungrin, M. D., C. Joshi, A. Nica et al. 2008. Reproducible, ultra high-throughput formation of multicellular organization from single cell suspension-derived human embryonic stem cell aggregates. *PLoS One* **3**(2): e1565.

van der Pol, L. and J. Tramper. 1998. Shear sensitivity of animal cells from a culture-medium perspective. *Trends Biotechnol* **16**(8): 323–8.

Van Handel, B., S. L. Prashad, N. Hassanzadeh-Kiabi et al. 2010. The first trimester human placenta is a site for terminal maturation of primitive erythroid cells. *Blood* **116**(17): 3321–30.

Vandervoort, J. and A. Ludwig. 2004. Preparation and evaluation of drug-loaded gelatin nanoparticles for topical ophthalmic use. *Eur J Pharm Biopharm* **57**(2): 251–61.

Vodyanik, M. A., J. A. Bork, J. A. Thomson et al. 2005. Human embryonic stem cell-derived CD34+ cells: Efficient production in the coculture with OP9 stromal cells and analysis of lymphohematopoietic potential. *Blood* **105**(2): 617–26.

Wakefield, L. M. and A. B. Roberts. 2002. TGF-beta signaling: Positive and negative effects on tumorigenesis. *Curr Opin Genet Dev* **12**(1): 22–9.

Wang, J., H. P. Zhao, G. Lin et al. 2005. *In vitro* hematopoietic differentiation of human embryonic stem cells induced by co-culture with human bone marrow stromal cells and low dose cytokines. *Cell Biol Int* **29**(8): 654–61.

Weitzer, G. 2006. Embryonic stem cell-derived embryoid bodies: An *in vitro* model of eutherian pregastrulation development and early gastrulation. *Handb Exp Pharmacol* (174): 21–51.

Wrenzycki, C., D. Herrmann, L. Keskintepe et al. 2001. Effects of culture system and protein supplementation on mRNA expression in pre-implantation bovine embryos. *Hum Reprod* 16(5): 893–901.

Xu, C., J. Q. He, T. J. Kamp et al. 2006. Human embryonic stem cell-derived cardiomyocytes can be maintained in defined medium without serum. *Stem Cells Dev* 15(6): 931–41.

Xu, C., M. S. Inokuma, J. Denham et al. 2001. Feeder-free growth of undifferentiated human embryonic stem cells. *Nat Biotechnol* 19(10): 971–4.

Xu, C., S. Police, N. Rao et al. 2002. Characterization and enrichment of cardiomyocytes derived from human embryonic stem cells. *Circ Res* 91(6): 501–8.

Xu, M. J., K. Tsuji, T. Ueda et al. 1998. Stimulation of mouse and human primitive hematopoiesis by murine embryonic aorta-gonad-mesonephros-derived stromal cell lines. *Blood* 92(6): 2032–40.

Xu, X. Q., R. Graichen, S. Y. Soo et al. 2008. Chemically defined medium supporting cardiomyocyte differentiation of human embryonic stem cells. *Differentiation* 76(9): 958–70.

Yagi, M., K. A. Ritchie, E. Sitnicka et al. 1999. Sustained *ex vivo* expansion of hematopoietic stem cells mediated by thrombopoietin. *Proc Natl Acad Sci USA* 96(14): 8126–31.

Yamaguchi, T. P. 2001. Heads or tails: Wnts and anterior-posterior patterning. *Curr Biol* 11(17): R713–24.

Yang, F., C. G. Williams, D. A. Wang et al. 2005. The effect of incorporating RGD adhesive peptide in polyethylene glycol diacrylate hydrogel on osteogenesis of bone marrow stromal cells. *Biomaterials* 26(30): 5991–8.

Yang, L., M. H. Soonpaa, E. D. Adler et al. 2008b. Human cardiovascular progenitor cells develop from a KDR+ embryonic-stem-cell-derived population. *Nature* 453(7194): 524–8.

Yang, S., H. Cai, H. Jin et al. 2008a. Hematopoietic reconstitution of CD34+ cells grown in static and stirred culture systems in NOD/SCID mice. *Biotechnol Lett* 30(1): 61–5.

Yoshimoto, M. and M. C. Yoder. 2009. Developmental biology: Birth of the blood cell. *Nature* 457(7231): 801–3.

Yu, X. and R. V. Bellamkonda. 2003. Tissue-engineered scaffolds are effective alternatives to autografts for bridging peripheral nerve gaps. *Tissue Eng* 9(3): 421–30.

Zandstra, P. W., C. Bauwens, T. Yin et al. 2003. Scalable production of embryonic stem cell-derived cardiomyocytes. *Tissue Eng* 9(4): 767–78.

Zeigler, B. M., D. Sugiyama, M. Chen et al. 2006. The allantois and chorion, when isolated before circulation or chorio-allantoic fusion, have hematopoietic potential. *Development* 133(21): 4183–92.

Zhang, J., C. Niu, L. Ye et al. 2003. Identification of the haematopoietic stem cell niche and control of the niche size. *Nature* 425(6960): 836–41.

Zovein, A. C., J. J. Hofmann, M. Lynch et al. 2008. Fate tracing reveals the endothelial origin of hematopoietic stem cells. *Cell Stem Cell* 3(6): 625–36.

zur Nieden, N. I., J. T. Cormier, D. E. Rancourt et al. 2007. Embryonic stem cells remain highly pluripotent following long term expansion as aggregates in suspension bioreactors. *J Biotechnol* 129(3): 421–32.

2

Cell Mechanobiology in Regenerative Medicine: Lessons from Cancer

Badriprasad
Ananthanarayanan
University of California, Berkeley

Sanjay Kumar
University of California, Berkeley

2.1 Introduction

The stem cell "niche" refers to the collective set of cell-extrinsic inputs that controls the functions of stem cells *in vivo*.[1,2] The key regulatory mechanisms within the niche include presentation of soluble and immobilized molecules such as growth factors and cytokines, direct interactions with other cells (e.g., stromal cells), and adhesion to the extracellular matrix (ECM). These diverse inputs are regulated and integrated in a temporally and spatially dynamic fashion to control self-renewal and differentiation, the two hallmark properties of stem cells. Traditionally, the field has approached this subject from a paradigm that is largely biochemical in nature, focusing on the regulatory roles of soluble and membrane-bound ligands on stem cell behavior. While it is clear that these inputs are indeed important, it is also increasingly being recognized that mechanical and other types of biophysical interactions between cells with their extracellular milieu can profoundly influence stem cell behavior. This idea is an extension of a broader awareness that many cell types can sense and apply forces to their surroundings,[3] and that the mechanical interactions of cells with their environment are critical regulators of function in physiology and disease, a concept now widely referred to as "cellular mechanobiology."[4,5] Early efforts in this area have demonstrated that, similar to other cell types in tissue, stem cells are also influenced by mechanical forces and that biophysical signaling can control stem cell self-renewal and differentiation.[6–8] These effects are mediated by intracellular signaling pathways that transduce force cues into biochemical signals that in turn drive fundamental cellular processes such as cell adhesion, motility, proliferation, and differentiation.[9,10]

Despite the growing interest in the mechanobiology of stem cells, our understanding of how these effects may be incorporated into a broader understanding of stem cell biology or leveraged to enhance stem cell-based therapies remains very limited. In addition, the mechanistic details of force transduction processes in stem cells are still incompletely understood. By contrast, there is a comparatively more advanced literature on the effects of mechanical signaling on a variety of other non-stem cells. In particular, it is now well accepted that dysfunctional interactions between cells and their ECM play a significant role in the initiation and progression of some solid tumors,[11,12] and that mechanical forces can influence malignant transformation, migration, and proliferation of cancer cells in culture.[13,14] Several recent studies have illuminated the role of mechanical signaling from native and engineered ECMs in the initiation and spread of cancer, such as malignant transformation,[15,16] migration,[17] and proliferation.[18] Indeed, it is possible to conceptualize the various stages in the progression of cancer in the form of a "force journey" in which mechanical interactions with the environment influence cellular behavior in concert with genetic and epigenetic cues.[13] This raises the possibility that one might draw upon an understanding of tumor cell mechanobiology to formulate instructive analogies to stem cell mechanobiology, and that this in turn might offer important clues about mechanisms and therapeutic applications.

While the biology of cancer and that of stem cells may appear at first sight to be unrelated, there are in fact several important similarities (Figure 2.1). First, many of the molecular mechanisms known to process force cues are not unique to tumor cells and indeed are critical to the function of many normal cell types, including stem and progenitor cells. These include integrin-mediated adhesion to the ECM, establishment and stabilization of cell structure by the cytoskeleton, generation of cell–ECM tractional forces by actomyosin complexes, and regulation of cytoskeletal assembly and mechanics by Rho-family GTPases.[10,19] Second, many of the processes that contribute to tumor growth, such as cell motility, ECM remodeling, and assembly of angiogenic vessels, are often critical to the success of tissue engineering and regenerative medicine strategies.[80] Finally, the hallmark ability of stem cells to undergo either self-renewal or differentiation bears direct mechanistic relevance to tumors inasmuch as tumor growth

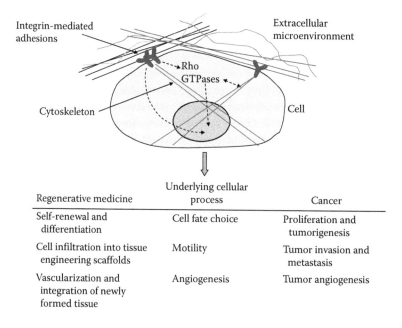

Regenerative medicine	Underlying cellular process	Cancer
Self-renewal and differentiation	Cell fate choice	Proliferation and tumorigenesis
Cell infiltration into tissue engineering scaffolds	Motility	Tumor invasion and metastasis
Vascularization and integration of newly formed tissue	Angiogenesis	Tumor angiogenesis

FIGURE 2.1 Similarities between cell–ECM mechanobiology of stem cells and cancer. A cell in its native microenvironment receives biophysical and biochemical inputs through integrin-mediated adhesions, initiating signaling cascades that direct the architecture and dynamics of the cellular cytoskeleton and in turn influence cellular contractility and force generation. These reciprocal relationships ultimately result in transcriptional programs effected by proteins such as the Rho family GTPases, thereby governing cell fate, motility, and angiogenesis. These fundamental cellular processes underlie phenomena of interest in regenerative medicine as well as in cancer.

frequently reflects profound dysregulation of cell-cycle progression, proliferation, differentiation, and death. This analogy has recently been articulated in a more literal way through the discovery of a privileged population of "cancer stem cells" within certain tumors, which often bear striking similarities to endogenous tissue stem cells.[20–24] The cancer stem cell concept argues that a subpopulation of cells within the tumor mass is largely responsible for sustaining tumor growth through continuous self-renewal, and that this process may be arrested if these cells can be directed toward an alternative fate choice (e.g., death or differentiation).

In this chapter, we seek to explore mechanistic and functional connections between tumor cell mechanobiology and stem cell mechanobiology, with the goal of using the former to guide understanding of the latter. We begin with a brief overview of the mechanobiology of stem cells and the molecular mechanisms that mediate the effects of mechanical signaling. We then focus on the mechanobiology of three critical cellular processes that have historically been investigated in the context of cancer but are equally applicable to stem cell biology: proliferation, motility, and angiogenesis. Finally, we offer a perspective on biomaterial systems that can enable investigation of stem cell and cancer mechanobiology in three-dimensional (3D) topologies, which are an important feature of many native tissue environments and are increasingly recognized to be critical to *in vivo* cell behavior.

2.2 Stem Cell Mechanobiology

Mammalian tissues exhibit a wide range of mechanical properties, ranging from soft tissues such as brain and fat to hard tissue such as cartilage and bone. In fact, there are often significant mechanical heterogeneities within a single tissue, as observed within the hippocampus of the brain.[25] The presence of these mechanical heterogeneities within the *in vivo* niche begs the question of whether they give rise to signals that can directly or indirectly modulate stem cell behavior, and this has recently begun to be addressed with the use of culture systems based on natural or synthetic polymeric matrices.[26–28] These material systems can be engineered to exhibit a wide range of elastic moduli, in contrast to traditionally used glass or plastic surfaces which are many orders of magnitude stiffer than most physiological tissues.

Several excellent reviews have covered the effects of mechanical signaling on stem cell fate,[6–8] so we will limit our focus to a few particularly illustrative examples. Dynamic mechanical loading is widely observed for mature tissues in the musculoskeletal system and vasculature, but has also been observed to be important in the early stages of development.[29] For example, application of force to the *Drosophila* embryo induces expression of *twist*, a gene central to the regulation of germ-layer formation and patterning.[30] Similarly, tensile forces in the cell cortex can promote the sorting of progenitor cells and organization of germ layers in the gastrulating zebrafish embryo.[31] At the cellular level, direct force application promotes myogenesis over adipogenesis in lung embryonic mesenchymal stem cells (MSCs),[32] downregulates pluripotency markers in mouse embryonic stem cells (mESCs),[33] and inhibits differentiation of human embryonic stem cells (hESCs).[34] Similarly, forces associated with shear flow, which have long been understood to be critical for the normal function of vascular endothelial and smooth muscle cells, are now recognized to also control the differentiation of stem cells into cardiovascular lineages[35] and the development of hematopoietic stem cells.[36,37]

The mechanical properties of the microenvironment have been shown to affect stem cell differentiation in dramatic ways even in the absence of directly applied forces. For example, when MSCs are shape-constrained through the use of micropatterned ECM islands and cultured in media permissive of multiple lineages, cells forced to adopt rounded shapes preferentially undergo adipogenesis, whereas cells allowed to spread more fully preferentially undergo osteogenesis.[38] Further, when MSCs are cultured on ECMs of varying stiffness under similar permissive media conditions, softer substrates (0.1–1 kPa) induce neurogenic differentiation, stiffer (8–17 kPa) substrates promote muscle formation, while the stiffest (25–40 kPa) substrates produce bone cells.[39] In other words, MSCs appear to differentiate into tissue types whose stiffness approximates that of the underlying ECM. In both cases, inhibition of actomyosin contractility abrogates ECM stiffness-dependent differences in MSC differentiation. More recently, ECM stiffness has been shown to regulate the proliferation of MSCs, with softer substrates inducing a quiescent state but not compromising the ability of cells to resume proliferation when transferred to stiff ECMs or to differentiate when treated with the appropriate factors.[40] Mechanosensitivity of stem cell differentiation has also been reported for tissues commonly regarded as protected from large external forces, such as the brain. For example, neural stem cells (NSCs) from the adult rat hippocampus differentiate optimally into neurons on soft substrates (~10 Pa), with stiffer substrates (~10 kPa) increasing glial differentiation.[41] This trend has subsequently been observed for hippocampal NSCs encapsulated in 3D alginate scaffolds[42] and for NSCs derived from other regions of the central nervous system,[43,44] although the precise relationship appears to depend on the tissue and the species source and the ECM ligand.

While the mechanistic details of the above effects remain to be completely elucidated, a large number of proteins and protein complexes have been implicated in the processing of force signals. The primary force sensors are often located in the plasma membrane—for example, G-protein-coupled receptors,[45] ion channels,[46,47] and integrins.[48] Indeed, the mechanosensitive growth and maturation of focal adhesions into structured complexes that contain a variety of cytoskeletal and signaling proteins represents one of the most important and well-studied ECM-mediated signaling pathways.[19,49,50] Another important class of proteins is the Rho family of GTPases, whose canonical members Rho, Rac, and Cdc42 serve as key control points for cytoskeletal assembly and dynamics.[51–53] These pathways directly influence the extent and nature of cell-generated forces, in part by regulating the assembly of actin stress fibers and bundles as well as the phosphorylation of nonmuscle myosin motor proteins that drive contraction of these structures.[54] Rho family proteins and actomyosin contractility have also been shown to mediate the mechanosensitive differentiation of MSCs and NSCs.[38,39,128] Together, these mechanosensitive pathways may contribute to the regulation of gene expression via transcription factors[55] as well as other indirect or epigenetic pathways[56] to direct, restrict, or impose selective pressure on stem cell fate choices.

2.3 Mechanobiology of Cell Proliferation

Self-renewal, the process by which a cell divides to generate daughter cells with developmental potentials that are indistinguishable from those of the mother cell, is one of the hallmark features of stem cells.[57] In other words, self-renewal involves mobilization of processes that promote proliferation concurrent with inhibition of differentiation into a less proliferative or terminally differentiated cell type. The factors that affect self-renewal of stem cells from different tissues and at different stages of development continue to be elucidated.[58] However, it is clearly recognized that the niche plays a central role in the maintenance of stem cells *in vivo*. It has been suggested that the subversion of these normal maintenance signals from the niche is one of the mechanisms through which cancer stem cells gain unlimited proliferative capacities.[59] Indeed, many of the signaling networks that are known to be essential for the self-renewal of stem cells, including the Notch, Wnt, and Hedgehog pathways, were originally identified as oncogenes based on their role in tumor formation.[60,61] This intimate connection between stemness and the proliferative properties of cancer raises the possibility that mechanisms identified as oncogenic in cancer might also facilitate stem cell self-renewal. Before exploring commonalities in signaling between the mechanobiology of tumor cell proliferation and the mechanobiology of stem cell self-renewal, we will discuss potential mechanisms that may underlie the mechanosensitivity of stem cell self-renewal.

There is evidence that some of the pathways that regulate self-renewal are sensitive to mechanical forces. The Wnt pathway is known to be important for the physiological adaptation of bone mass and structure to mechanical loading.[62] Both pulsatile fluid flow[63] and mechanical strain[64] have been shown to activate the Wnt/β-catenin pathway in bone cells, which results in nuclear translocation of β-catenin and increased proliferation. This pathway has also been implicated in tumorigenesis[65] and in controlling self-renewal of stem cells.[66] Similarly, mechanical forces have been shown to induce the expression of proteins of the Hedgehog family in smooth muscle cells[67] and chondrocytes.[68] The mechanosensitivity of these pathways has not yet been explored in the context of stem cell self-renewal.

There is a significant body of evidence supporting the role of mechanical forces in controlling proliferation, and it is becoming clear that several of these effects are communicated through cell–matrix focal adhesions. As we described earlier, these structures serve as organizing centers for both mechanotransductive and mitogenic signaling elements and grow and mature upon application of force. For example, focal adhesion kinase (FAK),[69] extracellular-signal-regulated kinase (ERK), and kinases of the Src family strongly promote proliferation and are all known to localize to focal adhesions.[49] Further, the Rho GTPases, previously mentioned for their role in organizing the cellular cytoskeleton, also play a direct role in controlling cell-cycle progression.[53,70,71] The effect of mechanical signaling on cell-cycle control was tested directly in a recent study in which cells from various tissues were cultured on variable-stiffness ECMs.[72] Compliant ECMs that mimic physiological tissue stiffness inhibited progression through the cell cycle (Figure 2.2), but highly stiff ECMs that mimic the stiffening associated with pathological matrix remodeling accelerated cell-cycle progression through various mechanisms including a FAK-Rac-cyclin D1 pathway. Rho GTPases have been shown to mediate the mechanosensitivity of mesenchymal stem cell differentiation in response to matrix elasticity[39] and cell shape.[38] Thus, mechanosensitive pathways known to be important in cancer and other cells may have direct roles in establishing self-renewal or directing differentiation.

Seminal work by Bissell and colleagues established that the tumor microenvironment plays a critical role in the formation and spread of tumors.[11,12,73] Later, Wang and colleagues showed that the stiffness of the ECM regulates the proliferative ability of normal cells, but that malignant transformation decreases this sensitivity to ECM mechanics, possibly allowing for anchorage-independent and uncontrolled proliferation.[74] This observation is reminiscent of the classical soft agar assay, in which cells are judged to be successfully transformed if they develop an ability to proliferate on soft, nonadhesive ECMs. The hypothesis that mechanics can mediate malignant transformation was tested directly in a landmark

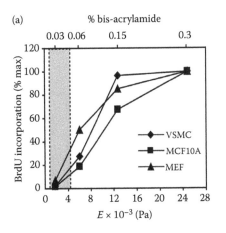
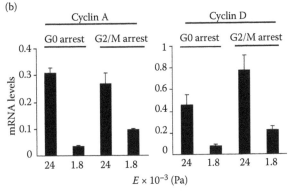

FIGURE 2.2 Mechanobiological control of cell-cycle progression. In this study, the effect of substrate stiffness on cell-cycle progression was assessed in mouse embryonic fibroblasts (MEFs), vascular smooth muscle cells (VSMCs), and MCF10A mammary epithelial cells. (a) Effect of substrate stiffness on cell proliferation. Increasing substrate stiffness results in a greater fraction of BrdU-positive cells for all cell types upon stimulation with mitogens. The shaded area highlights the range of elastic moduli measured in mouse mammary glands and arteries (data not shown). (b) Effect of substrate stiffness on expression of cell-cycle checkpoint genes. MEFs were synchronized at G0 (by 48 h serum starvation) or at G2/M (by treatment with 5 mg/mL nocodazole for 24 h) and then reseeded on hydrogels of varying stiffness and stimulated with 10% fetal bovine serum (FBS). Induction of cyclin A and cyclin D1 expression depended strongly on matrix stiffness regardless of whether cells entered G1 phase from G0 or G2/M, with higher stiffness substrates promoting increased cell-cycle progression. (Reproduced with permission from Klein, E.A. et al. *Current Biology* 2009, 19(18), 1511–1518.)

study by Weaver and colleagues, who showed that culturing nontumorigenic mammary epithelial cells on ECMs of tumor-like stiffness induces dysplasia, proliferation, and activation of oncogenic signaling pathways.[16] The recent finding that breast tumorigenesis is accompanied by crosslinking and stiffening of the collagenous matrix even in premalignant tissue verifies that this phenomenon is relevant to tumorigenesis *in vivo*. These effects are mediated by increased signaling through integrins and focal adhesions, and may be suppressed by the inhibition of lysyl oxidase (LOX).[15] A complementary set of studies with breast epithelial tumor cells in 3D collagen matrices has also elucidated the role of FAK, ERK, and Rho in the promotion of a proliferative and invasive phenotype in response to increased collagen density.[75,76] Our laboratory recently tested the link between ECM stiffness and the pathophysiology of malignant brain tumors *in vitro*.[18] When we cultured human glioblastoma multiforme (GBM) cells on

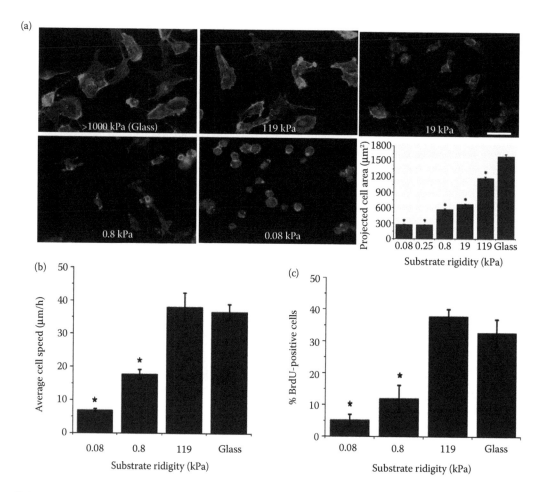

FIGURE 2.3 **(See color insert.)** Mechanobiological control of glioma cell behavior. The effect of mechanics on the morphology, motility, and proliferation of U373 MG glioblastoma multiforme tumor cells was assessed by plating cells on variable-stiffness polyacrylamide substrates coated with fibronectin. (a) Effect on cell morphology and adhesion. Cell morphology shows a steep dependence on substrate stiffness, with cells spreading extensively and forming well-defined focal adhesions and stress fibers on glass or stiff substrates, but not on softer substrates. Immunofluorescence images depict nuclear DNA (blue), F-actin (green), and the proliferation marker Ki67 (red). (b) Effect on motility. Increasing substrate stiffness increases the speed of random cell migration. (c) Effect on proliferation. Substrate stiffness also influences proliferation, with a greater fraction of BrdU-positive cells seen on stiffer substrates. (Reproduced with permission from Ulrich, T.A. et al. *Cancer Research* 2009, 69(10), 4167–4174.)

variable-stiffness ECMs coated with fibronectin, we found that ECM stiffness strongly regulates cellular morphology, motility, and proliferation (Figure 2.3). Increasing ECM stiffness resulted in a higher fraction of dividing cells, as determined by bromodeoxyuridine (BrdU) incorporation. Thus, proliferative signals generated by mechanosensitive pathways have been shown to influence the formation and progression of cancer, and bear investigation in the context of stem cell self-renewal. Mechanoregulation of self-renewal is important not just in niche-mediated maintenance of adult stem cell populations, but also for engineering stem cell therapies where control of cell fate is essential.

2.4 Mechanobiology of Cell Motility

Cell motility is a fundamental process that contributes to development, tissue homeostasis, wound healing, and a wide variety of pathological processes.[77,78] In embryonic development, movements of single cells and multicellular sheets contribute to segregation and patterning and establish the highly specified architecture of developing tissues.[77] Migration of progenitor cell populations is essential in tissues that undergo continuous regeneration during adult life such as the skin, intestinal epithelia, and the brain, where large-scale movements of neural progenitors along defined paths are observed.[78] Further, cell migration is essential during all phases of tissue repair and regeneration, including recruitment of leukocytes as part of the inflammatory response, reentry of cells into the wound area, and revascularization of the tissue.[79] Similarly, cell migration is essential for the success of regenerative therapies such as scaffold-based tissue engineering.[27,80] Indeed, cell infiltration into the scaffold has long been recognized as an important consideration in the design of tissue engineering scaffolds. This has spurred a significant interest in optimizing pore size within scaffolds, for example, for bone tissue engineering,[81] to allow sufficient cell penetration without compromising bulk mechanical properties. Similarly, significant efforts have been devoted to the development of synthetic matrices that can be proteolytically degraded by migrating cells.[82–84] Despite these advances in scaffold engineering, the field could benefit from a greater understanding of the mechanisms that govern cell motility in synthetic ECMs to efficiently design tissue engineering scaffolds for regenerative medicine.

Cell migration on two-dimensional (2D) substrates has been described as a physically integrated molecular process in which the cell undergoes cycles consisting of morphological polarization and membrane extension, attachment at the leading edge, contraction of the cell body, and finally detachment of the trailing edge.[85] In this mode of motility, known as mesenchymal motility, the cell must be able to physically exert force on the substratum through cell–matrix adhesions. This depends not only on the strength of these adhesions[86] but also on the mechanical compliance of the substrate, which determines the response to cell-applied forces. It has now been clearly established that the migration speed of a variety of cell types depends on the elasticity of the underlying substrate.[87–89] For example, we recently showed that the average speed of random migration of glioma cells significantly increases when the matrix stiffness is increased (Figure 2.3).[18] This trend was also observed for glioma cells cultured on variable-stiffness hydrogels composed of hyaluronic acid, thereby extending our previous observations to a brain-mimetic ECM platform.[129] Inhibition of nonmuscle myosin II-based contractility ablates this stiffness sensitivity and rescues motility on soft substrates, indicating a tight balance between protrusive and contractile forces within cells. The phenomenon of "durotaxis" describes cell motion in response to variations in substrate stiffness, with many cell types displaying a trend to migrate toward stiffer regions.[90,91] Therefore, engineering the mechanical properties of the matrix may enable better infiltration of stem cells into scaffolds for tissue engineering applications.

Several novel insights into the mechanisms of cell migration have been deduced from recent studies on tumor invasion and metastasis.[92,93] Perhaps the most intriguing of these is the recognition that tumor cells can exhibit several different modes of motility, differing not only in their average speeds but also in their requirement for cell–ECM adhesions, contractile force generation, and ECM remodeling via proteolysis. As tumor cells invade the surrounding matrix, they often exhibit mesenchymal motility, which is typically accompanied by pericellular proteolysis by secreted and membrane-associated enzymes such as matrix metalloproteases (MMPs). These enzymes can degrade the surrounding matrix to clear steric barriers against migration. However, in the absence of proteolytic abilities, or when proteolysis is specifically blocked by pharmacological agents, tumor cells have been observed to switch to an "amoeboid" form of motility in which cells depend primarily on contractile forces generated by the actomyosin cytoskeleton to extrude themselves through existing pores and channels in the ECM.[94–96] Amoeboid motility is often viewed as independent of protease activity and the strength of cell–matrix adhesions, permitting tumor cells to escape strategies directed against mesenchymal motility. These findings have obvious clinical relevance in therapies targeting cancer metastasis, but they are also relevant for tissue engineering. The exact nature of stem cell motility in tissue engineering scaffolds will dictate whether strong cell–ECM adhesions are required, or whether degradability by cellular proteases is an important design requirement. Further investigation of these questions should facilitate the formulation of more precise strategies for engineering stem cell behavior in synthetic scaffolds.

2.5 Mechanobiology of Angiogenesis

Vascularization is crucial for the viability of engineered tissue replacements.[97] The therapeutic potential of stem cells in medicine hinges on the ability to generate functional replacements of diseased cell types in the body; however, the efficacy of any stem cell-based therapy will ultimately depend on the extent of vascularization, innervation, and functional integration of the newly formed tissue. Since oxygen and nutrient supply and waste removal depend critically on the vasculature,[98] angiogenesis represents an important step in the success of regenerative therapies using stem cells. It is not surprising, therefore, that a significant amount of work in the development of scaffolds for tissue engineering has been focused on the controlled delivery of growth factors that promote angiogenesis.[27,99,100] Although soluble signaling via growth factors from the vascular-endothelial growth factor (VEGF)[101] and angiopoietin[102] families represent the primary mechanisms governing angiogenesis in mammalian tissue, it has also been recognized that solid-state biochemical and physical signals from the ECM play an important role.[103,104]

Angiogenesis is also clearly an important step in the progression of cancer.[105,106] As a tumor grows and spreads, it outstrips the capacity of diffusion to supply the oxygen and nutrients needed for continued proliferation and expansion. Some tumors acquire the ability to circumvent this limitation by directing the host vasculature to extend new blood vessels. This "angiogenic switch" has received increasing attention in recent years as a potential point for therapeutic intervention to limit the growth of tumors. Indeed, antiangiogenic interventions such as a monoclonal antibody against VEGF (e.g., bevacizumab, commercially marketed as Avastin) have shown clinical success in the treatment of colorectal cancer in combination with chemotherapy.[107] These successes have spurred interest in the diverse mechanisms that promote angiogenesis, including the role played by ECM-mediated mechanical signaling.[104]

Initial work in the mechanobiology of angiogenesis concerned the effects of mechanical signaling on the growth of endothelial cells. For example, it was found that fibronectin density governs cell shape and cell fate, directing proliferation when cells are spread on high fibronectin density substrates, but triggering apoptosis on rounded cells on low-density substrates.[108] The connection between cell shape and cell fate was established conclusively in a landmark study by Ingber, Whitesides, and colleagues, who used microcontact-printed fibronectin ECMs to control cell shape independently of matrix density and soluble factors, and showed that cell shape can independently drive proliferation, differentiation, and death.[109] Further work has focused on the development of microvasculature, such as the formation and structure of capillary networks, as a function of ECM density and stiffness. For example, it has been shown that the density of the collagenous matrix in which endothelial cells are cultured influences their ability to form branched capillaries with small lumens, resembling those found *in vivo*.[110] Similarly, the density of fibrin matrix surrounding endothelial cells cultured on beads has been shown to govern the extent of capillary network formation.[111] Both these results implicate cellular force generation due to actomyosin contractility as an important process through which cells sense and respond to mechanical forces in their environment. In addition to these angiogenic effects, mechanical signaling is also known to be important in force-dependent neovascularization via enlargement and elongation of existing blood vessels. For example, in an *in vivo* model of wound healing, neovascularization was found to depend on the ability of cells to stress and contract the collagenous matrix.[112]

Mechanistically, the transduction of mechanical force into angiogenic signals is known to partly follow the canonical routes of force transduction outlined previously, including the generation of cytoskeletal tension through the actomyosin apparatus and the activity of GTPases such as Rho.[113] In addition, it has recently been discovered that there may be direct crosstalk between force-mediated signaling and the classical VEGF signaling pathways that govern angiogenesis. In a recent study, it was determined that p190RhoGAP, an endogenous inhibitor of Rho GTPase activation, controls capillary network formation both *in vitro* and *in vivo* by sequestering transcription factors that govern sensitivity to VEGF via expression of the VEGFR2 receptor gene.[55] Further, p190RhoGAP activity may be decreased by increasing the stiffness of the substrate, resulting in increased Rho activation as well as promotion of VEGFR2 gene expression and VEGF-based angiogenesis. Thus, study of the mechanobiology of angiogenesis has revealed several interesting regulatory effects and their mechanisms. These studies can inform the design of material scaffolds and clinical protocols, which, by promoting angiogenesis and vascularization, might enable better integration of stem cell-derived engineered tissues *in vivo*.

2.6 Perspective: Three-Dimensional Material Systems for Investigating Mechanobiology

A large amount of the existing knowledge on cell–ECM interactions has been derived from *in vitro* studies using cells cultured on 2D surfaces. Although these studies have revealed a great deal about the mechanisms of cell adhesion, migration, and force transduction, it is becoming increasingly recognized that cells in their native 3D ECM exhibit behavior that is distinct from that seen in 2D.[114,115] For instance, cell–matrix adhesions in 3D display strikingly different morphology, effects on matrix organization, and protein recruitment patterns compared to those observed in 2D.[116] These fundamental differences in cell–ECM contacts result in a functionally different behavioral phenotype for cells in 3D matrices. This fact has been recognized for the last two decades in the context of the formation and growth of tumors,[11,117,118] and is beginning to be apparent in the context of stem cell self-renewal and differentiation. For example, hESCs cultured in a medium conditioned by fibroblast feeders were shown to undergo self-renewal in 3D scaffolds of crosslinked hyaluronic acid, but not on 2D surfaces of the same material.[119] Similarly, directed differentiation of mESCs into hematopoietic lineages has been shown to be more efficient in 3D culture.[120] Since mechanical communication between cells and the ECM is largely channeled through cell–ECM adhesions, it follows that force sensing and transduction and the concomitant effects on cellular physiology should also depend on the dimensionality of the matrix.[121] For example, we recently delineated the effects of one important aspect of 3D culture—cellular confinement in narrow spaces—by building a novel microfabricated polyacrylamide gel system, where tumor cells confined within narrow channels migrated faster than in wide channels or on flat surfaces of the same ECM stiffness, due to more efficient polarization of cell-generated traction forces.[130] Therefore, it is essential that cell–ECM mechanical signaling be explored in physiologically relevant 3D models.

Traditional approaches to study cell–ECM biology in 3D have focused on natural ECM proteins that form gels under physiological conditions, for example, collagen I and Matrigel. While these materials do partially recapitulate the rich biochemical milieu to which cells are exposed in native environments, they offer a fairly limited range of mechanical properties. Further, the mechanics, microstructure, and biochemistry of these gels are intimately linked, in that changing the bulk density of the gel-forming proteins simultaneously varies all the above properties, making it difficult to attribute observed differences in cell behavior unambiguously to chemical or mechanical stimuli. Further, many of these native biomaterials are inappropriate for stem cell-based regenerative medicine, because they are typically derived from animal sources and therefore suffer from batch-to-batch variability and pose unacceptable risks with respect to pathogenicity and immunogenicity. Therefore, there has been a significant drive toward the development of semisynthetic and synthetic 3D model ECMs that can be used to study cancer and stem cell biology and might potentially be appropriate for therapeutic use.[27,28,122–124] Several synthetic polymer systems have been developed that can be crosslinked to varying extents, and by inclusion of full-length proteins or short peptides, can mimic the native ECM and also permit independent variation of matrix stiffness and adhesive functionality. Taking a cue from the recent tissue engineering efforts,[125,126] we recently developed a system for studying cell–matrix mechanobiology in 3D based on mixtures of collagen I and agarose, a biologically inert polysaccharide that forms a filamentous meshwork and serves to stiffen collagen gels with modest effects on their fibrous architecture.[127] This hybrid system allows the study of cell mechanobiology in 3D while uncoupling the effects of matrix structure and mechanics from biochemistry. Studies of invasion of spheroids of glioma cells implanted in these gels revealed that increasing agarose concentrations created increasingly stiff gels but progressively slowed and eventually stopped invasion. This result was somewhat surprising, given that increasing stiffness was found to increase glioma cell motility on 2D surfaces (Figure 2.3). However, it appears that steric barriers created by the agarose meshwork present an obstacle to cell migration in 3D and limit the ability of the cells to contract and remodel collagen fibers, combining to prevent glioma invasion.[131] This study illustrates clearly that some aspects of cellular behavior, such as the dependence of motility on the porosity and degradability of the matrix, can only be captured in 3D environments. Therefore, the development of material systems that can increasingly mimic native 3D ECM while retaining independent control of various design parameters such as stiffness, porosity, biochemical functionality, and degradability is crucial for facilitating studies on the mechanobiology of stem cells and cancer.

2.7 Conclusions

Biophysical interactions of stem cells with the extracellular milieu in their native niches as well as in engineered tissue constructs represent an important class of inputs governing cell behavior. Some of the mechanisms by which cells detect and process these inputs are conserved among many cell types, including stem cells, normal cells, and tumor cells. Therefore, a comparative study of these mechanisms may allow us to leverage our knowledge of the mechanobiology of normal cells and cancer to accelerate our understanding of the processes that control stem cell fate and design more effective strategies for regenerative medicine.

Acknowledgments

We apologize to the many authors whose work could not be cited because of space limitations. Sanjay Kumar wishes to acknowledge the support of a UC Berkeley Stem Cell Center Seed Grant, the Arnold and Mabel Beckman Young Investigator Award, an NSF Research Award (CMMI-0727420), an NIH Physical Sciences in Oncology Center Grant (1U54CA143836), and the NIH Director's New Innovator Award (1DP2OD004213)—a part of the NIH Roadmap for Medical Research.

References

1. Fuchs, E.; Tumbar, T.; Guasch, G., Socializing with the neighbors: Stem cells and their niche. *Cell* 2004, 116(6), 769–778.
2. Moore, K. A.; Lemischka, I. R., Stem cells and their niches. *Science* 2006, 311(5769), 1880–1885.
3. Discher, D. E.; Janmey, P.; Wang, Y. L., Tissue cells feel and respond to the stiffness of their substrate. *Science* 2005, 310(5751), 1139–1143.
4. Ingber, D. E., Mechanobiology and diseases of mechanotransduction. *Annals of Medicine* 2003, 35(8), 564–577.
5. Vogel, V.; Sheetz, M., Local force and geometry sensing regulate cell functions. *Nature Reviews Molecular Cell Biology* 2006, 7(4), 265–275.

6. Keung, A. J.; Healy, K. E.; Kumar, S.; Schaffer, D. V., Biophysics and dynamics of natural and engineered stem cell microenvironments. *Wiley Interdisciplinary Reviews: Systems Biology and Medicine* 2009, 2(1), 49–64.

7. Discher, D. E.; Mooney, D. J.; Zandstra, P. W., Growth factors, matrices, and forces combine and control stem cells. *Science* 2009, 324(5935), 1673–1677.

8. Guilak, F.; Cohen, D. M.; Estes, B. T.; Gimble, J. M.; Liedtke, W.; Chen, C. S., Control of stem cell fate by physical interactions with the extracellular matrix. *Cell Stem Cell* 2009, 5(1), 17–26.

9. Ingber, D. E., Cellular mechanotransduction: Putting all the pieces together again. *FASEB J.* 2006, 20(7), 811–827.

10. Chen, C. S., Mechanotransduction—a field pulling together? *Journal of Cell Science* 2008, 121(20), 3285–3292.

11. Bissell, M. J.; Radisky, D., Putting tumours in context. *Nature Reviews Cancer* 2001, 1(1), 46–54.

12. Nelson, C. M.; Bissell, M. J., Of extracellular matrix, scaffolds, and signaling: Tissue architecture regulates development, homeostasis, and cancer. *Annual Review of Cell and Developmental Biology* 2006, (22), 287–309.

13. Kumar, S.; Weaver, V., Mechanics, malignancy, and metastasis: The force journey of a tumor cell. *Cancer and Metastasis Reviews* 2009, 28(1), 113–127.

14. Butcher, D. T.; Alliston, T.; Weaver, V. M., A tense situation: Forcing tumour progression. *Nature Reviews Cancer* 2009, 9(2), 108–122.

15. Levental, K. R.; Yu, H. M.; Kass, L.; Lakins, J. N.; Egeblad, M.; Erler, J. T.; Fong, S. F. T. et al., Matrix crosslinking forces tumor progression by enhancing integrin signaling. *Cell* 2009, 139(5), 891–906.

16. Paszek, M. J.; Zahir, N.; Johnson, K. R.; Lakins, J. N.; Rozenberg, G. I.; Gefen, A.; Reinhart-King, C. A. et al., Tensional homeostasis and the malignant phenotype. *Cancer Cell* 2005, 8(3), 241–254.

17. Zaman, M. H.; Trapani, L. M.; Sieminski, A. L.; MacKellar, D.; Gong, H.; Kamm, R. D.; Wells, A.; Lauffenburger, D. A.; Matsudaira, P., Migration of tumor cells in 3D matrices is governed by matrix stiffness along with cell-matrix adhesion and proteolysis. *Proceedings of the National Academy of Sciences of the United States of America* 2006, 103(29), 10889–10894.

18. Ulrich, T. A.; de Juan Pardo, E. M.; Kumar, S., The mechanical rigidity of the extracellular matrix regulates the structure, motility, and proliferation of glioma cells. *Cancer Research* 2009, 69(10), 4167–4174.

19. Geiger, B.; Spatz, J. P.; Bershadsky, A. D., Environmental sensing through focal adhesions. *Nature Reviews Molecular Cell Biology* 2009, 10(1), 21–33.

20. Al-Hajj, M.; Wicha, M. S.; Benito-Hernandez, A.; Morrison, S. J.; Clarke, M. F., Prospective identification of tumorigenic breast cancer cells. *Proceedings of the National Academy of Sciences of the United States of America* 2003, 100(7), 3983–3988.

21. Reya, T.; Morrison, S. J.; Clarke, M. F.; Weissman, I. L., Stem cells, cancer, and cancer stem cells. *Nature* 2001, 414(6859), 105–111.

22. Singh, S. K.; Hawkins, C.; Clarke, I. D.; Squire, J. A.; Bayani, J.; Hide, T.; Henkelman, R. M.; Cusimano, M. D.; Dirks, P. B., Identification of human brain tumour initiating cells. *Nature* 2004, 432(7015), 396–401.

23. Visvader, J. E.; Lindeman, G. J., Cancer stem cells in solid tumours: Accumulating evidence and unresolved questions. *Nature Reviews Cancer* 2008, 8(10), 755–768.

24. Bonnet, D.; Dick, J. E., Human acute myeloid leukemia is organized as a hierarchy that originates from a primitive hematopoietic cell. *Nature Medicine* 1997, 3(7), 730–737.

25. Elkin, B. S.; Azeloglu, E. U.; Costa, K. D.; Morrison, B., Mechanical heterogeneity of the rat hippocampus measured by atomic force microscope indentation. *Journal of Neurotrauma* 2007, 24(5), 812–822.

26. Saha, K.; Pollock, J. F.; Schaffer, D. V.; Healy, K. E., Designing synthetic materials to control stem cell phenotype. *Current Opinion in Chemical Biology* 2007, 11(4), 381–387.

27. Lutolf, M. P.; Hubbell, J. A., Synthetic biomaterials as instructive extracellular microenvironments for morphogenesis in tissue engineering. *Nature Biotechnology* 2005, 23(1), 47–55.

28. Lutolf, M. P.; Gilbert, P. M.; Blau, H. M., Designing materials to direct stem-cell fate. *Nature* 2009, 462(7272), 433–441.

29. Ghosh, K.; Ingber, D. E., Micromechanical control of cell and tissue development: Implications for tissue engineering. *Advanced Drug Delivery Reviews* 2007, 59(13), 1306–1318.

30. Farge, E., Mechanical induction of twist in the Drosophila foregut/stomodeal primordium. *Current Biology* 2003, 13(16), 1365–1377.

31. Krieg, M.; Arboleda-Estudillo, Y.; Puech, P. H.; Kafer, J.; Graner, F.; Muller, D. J.; Heisenberg, C. P., Tensile forces govern germ-layer organization in zebrafish. *Nature Cell Biology* 2008, 10(4), 429–436.

32. Yang, Y.; Beqaj, S.; Kemp, P.; Ariel, I.; Schuger, L., Stretch-induced alternative splicing of serum response factor promotes bronchial myogenesis and is defective in lung hypoplasia. *Journal of Clinical Investigation* 2000, 106(11), 1321–1330.

33. Chowdhury, F.; Na, S.; Li, D.; Poh, Y.C.; Tanaka, T. S.; Wang, F.; Wang, N., Material properties of the cell dictate stress-induced spreading and differentiation in embryonic stem cells. *Nature Materials* 2010, 9(1), 82–88.

34. Saha, S.; Lin, J.; De Pablo, J. J.; Palecek, S. P., Inhibition of human embryonic stem cell differentiation by mechanical strain. *Journal of Cellular Physiology* 2006, 206(1), 126–137.

35. Illi, B.; Scopece, A.; Nanni, S.; Farsetti, A.; Morgante, L.; Biglioli, P.; Capogrossi, M. C.; Gaetano, C., Epigenetic histone modification and cardiovascular lineage programming in mouse embryonic stem cells exposed to laminar shear stress. *Circulation Research* 2005, 96(5), 501–508.

36. North, T. E.; Goessling, W.; Peeters, M.; Li, P. L.; Ceol, C.; Lord, A. M.; Weber, G. J. et al., Hematopoietic stem cell development is dependent on blood flow. *Cell* 2009, 137(4), 736–748.

37. Adamo, L.; Naveiras, O.; Wenzel, P. L.; McKinney-Freeman, S.; Mack, P. J.; Gracia-Sancho, J.; Suchy-Dicey, A. et al., Biomechanical forces promote embryonic haematopoiesis. *Nature* 2009, 459(7250), 1131–U120.

38. McBeath, R.; Pirone, D. M.; Nelson, C. M.; Bhadriraju, K.; Chen, C. S., Cell shape, cytoskeletal tension, and RhoA regulate stem cell lineage commitment. *Developmental Cell* 2004, 6(4), 483–495.

39. Engler, A. J.; Sen, S.; Sweeney, H. L.; Discher, D. E., Matrix elasticity directs stem cell lineage specification. *Cell* 2006, 126(4), 677–689.

40. Winer, J. P.; Janmey, P. A.; McCormick, M. E.; Funaki, M., Bone marrow-derived human mesenchymal stem cells become quiescent on soft substrates but remain responsive to chemical or mechanical stimuli. *Tissue Engineering Part A* 2009, 15(1), 147–154.

41. Saha, K.; Keung, A. J.; Irwin, E. F.; Li, Y.; Little, L.; Schaffer, D. V.; Healy, K. E., Substrate modulus directs neural stem cell behavior. *Biophysical Journal* 2008, 95(9), 4426–4438.

42. Banerjee, A.; Arha, M.; Choudhary, S.; Ashton, R. S.; Bhatia, S. R.; Schaffer, D. V.; Kane, R. S., The influence of hydrogel modulus on the proliferation and differentiation of encapsulated neural stem cells. *Biomaterials* 2009, 30(27), 4695–4699.

43. Seidlits, S. K.; Khaing, Z. Z.; Petersen, R. R.; Nickels, J. D.; Vanscoy, J. E.; Shear, J. B.; Schmidt, C. E., The effects of hyaluronic acid hydrogels with tunable mechanical properties on neural progenitor cell differentiation. *Biomaterials* 2010, 31(14), 3930–3940.

44. Leipzig, N. D.; Shoichet, M. S., The effect of substrate stiffness on adult neural stem cell behavior. *Biomaterials* 2009, 30(36), 6867–6878.

45. Chachisvilis, M.; Zhang, Y. L.; Frangos, J. A., G protein-coupled receptors sense fluid shear stress in endothelial cells. *Proceedings of the National Academy of Sciences of the United States of America* 2006, 103(42), 15463–15468.

46. Martinac, B., Mechanosensitive ion channels: Molecules of mechanotransduction. *Journal of Cell Science* 2004, 117(12), 2449–2460.

47. Kung, C., A possible unifying principle for mechanosensation. *Nature* 2005, 436(7051), 647–654.

48. Katsumi, A.; Orr, A. W.; Tzima, E.; Schwartz, M. A., Integrins in mechanotransduction. *Journal of Biological Chemistry* 2004, 279(13), 12001–12004.

49. Berrier, A. L.; Yamada, K. M., Cell-matrix adhesion. *Journal of Cellular Physiology* 2007, 213(3), 565–573.

50. Chen, C. S.; Tan, J.; Tien, J., Mechanotransduction at cell-matrix and cell-cell contacts. *Annual Review of Biomedical Engineering* 2004, 6, 275–302.

51. Nobes, C. D.; Hall, A., Rho, Rac, and Cdc42 GTPases regulate the assembly of multimolecular focal complexes associated with actin stress fibers, lamellipodia, and filopodia. *Cell* 1995, 81(1), 53–62.

52. Burridge, K.; Wennerberg, K., Rho and Rac take center stage. 2004, 116(2), 167–179.

53. Etienne-Manneville, S.; Hall, A., Rho GTPases in cell biology. *Nature* 2002, 420(6916), 629–635.

54. Pellegrin, S.; Mellor, H., Actin stress fibres. *Journal of Cell Science* 2007, 120(20), 3491–3499.

55. Mammoto, A.; Connor, K. M.; Mammoto, T.; Yung, C. W.; Huh, D.; Aderman, C. M.; Mostoslavsky, G.; Smith, L. E. H.; Ingber, D. E., A mechanosensitive transcriptional mechanism that controls angiogenesis. *Nature* 2009, 457(7233), 1103–U57.

56. Le Beyec, J.; Xu, R.; Lee, S. Y.; Nelson, C. M.; Rizki, A.; Alcaraz, J.; Bissell, M. J., Cell shape regulates global histone acetylation in human mammary epithelial cells. *Experimental Cell Research* 2007, 313(14), 3066–3075.

57. Molofsky, A. V.; Pardal, R.; Morrison, S. J., Diverse mechanisms regulate stem cell self-renewal. *Current Opinion in Cell Biology* 2004, 16(6), 700–707.

58. Morrison, S. J.; Spradling, A. C., Stem cells and niches: Mechanisms that promote stem cell maintenance throughout life. *Cell* 2008, 132(4), 598–611.

59. Li, L. H.; Neaves, W. B., Normal stem cells and cancer stem cells: The niche matters. *Cancer Research* 2006, 66(9), 4553–4557.

60. Pardal, R.; Clarke, M. F.; Morrison, S. J., Applying the principles of stem-cell biology to cancer. *Nature Reviews Cancer* 2003, 3(12), 895–902.

61. Al-Hajj, M.; Clarke, M. F., Self-renewal and solid tumor stem cells. *Oncogene* 2004, 23(43), 7274–7282.

62. Bonewald, L. F.; Johnson, M. L., Osteocytes, mechanosensing and Wnt signaling. *Bone* 2008, 42(4), 606–615.

63. Santos, A.; Bakker, A. D.; Zandieh-Doulabi, B.; de Blieck-Hogervorst, J. M. A.; Klein-Nulend, J., Early activation of the beta-catenin pathway in osteocytes is mediated by nitric oxide, phosphatidyl inositol-3 kinase/Akt, and focal adhesion kinase. *Biochemical and Biophysical Research Communications* 2010, 391(1), 364–369.

64. Armstrong, V. J.; Muzylak, M.; Sunters, A.; Zaman, G.; Saxon, L. K.; Price, J. S.; Lanyon, L. E., Wnt/beta-catenin signaling is a component of osteoblastic bone cell early responses to load-bearing and requires estrogen receptor alpha. *Journal of Biological Chemistry* 2007, 282(28), 20715–20727.

65. Taipale, J.; Beachy, P. A., The Hedgehog and Wnt signaling pathways in cancer. *Nature* 2001, 411(6835), 349–354.

66. Reya, T.; Duncan, A. W.; Ailles, L.; Domen, J.; Scherer, D. C.; Willert, K.; Hintz, L.; Nusse, R.; Weissman, I. L., A role for Wnt signalling in self-renewal of haematopoietic stem cells. *Nature* 2003, 423(6938), 409–414.

67. Morrow, D.; Sweeney, C.; Birney, Y. A.; Guha, S.; Collins, N.; Cummins, P. M.; Murphy, R.; Walls, D.; Redmond, E. M.; Cahill, P. A., Biomechanical regulation of hedgehog signaling in vascular smooth muscle cells *in vitro* and *in vivo*. *American Journal of Physiology-Cell Physiology* 2007, 292(1), C488–C496.

68. Wu, Q. Q.; Zhang, Y.; Chen, Q., Indian hedgehog is an essential component of mechanotransduction complex to stimulate chondrocyte proliferation. *Journal of Biological Chemistry* 2001, 276(38), 35290–35296.

69. Zhao, J.; Guan, J. L., Signal transduction by focal adhesion kinase in cancer. *Cancer and Metastasis Reviews* 2009, 28(1–2), 35–49.

70. Welsh, C. F.; Roovers, K.; Villanueva, J.; Liu, Y. Q.; Schwartz, M. A.; Assoian, R. K., Timing of cyclin D1 expression within G1 phase is controlled by Rho. *Nature Cell Biology* 2001, 3(11), 950–957.

71. Olson, M. F.; Ashworth, A.; Hall, A., An essential role for Rho, Rac, and Cdc42 GTPases in cell-cycle progression through G(1). *Science* 1995, 269(5228), 1270–1272.

72. Klein, E. A.; Yin, L. Q.; Kothapalli, D.; Castagnino, P.; Byfield, F. J.; Xu, T. N.; Leventhal, I.; Hawthorne, E.; Janmey, P. A.; Assoian, R. K., Cell-cycle control by physiological matrix elasticity and *in vivo* tissue stiffening. *Current Biology* 2009, 19(18), 1511–1518.

73. Boudreau, N.; Sympson, C. J.; Werb, Z.; Bissell, M. J., Suppression of ICE and apoptosis in mammary epithelial cells by extracellular matrix. *Science* 1995, 267(5199), 891–893.

74. Wang, H. B.; Dembo, M.; Wang, Y. L., Substrate flexibility regulates growth and apoptosis of normal but not transformed cells. *American Journal of Physiology-Cell Physiology* 2000, 279(5), C1345–C1350.

75. Provenzano, P. P.; Inman, D. R.; Eliceiri, K. W.; Keely, P. J., Matrix density-induced mechanoregulation of breast cell phenotype, signaling and gene expression through a FAK-ERK linkage. *Oncogene* 2009, 28(49), 4326–4343.

76. Provenzano, P. P.; Inman, D. R.; Eliceiri, K. W.; Knittel, J. G.; Yan, L.; Rueden, C. T.; White, J. G.; Keely, P. J., Collagen density promotes mammary tumor initiation and progression. *BMC Medicine* 2008, 6, 11.

77. Locascio, A.; Nieto, M. A., Cell movements during vertebrate development: Integrated tissue behaviour versus individual cell migration. *Current Opinion in Genetics & Development* 2001, 11(4), 464–469.

78. Hatten, M. E., Central nervous system neuronal migration. *Annual Review of Neuroscience* 1999, 22, 511–539.

79. Martin, P., Wound healing—aiming for perfect skin regeneration. *Science* 1997, 276(5309), 75–81.

80. Griffith, L. G.; Naughton, G., Tissue engineering—current challenges and expanding opportunities. *Science* 2002, 295(5557), 1009–1014.

81. Karageorgiou, V.; Kaplan, D., Porosity of 3D biomaterial scaffolds and osteogenesis. *Biomaterials* 2005, 26(27), 5474–5491.

82. Luo, Y.; Shoichet, M. S., A photolabile hydrogel for guided three-dimensional cell growth and migration. *Nature Materials* 2004, 3(4), 249–253.

83. Lutolf, M. P.; Lauer-Fields, J. L.; Schmoekel, H. G.; Metters, A. T.; Weber, F. E.; Fields, G. B.; Hubbell, J. A., Synthetic matrix metalloproteinase-sensitive hydrogels for the conduction of tissue regeneration: Engineering cell-invasion characteristics. *Proceedings of the National Academy of Sciences of the United States of America* 2003, 100(9), 5413–5418.

84. Mann, B. K.; Gobin, A. S.; Tsai, A. T.; Schmedlen, R. H.; West, J. L., Smooth muscle cell growth in photopolymerized hydrogels with cell adhesive and proteolytically degradable domains: Synthetic ECM analogs for tissue engineering. *Biomaterials* 2001, 22(22), 3045–3051.

85. Lauffenburger, D. A.; Horwitz, A. F., Cell migration: A physically integrated molecular process. *Cell* 1996, 84(3), 359–369.

86. Palecek, S. P.; Loftus, J. C.; Ginsberg, M. H.; Lauffenburger, D. A.; Horwitz, A. F., Integrin-ligand binding properties govern cell migration speed through cell-substratum adhesiveness. *Nature* 1997, 385(6616), 537–540.

87. Pelham, R. J.; Wang, Y. L., Cell locomotion and focal adhesions are regulated by substrate flexibility. *Proceedings of the National Academy of Sciences of the United States of America* 1997, 94(25), 13661–13665.

88. Peyton, S. R.; Putnam, A. J., Extracellular matrix rigidity governs smooth muscle cell motility in a biphasic fashion. *Journal of Cellular Physiology* 2005, 204(1), 198–209.

89. Engler, A.; Bacakova, L.; Newman, C.; Hategan, A.; Griffin, M.; Discher, D., Substrate compliance versus ligand density in cell on gel responses. *Biophysical Journal* 2004, 86(1), 617–628.

90. Lo, C. M.; Wang, H. B.; Dembo, M.; Wang, Y. L., Cell movement is guided by the rigidity of the substrate. *Biophysical Journal* 2000, 79(1), 144–152.

91. Wong, J. Y.; Velasco, A.; Rajagopalan, P.; Pham, Q., Directed movement of vascular smooth muscle cells on gradient-compliant hydrogels. *Langmuir* 2003, 19(5), 1908–1913.

92. Friedl, P.; Wolf, K., Tumour-cell invasion and migration: Diversity and escape mechanisms. *Nature Reviews Cancer* 2003, 3(5), 362–374.

93. Sahai, E., Mechanisms of cancer cell invasion. *Current Opinion in Genetics & Development* 2005, 15(1), 87–96.

94. Sahai, E.; Marshall, C. J., Differing modes of tumour cell invasion have distinct requirements for Rho/ROCK signalling and extracellular proteolysis. *Nature Cell Biology* 2003, 5(8), 711–719.

95. Wolf, K.; Mazo, I.; Leung, H.; Engelke, K.; von Andrian, U. H.; Deryugina, E. I.; Strongin, A. Y.; Brocker, E. B.; Friedl, P., Compensation mechanism in tumor cell migration: Mesenchymal-amoeboid transition after blocking of pericellular proteolysis. *Journal of Cell Biology* 2003, 160(2), 267–277.

96. Friedl, P.; Wolf, K., Proteolytic interstitial cell migration: A five-step process. *Cancer and Metastasis Reviews* 2009, 28(1–2), 129–135.

97. Laschke, M. W.; Harder, Y.; Amon, M.; Martin, I.; Farhadi, J.; Ring, A.; Torio-Padron, N. et al., Angiogenesis in tissue engineering: Breathing life into constructed tissue substitutes. *Tissue Engineering* 2006, 12(8), 2093–2104.

98. Carmeliet, P., Angiogenesis in health and disease. *Nature Medicine* 2003, 9(6), 653–660.

99. Zisch, A. H.; Lutolf, M. P.; Hubbell, J. A., Biopolymeric delivery matrices for angiogenic growth factors. *Cardiovascular Pathology* 2003, 12(6), 295–310.

100. Richardson, T. P.; Peters, M. C.; Ennett, A. B.; Mooney, D. J., Polymeric system for dual growth factor delivery. *Nature Biotechnology* 2001, 19(11), 1029–1034.

101. Ferrara, N.; Gerber, H. P.; LeCouter, J., The biology of VEGF and its receptors. *Nature Medicine* 2003, 9(6), 669–676.

102. Davis, S.; Yancopoulos, G. D., The angiopoietins: Yin and yang in angiogenesis. In *Vascular Growth Factors and Angiogenesis*, Lena Claesson-Welsh (Ed.), 1999; Vol. 237, pp. 173–185.

103. Ingber, D. E.; Folkman, J., How does extracellular-matrix control capillary morphogenesis. *Cell* 1989, 58(5), 803–805.

104. Ingber, D. E., Mechanical signalling and the cellular response to extracellular matrix in angiogenesis and cardiovascular physiology. *Circulation Research* 2002, 91(10), 877–887.

105. Carmeliet, P.; Jain, R. K., Angiogenesis in cancer and other diseases. *Nature* 2000, 407(6801), 249–257.

106. Folkman, J.; Bach, M.; Rowe, J. W.; Davidoff, F.; Lambert, P.; Hirsch, C.; Goldberg, A.; Hiatt, H. H.; Glass, J.; Henshaw, E., Tumor angiogenesis—therapeutic implications. *New England Journal of Medicine* 1971, 285(21), 1182–1186

107. Ferrara, N., Vascular endothelial growth factor: Basic science and clinical progress. *Endocrine Reviews* 2004, 25(4), 581–611.

108. Ingber, D. E., Fibronectin controls capillary endothelial-cell growth by modulating cell-shape. *Proceedings of the National Academy of Sciences of the United States of America* 1990, 87(9), 3579–3583.

109. Chen, C. S.; Mrksich, M.; Huang, S.; Whitesides, G. M.; Ingber, D. E., Geometric control of cell life and death. *Science* 1997, 276(5317), 1425–1428.

110. Sieminski, A. L.; Hebbel, R. P.; Gooch, K. J., The relative magnitudes of endothelial force generation and matrix stiffness modulate capillary morphogenesis *in vitro*. *Experimental Cell Research* 2004, 297(2), 574–584.

111. Kniazeva, E.; Putnam, A. J., Endothelial cell traction and ECM density influence both capillary morphogenesis and maintenance in 3-D. *American Journal of Physiology-Cell Physiology* 2009, 297(1), C179–C187.

112. Kilarski, W. W.; Samolov, B.; Petersson, L.; Kvanta, A.; Gerwins, P., Biomechanical regulation of blood vessel growth during tissue vascularization. *Nature Medicine* 2009, 15(6), 657–664.

113. Moore, K. A.; Polte, T.; Huang, S.; Shi, B.; Alsberg, E.; Sunday, M. E.; Ingber, D. E., Control of basement membrane remodeling and epithelial branching morphogenesis in embryonic lung by Rho and cytoskeletal tension. *Developmental Dynamics* 2005, 232(2), 268–281.

114. Yamada, K. M.; Cukierman, E., Modeling tissue morphogenesis and cancer in 3D. *Cell* 2007, 130(4), 601–610.

115. Cukierman, E.; Pankov, R.; Yamada, K. M., Cell interactions with three-dimensional matrices. *Current Opinion in Cell Biology* 2002, 14(5), 633–639.

116. Cukierman, E.; Pankov, R.; Stevens, D. R.; Yamada, K. M., Taking cell-matrix adhesions to the third dimension. *Science* 2001, 294(5547), 1708–1712.

117. Jacks, T.; Weinberg, R. A., Taking the study of cancer cell survival to a new dimension. *Cell* 2002, 111(7), 923–925.
118. Debnath, J.; Brugge, J. S., Modelling glandular epithelial cancers in three-dimensional cultures. *Nature Reviews Cancer* 2005, 5(9), 675–688.
119. Gerecht, S.; Burdick, J. A.; Ferreira, L. S.; Townsend, S. A.; Langer, R.; Vunjak-Novakovic, G., Hyaluronic acid hydrogel for controlled self-renewal and differentiation of human embryonic stem cells. *Proceedings of the National Academy of Sciences of the United States of America* 2007, 104(27), 11298–11303.
120. Liu, H.; Roy, K., Biomimetic three-dimensional cultures significantly increase hematopoietic differentiation efficacy of embryonic stem cells. *Tissue Engineering* 2005, 11(1–2), 319–330.
121. Pedersen, J. A.; Swartz, M. A., Mechanobiology in the third dimension. *Annals of Biomedical Engineering* 2005, 33(11), 1469–1490.
122. Griffith, L. G.; Swartz, M. A., Capturing complex 3D tissue physiology *in vitro*. *Nature Reviews Molecular Cell Biology* 2006, 7(3), 211–224.
123. Hutmacher, D. W., Biomaterials offer cancer research the third dimension. *Nature Materials* 2010, 9(2), 90–93.
124. Tibbitt, M. W.; Anseth, K. S., Hydrogels as extracellular matrix mimics for 3D cell culture. *Biotechnology and Bioengineering* 2009, 103(4), 655–663.
125. Balgude, A. P.; Yu, X.; Szymanski, A.; Bellamkonda, R. V., Agarose gel stiffness determines rate of DRG neurite extension in 3D cultures. *Biomaterials* 2001, 22(10), 1077–1084.
126. Batorsky, A.; Liao, J. H.; Lund, A. W.; Plopper, G. E.; Stegemann, J. P., Encapsulation of adult human mesenchymal stem cells within collagen-agarose microenvironments. *Biotechnology and Bioengineering* 2005, 92(4), 492–500.
127. Ulrich, T. A.; Jain, A.; Tanner, K.; MacKay, J. L.; Kumar, S., Probing cellular mechanobiology in three-dimensional culture with collagen-agarose matrices. *Biomaterials* 2010, 31(7), 1875–1884.
128. Keung, A. J.; de Juan Pardo, E. M.; Schaffer, D. V.; Kumar, S., Rho GTPases mediate the mechanosensitive lineage commitment of neural stem cells. *Stem Cells* 2011, 29(11), 1886–1897.
129. Ananthanarayanan, B.; Kim, Y.; Kumar, S., Elucidating the mechanobiology of malignant brain tumors using a brain matrix-mimetic hyaluronic acid hydrogel platform. *Biomaterials* 2011, 32(31), 7913–7923.
130. Pathak, A.; Kumar, S., Independent regulation of tumor cell migration by matrix stiffness and confinement. *Proceedings of the National Academy of Sciences* 2012, 109(26), 10334–10339.
131. Ulrich, T. A.; Lee, T. G.; Shon, H. K.; Moon, D. W.; Kumar, S., Microscale mechanisms of agarose-induced disruption of collagen remodeling. *Biomaterials* 2011, 32(24), 5633–5642.

3

Systems-Engineering Principles in Signal Transduction and Cell-Fate Choice

Karin J. Jensen
University of Virginia

Anjun K. Bose
University of Virginia

Kevin A. Janes
University of Virginia

3.1 Introduction

Tissue, organs, and organisms arise because of the choices made by individual cells during development (Gilbert 2006). Stem cells and other progenitors are constantly faced with decisions about when and whether to proliferate, differentiate, or die (Weissman 2000). Proper cell-fate decisions are critical for tissue morphogenesis and physiology (Meier et al. 2000; Srivastava 2006). Inappropriate decision-making leads to developmental abnormalities, degenerative diseases, and cancer (Chien and Karsenty 2005; Hanahan and Weinberg 2000; Zelzer and Olsen 2003).

How do cells choose one fate or another? Sometimes, a cell is "hard-wired" to make predetermined decisions based on the genes and proteins that it expresses (Ingham et al. 1991). But more commonly, the local cellular environment provides "cues" that influence which choice a cell will make (Figure 3.1a). Cells respond to a remarkable array of cues: diffusible proteins, metabolites and other small molecules, matrix proteins, mechanical forces, radiation, osmolarity, and many others. These diverse stimuli are presented in a time-dependent and combinatorial manner (Janes et al. 2005), which further complicates the challenge of interpreting a set of environmental cues and responding correctly.

The process by which cells receive these inputs and relay information to the cell interior is called signal transduction (Downward 2001). Signal transduction occurs via a large group of cellular enzymes and binding proteins, which became interconnected during evolution to form networks (Jordan et al. 2000). Unlike genome sequences and certain subcellular proteomes (Au et al. 2007), signaling networks cannot be exhaustively characterized (Albeck et al. 2006). The state of a cell's network depends on the environmental cues, the time since these cues were introduced, and the particular wiring of signaling proteins in that cell type. Therefore, just as the blueprint of one chemical factory does not immediately

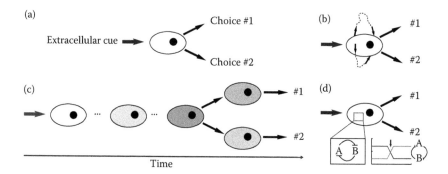

FIGURE 3.1 **(See color insert.)** Recurring systems-level themes in cell-fate decisions. (a) Cell signaling connects extracellular cues with cellular decision-making. Cues are transduced through the cell by networks of signaling pathways (not shown), which together coordinate cell-fate choices. (b) Autocrine circuits are a component of cell-signal processing. Regulated release of intrinsic autocrine cues provides microenvironment-dependent feedback to reinforce cell fate. (c) Signaling dynamics allow time-dependent evolution of cell state toward key decision points. The rate and trajectory of these signaling events ultimately determine the cell-fate choice. (d) Modeling the role of network topology in cell decisions. Certain signaling architectures, such as double-negative feedback (shown), can give rise to bistable networks that can flip in response to a cue (red).

allow an engineer to make predictions about refineries in general, the specifics of one signaling network do not readily translate to all others.

Despite this challenge, there are recurring themes in signal transduction that are shared by many networks. For example, signaling proteins are highly modular in function and are comprised of evolutionarily conserved domains that provide binding and enzymatic properties (Pawson 2004). This is analogous to engineered networks—every chemical refinery may be different, but each is built from the same fundamental parts list (chillers, evaporators, reactors, etc.).

Several excellent reviews of signaling-protein modules are available in the literature (Seet et al. 2006; Yaffe 2002; Yaffe and Elia 2001); therefore, we do not cover the topic here. We instead take a higher-level view in this chapter, focusing on systems principles of signaling networks that have repeatedly been shown to impact cell-fate choice. Two biological themes stood out because of their recognized importance and complexity: (1) inducible autocrine–paracrine factors and (2) signaling dynamics, and we dedicate sections to each (Figure 3.1b and c). At the end of the chapter, we discuss how computational models have played an important role in defining network-level principles of signal transduction (Figure 3.1d). Many of the published examples we discuss are theoretical. However, as techniques for measuring signal-transduction networks become more advanced, we predict that models will soon become essential for understanding signaling data more fully (Janes and Yaffe 2006b). Only by the combination of modeling and experiment do we stand a chance at understanding the coordination of cell choice.

3.2 Autocrine and Paracrine Signaling in Cell-Fate Determination

Extracellular stimuli initiate cascades of intracellular signaling activity. The binding of an extracellular ligand to its target cell-surface receptor activates intracellular signaling pathways that ultimately dictate the cell's response. Signals are transduced along enzymatic pathways through a complex network of interactions from the cell surface to the nucleus, leading to changes in gene transcription and cell behavior.

Cells receive extracellular cues from ligands circulating in the bloodstream as well as those released from neighboring cells. Autocrine signaling occurs when the cell secretes a ligand that eventually binds to a receiver on the same cell's surface to initiate downstream intracellular signaling (Figure 3.1b). A major source of crosstalk between signal transduction pathways is through extracellular autocrine signaling (Lauffenburger et al. 1995). Autocrine circuits can provide positive or negative feedback to individual transduction pathways. Perturbations in autocrine feedback can greatly affect the magnitude and duration of signaling across multiple pathways, ultimately influencing cell-fate determination (Janes et al. 2006a).

Cell-fate choices are prompted by the initial fast activation of signaling cascades within seconds to minutes and then reinforced by the slow evolution of cell state from hours to days. In general, the early steps in this progression are posttranslational and the late steps are predominantly transcriptional. Autocrine signaling lies at the interface between fast and slow timescales (minutes to hours) and provides a mechanism for bridging them. For example, many ligands induce the transcription of autocrine factors that can then induce a second phase of signaling (Busch et al. 2008; Janes et al. 2006a; Schulze et al. 2004).

Nevertheless, this disparity of timescales must be reconciled to define how signaling networks should be measured if the goal is to predict cellular outcomes. Theoretical studies have shown that in complex systems with multiple timescales, the long-term system behavior can be approximated by the slowest developing element. Busch and coworkers applied this principle to the long-term decision-making of keratinocytes in response to the migration-inducing cue, the hepatocyte growth factor (HGF) (Busch et al. 2008). Cell migration occurs through reversible protein modifications, macromolecular assembly/disassembly cycles, and gene transcription (Lauffenburger and Horwitz 1996). The timescale over which a cell decides to migrate or remain at rest is several hours; therefore, the slowest evolving process for such a choice is gene expression. Indeed, the authors found that HGF-induced migration requires new gene transcription because inhibition with actinomycin D treatment blocked the keratinocyte choice to migrate. Transcriptional profiling of HGF-stimulated keratinocytes revealed that expression of the epidermal growth factor (EGF)-family ligand *HBEGF* was strongly and specifically upregulated (Busch et al. 2008). The authors went on to show that HBEGF-induced mitogen-activated protein kinase (MAPK) activation through the EGF receptor (EGFR) was essential for achieving a sustained migratory response to HGF stimulation ($t > 1.5$ h).

Cells are exposed to a cocktail of extracellular stimuli *in vivo* that simultaneously activate multiple pathways and act synergistically or antagonistically. Autocrine factors add to this cocktail, and it has recently been found that they can reconfigure the mix of extracellular stimuli over time. Janes et al. studied the time-dependent release of autocrine factors in the response of epithelial cells to tumor necrosis factor (TNF) (Janes et al. 2006a), an inflammatory cytokine that induces both pro- and anti-apoptotic signals. The authors found that TNF stimulation leads to the secretion of three autocrine agents: transforming growth factor-α (TGF-α), interleukin-1α (IL-1α), and IL-1 receptor antagonist (IL-1ra) (Figure 3.2). The induced autocrine signaling is sequential, with TGF-α released within minutes and contributing to early MAPK activation, IL-1α released late to cause sustained activation of the nuclear factor-κB (NF-κB) pathway, and IL-1ra released thereafter. Importantly, the authors also found an underlying logic that interconnected the different autocrine factors: TNF-induced TGF-α was required for the subsequent release of IL-1α, and IL-1ra fed back negatively on cells to limit the duration of autocrine IL-1α signaling. The combined effects of the different autocrine factors ultimately shaped the epithelial apoptotic response to TNF stimulation. This work put forth the idea that autocrine ligands were not isolated loops but instead could be wired together as extracellular cascades that engaged a series of receptor pathways following an individual stimulus.

Our introduction to this chapter touched on the challenge of making general statements about signal transduction and signaling networks. Is the same true for autocrine networks? Cosgrove and coworkers asked this question by reexamining the TNF-induced autocrine cascade in hepatocytes (Cosgrove et al. 2008), which normally interpret TNF as a proliferative cue rather than an apoptotic one (Beg et al. 1995). Incredibly, much of the autocrine cascade identified in other epithelia remained true in hepatocytes— TNF stimulated the autocrine release of TGF-α, IL-1, and IL-1ra, and TGF-α signaling was required for

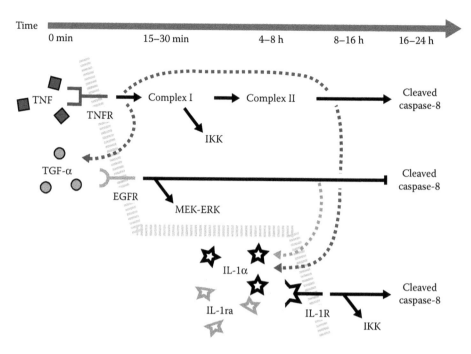

FIGURE 3.2 **(See color insert.)** A time-dependent autocrine cascade induced by TNF. TNF drives the early release of TGF-α, which cooperates with TNF to induce IL-1α at intermediate times. Later, TNF induces the release of IL-1ra to inhibit IL-1α signaling. TNF induces early IKK signaling through the TNF receptor Complex I (Micheau and Tschopp 2003), whereas TGF-α induces early MEK–ERK signaling and IL-1α induces late IKK signaling. TNF and IL-1α promote cleavage and activation of the initiator apoptotic enzyme, caspase-8, whereas TGF-α inhibits caspase-8 cleavage. The plasma membrane (gray) is staggered to show the sequence of induced autocrine factors.

IL-1 release and signaling, which was later attenuated by IL-1ra (Cosgrove et al. 2008; Janes et al. 2006a). The major difference was how each autocrine factor affected hepatocyte response. TNF and TGF-α promoted proliferation, whereas IL-1 inhibited proliferation. Strikingly, when hepatocytes were preinfected with adenovirus to promote TNF-induced apoptosis (Miller-Jensen et al. 2006), Cosgrove et al. found that the TNF-induced autocrine cascade functionally converged to what had been previously identified (Janes et al. 2006a). It is tempting to speculate that greater commonalities between cell types may be found by examining inducible autocrine signatures than by comparing their immediate early signaling patterns, which are highly cell-type specific (Miller-Jensen et al. 2007).

The preceding examples illustrate how autocrine factors coordinate signaling and cell-fate choices over time. During development, spatial control of signaling is achieved by morphogens, which are produced by neighboring cells to organize gradients of cues that signal in a paracrine fashion. Cells are assigned a positional value depending on the morphogen concentration at that particular location and switch genes on or off to specify distinct cell fates depending on the individual cell's morphogen threshold level (Gurdon and Bourillot 2001).

Morphogen signaling is an area where computational modeling and engineering analysis has been particularly valuable. *In silico* models of morphogen transport and interaction with cell receptors have been created to understand how morphogens distribute through tissue and affect cell patterning. Morphogen gradients form a "source" region of high concentration to a "sink" region of low concentration. In principle, concentration gradients across tissues can be established and maintained in multiple ways. Consequently, there have been conflicting theories about the mechanism of morphogen transport across tissue. Lander and coworkers developed a reaction–diffusion model to test different mechanisms of morphogen transport and then compared these results with published data *in vivo* (Lander et al.

2002). Their computational model revealed that a surprising number of experimental observations could be explained by diffusion. The authors further showed that nondiffusive mechanisms, such as transcytosis of morphogen ligands, would have to occur at impossibly fast rates to accommodate the rapid establishment of gradients that develop in embryos. Several perturbation experiments, which at first would seem to exclude a diffusive mechanism, were shown by the authors to be compatible with diffusion when viewed in the context of the computational model.

Morphogen gradients are established substantially before cell-fate choices are apparent. Therefore, gradients must be stably maintained to act as effective cues for target cells (Strigini and Cohen 1999). For long-range gradients to be stable, morphogen concentrations must decay rapidly near their source and then at a much slower rate over the remainder of the field. The simplest models of morphogen dynamics characterize the degradation rate as being linearly proportional to the morphogen concentration, which leads to an exponential concentration distribution at steady state. The problem is that this gradient is not stable to fluctuations in morphogen synthesis, which could vary because of environmental conditions or genetic variations. Eldar and coworkers showed through simple reaction–diffusion models that gradients could be robust to such fluctuations when morphogens underwent enhanced degradation as a function of concentration (Eldar et al. 2003). From a dynamical-systems perspective, stable gradients form because the solution to the governing equation becomes a power-law distribution, which is independent of morphogen synthesis rate. The authors identified multiple examples for enhanced degradation of morphogens, suggesting it may represent an evolutionary strategy for stabilizing developmental variation.

Autocrine factors that may not fit the strict definition of a morphogen can also play important roles in tissue patterning (Gurdon and Bourillot 2001). This process occurs through the diffusion of autocrine ligands that selectively amplify or inhibit key signaling pathways. Shvartsman et al. used computational approaches to study MAPK pathway activation during *Drosophila* oogenesis (Shvartsman et al. 2002). Localized peaks of MAPK activation are critical for the development of respiratory appendages on the fly eggshell. MAPK activity is positively regulated by the autocrine ligand Spitz (a TGF-α-like molecule) and negatively regulated by the EGFR inhibitor, Argos. MAPK activation leads to the expression of *rhomboid*, a protease that cleaves Spitz into its mature active form, providing a positive feedback loop to the system. The authors found that the spatial range of Argos (defined by its effective diffusivity) must substantially exceed the range of Spitz to recapitulate the two peaks of MAPK activation observed *in vivo*. This work and that of Lander et al. nicely illustrate how computational models can be useful for assessing the sufficiency of mechanisms implicated in the developmental pattern formation (Lander et al. 2002).

3.3 Signaling Dynamics in Cell-Fate Determination

One key dimension for encoding signaling information is time (Figure 3.1c). MAPKs are a classic illustration of how time-dependent signals control distinct cell-fate choices (Werlen et al. 2003). MAPK signaling pathways are cascades of three or more kinases that are sequentially phosphorylated by the preceding kinase and terminate with a final dual threonine–tyrosine phosphorylation that creates a fully active MAPK. MAPKs play a role in diverse cellular functions, including differentiation, death, and proliferation (Pearson et al. 2001). The Ras–Raf–MEK–ERK MAPK pathway often regulates proliferation but can promote different cellular outcomes depending on the characteristics of the signal and the context in which it was received (Marshall 1995).

ERK mediates a particularly divergent cell-fate choice in PC12 pheochromocytoma cells, which have served as a prototype for studying how signaling dynamics control cellular outcomes (Marshall 1995). In PC12 cells, EGF and nerve growth factor (NGF) both activate ERK, but they do so with different kinetics: EGF induces transient ERK activity that returns to basal levels within 20 min, whereas NGF induces sustained ERK activity that persists for hours (Marshall 1995). The differences in ERK dynamics are critical to the PC12 cell-fate choice. Transient EGF-induced ERK activation leads to PC12 proliferation, whereas sustained NGF-induced ERK activation causes differentiation into cells with many similar properties to sympathetic neurons (Cowley et al. 1994).

The PC12 cell-fate dichotomy first raises questions about how different ERK dynamics can be achieved with the same MAPK signaling module. Sasagawa and coworkers showed through quantitative modeling and experiments that activators upstream of the MAPK module play a key role in initiating transient-versus-sustained ERK kinetics (Sasagawa et al. 2005). EGF and NGF stimuli both activate ERK through MAPK/ERK kinase (MEK), but their paths to MEK activation are divergent and subject to different feedback mechanisms (Figure 3.3). EGF binds the EGF receptor, whose tyrosine phosphorylation recruits son of sevenless (SOS), a Ras guanine nucleotide exchange factor (GEF), by the SH2-domain-containing adaptor proteins Shc and Grb2 (Figure 3.3a). SOS activates Ras by exchanging bound GDP for GTP, and active Ras-GTP activates c-Raf that then phosphorylates MEK. Upon EGF stimulation, ERK indirectly downregulates itself by phosphorylating and inactivating SOS, leading to lower levels of Ras activation (Sasagawa et al. 2005). In contrast, NGF stimulation leads to the preferential activation of the small G protein Rap1 over Ras (Vaudry et al. 2002) (Figure 3.3b). Rap1 in turn activates a separate Raf isoform (B-Raf) that phosphorylates MEK leading to ERK activation (Sasagawa et al. 2005). Importantly, in bypassing the requirement for SOS, NGF-induced ERK activity is not subject to the negative feedback intrinsic to EGF signaling and thus signals in a sustained fashion. Selective recruitment of signaling molecules to activated receptors is a recurrent mechanism for achieving specificity in signal transduction (Jones et al. 2006; Yaffe 2002).

Interestingly, the use of distinct small G proteins in response to EGF or NGF also allows the ERK pathway to differentially sense how each ligand is presented (Sasagawa et al. 2005). Transient Ras-mediated ERK signaling depends on sudden spikes of ligand because slower additions of growth factor allow the ERK–SOS negative feedback to dampen most of the MAPK signal. Conversely, because the

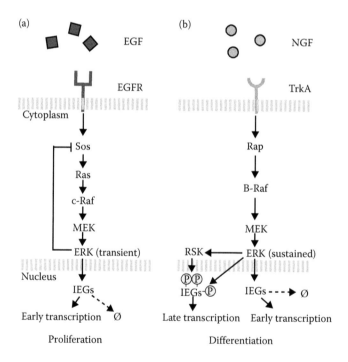

FIGURE 3.3 **(See color insert.)** Changes in ERK activity kinetics and gene expression induced by two different extracellular cues. (a) EGF induces transient ERK activity and proliferation in PC12 cells. ERK deactivation is achieved by negative feedback of SOS. Activated ERK induces IEGs that are rapidly degraded without a sustained ERK signal. (b) NGF induces sustained ERK activity and differentiation in PC12 cells. NGF-induced ERK activity does not require SOS and thus is not subject to ERK-mediated negative feedback. Prolonged ERK activation causes hyperphosphorylation of IEGs through ERK and RSK, which stabilize IEGs and lead to prolonged gene expression.

overall NGF-induced MAPK signal is not subject to feedback inhibition, sustained Rap1-mediated ERK activation depends on the final concentration of ligand rather than its rate of change. Therefore, the ERK pathway acts not only as a ligand discriminator but also as a differential or steady-state transducer of ligand presentation.

The work of Sasagawa et al. illustrates the important role that network "wiring" plays in determining downstream signaling dynamics (Sasagawa et al. 2005). More recently, Santos and coworkers directly tested the importance of network topology in the PC12 culture model (Santos et al. 2007). The authors focused on an NGF-stimulated positive-feedback loop involving protein kinase C (PKC) signaling, which the authors speculated would be activated by NGF but not by EGF. After creating or eliminating feedback with PKC activators or inhibitors, Santos et al. showed that ERK activation kinetics could be reversed, causing EGF to promote differentiation and NGF to promote proliferation. The modularity of signaling domains (see Section 3.1) makes it easy to envision how feedback interactions could have been manipulated over the course of evolution to provide complex input–output characteristics to networks (Bashor et al. 2008).

If networks are configured to elicit distinct time-dependent signaling profiles, how then are these profiles transduced into different cell-fate choices? One possible mechanism for interpreting transient-versus-sustained ERK signals was discovered in studies of 3T3 fibroblasts and their decision to enter the cell cycle (Murphy et al. 2002). In 3T3 cells, platelet-derived growth factor (PDGF) induces sustained ERK activity and promotes S phase entry, whereas EGF induces transient ERK activity and cells remain quiescent. Pioneering work by Murphy and coworkers revealed that immediate-early genes (IEGs) induced by early-phase ERK activity were themselves ERK substrates (Murphy et al. 2002) (Figure 3.3). IEGs are normally unstable, but the authors showed that hyperphosphorylation by ERK and the downstream kinase p90-ribosomal S6 kinase (RSK) prolonged the IEG protein expression in cells (Figure 3.3b). Transient ERK activation has declined before the IEG products accumulate, which prevents hyperphosphorylation and stabilization (Figure 3.3a). IEGs often encode for transcription factors or repressors; thus, their stability is key for the overall transcriptional response to growth-factor stimuli (Amit et al. 2007). Many IEGs have now been shown to possess the protein domains required for acting as "ERK sensors" and together may be important for controlling the G_1–S transition (Murphy et al. 2004).

Unbiased approaches have also been used in the PC12 model to identify novel mechanisms of ERK feedback and suggest candidate ERK effectors. von Kriegsheim et al. used quantitative mass spectrometry to observe dynamic ERK signaling complexes in EGF- or NGF-stimulated PC12 cells (von Kriegsheim et al. 2009). Dozens of proteins were identified whose association with ERK increased or decreased during sustained NGF-induced activation. Aside from increased binding to transcription factors that likely act as ERK substrates (ERF, TRPS1), the authors also identified a sustained decrease in binding of NF1 (a Ras-GAP) and PEA-15 (a cytoplasmic binding partner) upon NGF stimulation. This work illustrates that the NGF-induced choice to differentiate involves a multivariate series of signaling events that center on the generation and interpretation of a dynamically encoded ERK signal.

The stress-activated c-jun N-terminal kinase (JNK) MAPK has also been suggested to exert time-dependent control of cell outcomes, such as apoptosis. The involvement of JNK signaling in apoptosis remains controversial (Davis 2000; Varfolomeev and Ashkenazi 2004), but for TNF, transient JNK activation frequently correlates with survival and sustained JNK with cell death (Guo et al. 1998). In fibroblasts, JNK activation kinetics are dependent on the generation of reactive oxygen species (ROS) because ROS can potently deactivate JNK phosphatases to cause sustained JNK activation and cell death by necrosis (Kamata et al. 2005; Ventura et al. 2004). Prosurvival JNK signaling has been linked to phosphorylation of the junD transcription factor (Lamb et al. 2003). Conversely, prodeath JNK signaling involves phosphorylation of Bcl-2 family members (Deng et al. 2003; Lei and Davis 2003). Unlike for ERK in PC12 cells, however, it has yet to be determined in a single experimental model how different JNK activation dynamics might exert opposing control on apoptosis.

Signaling dynamics are also important for pathways other than the MAPK cascades. Janes et al. examined the importance of the phosphatidylinositol-3-kinase (PI3K)–Akt pathway in human colon adenocarcinoma cells (Janes et al. 2003), which can withstand TNF treatment upon insulin costimulation. Insulin was found to induce two distinct phases of Akt activity, an early phase ($t = 0$–4 h after stimulation) and a late phase ($t = 4$–24 h). By inhibiting PI3K with reversible small molecules, the authors found that late-phase PI3K–Akt was the critical antiapoptotic signal for inhibiting TNF-induced apoptosis. As high-throughput techniques for quantifying signaling networks continue to expand (Albeck et al. 2006), we predict that other pathways will be found to encode information in their temporal activation pattern.

Cells can also elicit and interpret more complex dynamics beyond "transient" and "sustained" activation patterns. Repeated spikes in calcium signaling have been appreciated for decades (Berridge 1990; Lipp et al. 1997), and oscillations have more recently been reported in several other pathways. Asynchronous oscillations require real-time single-cell reporters to view the time-dependent location and abundance of proteins in individual cells. By fusing green fluorescent protein (GFP) to the p65 subunit of the NF-κB transcription factor, Nelson and coworkers identified oscillations in the nucleo-cytoplasmic localization of NF-κB in response to TNF (Nelson et al. 2004). The duration, amplitude, and damping of NF-κB oscillations varied across different cell lines and could account for differential NF-κB-dependent gene expression observed in these cells (Ashall et al. 2009). Expression levels of the tumor suppressor p53 have also been shown to oscillate in response to DNA damage. Interestingly, it is not the frequency or amplitude of p53 oscillations that indicates the level of DNA damage, but rather the number of spikes (Lahav et al. 2004). One challenge for studying asynchronous oscillations is the requirement for tagged fluorescent reporters. Overexpression of tagged reporters may unintentionally disrupt the balance of different signaling proteins and cause artifactual behavior of the network (Barken et al. 2005). Newer approaches that fuse reporters to endogenous proteins can overcome this limitation and provide a more faithful readout of protein levels and locations (Cohen et al. 2008).

Signaling dynamics influence cell decisions and also the timing of when decisions are made. Asynchronies in timing will cause cell-to-cell variations to emerge in clonal cell populations. Spencer and coworkers studied this phenomenon in response to TNF-related apoptosis-inducing ligand (TRAIL) (Spencer et al. 2009). Upon TRAIL stimulation and the resulting inhibition of protein synthesis, epithelial cells will reproducibly die, but the time course over which cells apoptose ranges from 45 min to 12 h (see Albeck et al. 2008b). The authors examined the heritability of this delay time (T_d) by live-cell tracking of sister cells after TRAIL exposure. T_d was strongly correlated in sisters that had recently divided, but this correlation disappeared as the time since division increased. The authors linked the covariation of T_d to the inheritance of signaling proteins after mitosis. Shortly after division, sisters would have roughly equal copies of key signaling proteins, but this correlation would fade as proteins were heterogeneously degraded and replenished in single cells. Cell-to-cell expression variation has received increasing attention of late (Feinerman et al. 2008; Janes et al. 2010; Raj and van Oudenaarden 2008), and the work of Spencer et al. (2009) suggests that such differences could be an important mechanism for asynchronous decision-making.

3.4 Computational Modeling of Cell-Fate Determination

In the preceding sections, we discussed how dynamics and the microenvironment are important considerations for signal transduction and cell-fate choice. The biology is obviously complicated and our knowledge is always evolving. Notwithstanding this uncertainty, it is possible to construct computational models of signaling that are firmly grounded in experimental observations. Many such models have made important contributions toward conceptually advancing our understanding of network function (Figure 3.1d).

The intricate signaling networks inside cells have often been compared to the wiring of electrical circuits (Lazebnik 2002). However, signal transmission by intracellular networks bears a closer resemblance to a transistor radio (Lazebnik 2002), with a mixture of analog and digital-like components, than it does to a modern, all-digital integrated circuit. Early work by McAdams and Shapiro on the bacteriophage λ lysis-lysogeny decision circuit revealed an explicit requirement for graded biochemical reactions on the phage "logic board" (McAdams and Shapiro 1995). More recently, Altan-Bonnet and Germain arrived at the converse requirement for digital signaling components in a biochemical-reaction model of T cell antigen discrimination (Altan-Bonnet and Germain 2005). Therefore, although signaling networks are undoubtedly built from digital information (i.e., DNA (Hood and Galas 2003)), it appears that network function is more accurately captured by hybrid models, which can accommodate analog- and digital-like elements (Aldridge et al. 2009; Amonlirdviman et al. 2005).

If cell-decision networks are built from parts with analog or digital input–output behaviors, how can we identify and classify them? One strategy is to build models exclusively from analog or digital components and see where the predictions fail (Altan-Bonnet and Germain 2005; McAdams and Shapiro 1995). Janes et al. (2008) recently introduced a complementary approach, called model-breakpoint analysis that starts with a predictive model and perturbs the underlying assumptions until the predictions fail. Using a data-driven partial least squares model of cytokine-induced apoptosis (Janes et al. 2005), the authors took the analog measurements on which the model was based and then discretized them to simulate digital elements. With data from perturbations of the TNF-induced autocrine cascade (Janes et al. 2006a), the analysis suggested that autocrine TGF-α was interpreted as an analog stimulus, whereas autocrine IL-1α was interpreted as a digital stimulus (Janes et al. 2008). Interestingly, the sequential release of TGF-α (early) and IL-1α (late) parallels the transition from analog decision-making at early times to digital cell-death execution at late times. This suggests that an analog-to-digital conversion of signaling is taking place at some point between cytokine receptors and the effector proteins that mediate apoptosis (see below).

A long-standing question of both theoretical and practical interest is how intracellular biochemical reactions, which often are reversible and graded, can lead to binary cell-fate choices (Ferrell 1996). From a dynamical-systems perspective, the bifurcation of cell "trajectories" into one of two states would argue that cell-decision networks must be bistable. The most straightforward way of achieving bistability is through positive (or double-negative) feedback (Gardner et al. 2000). Indeed, this architecture has been shown to be important for some cell-fate switches, such as the decision of *Xenopus* oocytes to mature in response to hormone (Ferrell and Machleder 1998). More recently, however, Brandman and coworkers showed that a single positive feedback loop may not be sufficient in situations where cell-fate switches must be made rapidly and reliably amidst biological fluctuations (Brandman et al. 2005). Through simulations, the authors showed that fast single-feedback networks switched unpredictably between states when stochastic fluctuations in the levels of signaling molecules were considered. Random cell-fate switching was suppressed when a fast positive-feedback loop was interconnected with a slow positive-feedback loop. Yet, such an architecture could respond reliably when a fate-inducing stimulus was added to the network. Interlinked positive feedback is found in many cell-decision networks, suggesting that evolution has converged toward this simple architecture for robust cell-state changes.

Feedback may even trump "irreversible" chemical reactions in their importance for enforcing cell-state transitions. Novak and coworkers noted that irreversible modifications, such as protein cleavage, are usually balanced by a reciprocal irreversible process (protein synthesis) that negate the overall irreversibility (Novak et al. 2007). For example, during the cell cycle, proteolytic degradation of cyclin-dependent kinase inhibitors (CKIs) is thought to confer irreversibility to the entry into S phase. However, from a systems perspective, the irreversibility of the G_1–S transition is not because of the degradation process itself, but because active cyclin-dependent kinases (CDKs) downregulate the expression of CKIs. Thus, it is the interplay between feedback mechanisms (CKIs inhibit CDK activity and CDKs inhibit *CKI* expression—double-negative feedback) that creates the irreversible transition. CKI proteolysis is required for irreversible S phase entry, but it will not be sufficient without CDK-mediated downregulation of *CKI* transcription to close the feedback loop.

Ironically, for cell-fate changes that are never repeated, neither bistability nor positive feedback is required. Apoptosis is a classic example of such a cell-fate choice that must be executed decisively once (Meier et al. 2000). There is often a huge disparity in timescales between cell-death processing and cell-death execution. In the presence of an extrinsic death cue, a cell will remain viable for 8–12 h but then dismantle its constituents within 30 min (Rehm et al. 2002). Albeck et al. (2008a) investigated the molecular mechanisms that control this rapid execution phase during the apoptotic response to TRAIL. The authors built a mass-action kinetic model of the relevant TRAIL-induced signaling network, consisting of three main components: (1) a ligand–receptor–adaptor module; (2) a proteolytic cascade of caspases, which execute the apoptotic phenotype; and (3) a mitochondrial circuit that feeds forward to promote effector–caspase activation. Extensive training and validation of the model through quantitative Western blotting, flow cytometry, and live-cell imaging revealed that positive feedback through the caspase cascade was not required for the "snap-action" execution of cell death. Instead, snap-action behavior was captured by a model that explicitly incorporated release of the prodeath factors Smac and cytochrome c from the mitochondria. In normal cells, Smac and cytochrome c are localized to the mitochondrial outer membrane, creating a steep gradient in concentration relative to the cytosol. The Smac–cytochrome c gradient is relieved when the mitochondrial outer membrane is permeabilized by the prodeath Bcl-2 protein, Bax. Cytosolic Smac and cytochrome c independently promote effector–caspase activation, and their rapid translocation from permeabilized mitochondria acutely amplifies the apoptotic response. Therefore, by sequestering effector proteins in subcellular organelles, a cell can achieve switch-like behavior without positive feedback.

A corollary to the mechanisms of cell-death execution is that a dying cell does not need to reach a stable apoptotic state to be dead. This creates a logistical problem for many standard dynamical-systems approaches (nullclines, bifurcation analysis, etc.) that are based on steady-state solutions to the governing equations. Aldridge and coworkers applied finite-time approaches (that do not require $d/dt = 0$) to define the separation of cellular trajectories in a model of caspase activation (Aldridge et al. 2006). These types of approaches may be particularly relevant for other irreversible cell-fate choices, such as during the mitotic-spindle checkpoint.

3.5 Conclusions and Future Directions

Throughout this chapter, we presented cell-fate decision-making as a process that is deterministic. However, in some circumstances, cells decide their fate stochastically by making a random, binary choice between two states (Losick and Desplan 2008). The mechanisms of stochastic decision processes have often been linked to intrinsic noise in small numbers of key signaling or effector molecules (Arkin et al. 1998; Elowitz et al. 2002; Raj et al. 2010; Weinberger et al. 2005). However, we predict that "regulation of randomness" will be an active area of research going forward, especially in areas such as cell reprogramming. The induction of pluripotent stem cells with cocktails of exogenous transcription factors is a reproducibly rare decision made by somatic cells (Takahashi and Yamanaka 2006). Early modeling work has implicated the cell-division rate as a key factor in the reprogramming process (Hanna et al. 2009), but these models will refine as our understanding of the biology accelerates.

Biomedical engineers have much to offer in the area of signal transduction and cell-fate choice. They also have much to gain from it. Although there is complexity and uncertainty in the molecular mechanisms, an underlying logic awaits those that seek to identify it. By modeling intracellular and extracellular "circuits," we can gauge our current reasoning by the accuracy of our predictions. This is the first step toward eventually manipulating these outcomes for therapeutic or biotechnological applications (Willerth and Sakiyama-Elbert 2009).

References

Albeck, J.G., Burke, J.M., Aldridge, B.B. et al. 2008a. Quantitative analysis of pathways controlling extrinsic apoptosis in single cells. *Mol Cell* 30:11–25.
Albeck, J.G., Burke, J.M., Spencer, S.L., Lauffenburger, D.A., and Sorger, P.K. 2008b. Modeling a snap-action, variable-delay switch controlling extrinsic cell death. *PLoS Biol* 6:2831–2852.

Albeck, J.G., MacBeath, G., White, F.M. et al. 2006. Collecting and organizing systematic sets of protein data. *Nat Rev Mol Cell Biol* 7:803–812.

Aldridge, B.B., Haller, G., Sorger, P.K., and Lauffenburger, D.A. 2006. Direct Lyapunov exponent analysis enables parametric study of transient signalling governing cell behaviour. *Syst Biol (Stevenage)* 153:425–432.

Aldridge, B.B., Saez-Rodriguez, J., Muhlich, J.L., Sorger, P.K., and Lauffenburger, D.A. 2009. Fuzzy logic analysis of kinase pathway crosstalk in TNF/EGF/insulin-induced signaling. *PLoS Comput Biol* 5: e1000340.

Altan-Bonnet, G. and Germain, R.N. 2005. Modeling T cell antigen discrimination based on feedback control of digital ERK responses. *PLoS Biol* 3:e356.

Amit, I., Citri, A., Shay, T. et al. 2007. A module of negative feedback regulators defines growth factor signaling. *Nat Genet* 39:503–512.

Amonlirdviman, K., Khare, N.A., Tree, D.R. et al. 2005. Mathematical modeling of planar cell polarity to understand domineering nonautonomy. *Science* 307:423–426.

Arkin, A., Ross, J., and McAdams, H.H. 1998. Stochastic kinetic analysis of developmental pathway bifurcation in phage lambda-infected Escherichia coli cells. *Genetics* 149:1633–1648.

Ashall, L., Horton, C.A., Nelson, D.E. et al. 2009. Pulsatile stimulation determines timing and specificity of NF-kappaB-dependent transcription. *Science* 324:242–246.

Au, C.E., Bell, A.W., Gilchrist, A. et al. 2007. Organellar proteomics to create the cell map. *Curr Opin Cell Biol* 19:376–385.

Barken, D., Wang, C.J., Kearns, J. et al. 2005. Comment on "Oscillations in NF-kappaB signaling control the dynamics of gene expression". *Science* 308:52; author reply 52.

Bashor, C.J., Helman, N.C., Yan, S., and Lim, W.A. 2008. Using engineered scaffold interactions to reshape MAP kinase pathway signaling dynamics. *Science* 319:1539–1543.

Beg, A.A., Sha, W.C., Bronson, R.T., Ghosh, S., and Baltimore, D. 1995. Embryonic lethality and liver degeneration in mice lacking the RelA component of NF-kappa B. *Nature* 376:167–170.

Berridge, M.J. 1990. Calcium oscillations. *J Biol Chem* 265:9583–9586.

Brandman, O., Ferrell, J.E., Jr., Li, R., and Meyer, T. 2005. Interlinked fast and slow positive feedback loops drive reliable cell decisions. *Science* 310:496–498.

Busch, H., Camacho-Trullio, D., Rogon, Z. et al. 2008. Gene network dynamics controlling keratinocyte migration. *Mol Syst Biol* 4:199.

Chien, K.R. and Karsenty, G. 2005. Longevity and lineages: Toward the integrative biology of degenerative diseases in heart, muscle, and bone. *Cell* 120:533–544.

Cohen, A.A., Geva-Zatorsky, N., Eden, E. et al. 2008. Dynamic proteomics of individual cancer cells in response to a drug. *Science* 322:1511–1516.

Cosgrove, B.D., Cheng, C., Pritchard, J.R. et al. 2008. An inducible autocrine cascade regulates rat hepatocyte proliferation and apoptosis responses to tumor necrosis factor-alpha. *Hepatology* 48:276–288.

Cowley, S., Paterson, H., Kemp, P., and Marshall, C.J. 1994. Activation of MAP kinase kinase is necessary and sufficient for PC12 differentiation and for transformation of NIH 3T3 cells. *Cell* 77:841–852.

Davis, R.J. 2000. Signal transduction by the JNK group of MAP kinases. *Cell* 103:239–252.

Deng, Y., Ren, X., Yang, L., Lin, Y., and Wu, X. 2003. A JNK-dependent pathway is required for TNFalpha-induced apoptosis. *Cell* 115:61–70.

Downward, J. 2001. The ins and outs of signalling. *Nature* 411:759–762.

Eldar, A., Rosin, D., Shilo, B.Z., and Barkai, N. 2003. Self-enhanced ligand degradation underlies robustness of morphogen gradients. *Dev Cell* 5:635–646.

Elowitz, M.B., Levine, A.J., Siggia, E.D., and Swain, P.S. 2002. Stochastic gene expression in a single cell. *Science* 297:1183–1186.

Feinerman, O., Veiga, J., Dorfman, J.R., Germain, R.N., and Altan-Bonnet, G. 2008. Variability and robustness in T cell activation from regulated heterogeneity in protein levels. *Science* 321:1081–1084.

Ferrell, J.E., Jr. 1996. Tripping the switch fantastic: How a protein kinase cascade can convert graded inputs into switch-like outputs. *Trends Biochem Sci* 21:460–466.

Ferrell, J.E., Jr., and Machleder, E.M. 1998. The biochemical basis of an all-or-none cell fate switch in Xenopus oocytes. *Science* 280:895–898.

Gardner, T.S., Cantor, C.R., and Collins, J.J. 2000. Construction of a genetic toggle switch in *Escherichia coli*. *Nature* 403:339–342.

Gilbert, S.F. 2006. *Developmental Biology*. Sunderland: Sinauer Associates.

Guo, Y.L., Baysal, K., Kang, B., Yang, L.J., and Williamson, J.R. 1998. Correlation between sustained c-Jun N-terminal protein kinase activation and apoptosis induced by tumor necrosis factor-alpha in rat mesangial cells. *J Biol Chem* 273:4027–4034.

Gurdon, J.B. and Bourillot, P.Y. 2001. Morphogen gradient interpretation. *Nature* 413:797–803.

Hanahan, D. and Weinberg, R.A. 2000. The hallmarks of cancer. *Cell* 100:57–70.

Hanna, J., Saha, K., Pando, B. et al. 2009. Direct cell reprogramming is a stochastic process amenable to acceleration. *Nature* 462:595–601.

Hood, L. and Galas, D. 2003. The digital code of DNA. *Nature* 421:444–448.

Ingham, P.W., Taylor, A.M., and Nakano, Y. 1991. Role of the *Drosophila* patched gene in positional signalling. *Nature* 353:184–187.

Janes, K.A., Albeck, J.G., Gaudet, S. et al. 2005. A systems model of signaling identifies a molecular basis set for cytokine-induced apoptosis. *Science* 310:1646–1653.

Janes, K.A., Albeck, J.G., Peng, L.X. et al. 2003. A high-throughput quantitative multiplex kinase assay for monitoring information flow in signaling networks: Application to sepsis-apoptosis. *Mol Cell Proteomics* 2:463–473.

Janes, K.A., Gaudet, S., Albeck, J.G. et al. 2006a. The response of human epithelial cells to TNF involves an inducible autocrine cascade. *Cell* 124:1225–1239.

Janes, K.A., Reinhardt, H.C., and Yaffe, M.B. 2008. Cytokine-induced signaling networks prioritize dynamic range over signal strength. *Cell* 135:343–354.

Janes, K.A., Wang, C.C., Holmberg, K.J., Cabral, K., and Brugge, J.S. 2010. Identifying single-cell molecular programs by stochastic profiling. *Nat Methods* 7:311–317.

Janes, K.A. and Yaffe, M.B. 2006b. Data-driven modelling of signal-transduction networks. *Nat Rev Mol Cell Biol* 7:820–828.

Jones, R.B., Gordus, A., Krall, J.A., and MacBeath, G. 2006. A quantitative protein interaction network for the ErbB receptors using protein microarrays. *Nature* 439:168–174.

Jordan, J.D., Landau, E.M., and Iyengar, R. 2000. Signaling networks: The origins of cellular multitasking. *Cell* 103:193–200.

Kamata, H., Honda, S., Maeda, S. et al. 2005. Reactive oxygen species promote TNFalpha-induced death and sustained JNK activation by inhibiting MAP kinase phosphatases. *Cell* 120:649–661.

Lahav, G., Rosenfeld, N., Sigal, A. et al. 2004. Dynamics of the p53-Mdm2 feedback loop in individual cells. *Nat Genet* 36:147–150.

Lamb, J.A., Ventura, J.J., Hess, P., Flavell, R.A., and Davis, R.J. 2003. JunD mediates survival signaling by the JNK signal transduction pathway. *Mol Cell* 11:1479–1489.

Lander, A.D., Nie, Q., and Wan, F.Y. 2002. Do morphogen gradients arise by diffusion? *Dev Cell* 2:785–796.

Lauffenburger, D.A., Forsten, K.E., Will, B., and Wiley, H.S. 1995. Molecular/cell engineering approach to autocrine ligand control of cell function. *Ann Biomed Eng* 23:208–215.

Lauffenburger, D.A. and Horwitz, A.F. 1996. Cell migration: A physically integrated molecular process. *Cell* 84:359–369.

Lazebnik, Y. 2002. Can a biologist fix a radio? Or, what I learned while studying apoptosis. *Cancer Cell* 2:179–182.

Lei, K. and Davis, R.J. 2003. JNK phosphorylation of Bim-related members of the Bcl2 family induces Bax-dependent apoptosis. *Proc Natl Acad Sci USA* 100:2432–2437.

Lipp, P., Thomas, D., Berridge, M.J., and Bootman, M.D. 1997. Nuclear calcium signalling by individual cytoplasmic calcium puffs. *EMBO J* 16:7166–7173.

Losick, R. and Desplan, C. 2008. Stochasticity and cell fate. *Science* 320:65–68.

Marshall, C.J. 1995. Specificity of receptor tyrosine kinase signaling: Transient versus sustained extracellular signal-regulated kinase activation. *Cell* 80:179–185.

McAdams, H.H. and Shapiro, L. 1995. Circuit simulation of genetic networks. *Science* 269:650–656.

Meier, P., Finch, A., and Evan, G. 2000. Apoptosis in development. *Nature* 407:796–801.

Micheau, O. and Tschopp, J. 2003. Induction of TNF receptor I-mediated apoptosis via two sequential signaling complexes. *Cell* 114:181–190.

Miller-Jensen, K., Janes, K.A., Brugge, J.S., and Lauffenburger, D.A. 2007. Common effector processing mediates cell-specific responses to stimuli. *Nature* 448:604–608.

Miller-Jensen, K., Janes, K.A., Wong, Y.L., Griffith, L.G., and Lauffenburger, D.A. 2006. Adenoviral vector saturates Akt pro-survival signaling and blocks insulin-mediated rescue of tumor-necrosis-factor-induced apoptosis. *J Cell Sci* 119:3788–3798.

Murphy, L.O., MacKeigan, J.P., and Blenis, J. 2004. A network of immediate early gene products propagates subtle differences in mitogen-activated protein kinase signal amplitude and duration. *Mol Cell Biol* 24:144–153.

Murphy, L.O., Smith, S., Chen, R.H., Fingar, D.C., and Blenis, J. 2002. Molecular interpretation of ERK signal duration by immediate early gene products. *Nat Cell Biol* 4:556–564.

Nelson, D.E., Ihekwaba, A.E., Elliott, M. et al. 2004. Oscillations in NF-kappaB signaling control the dynamics of gene expression. *Science* 306:704–708.

Novak, B., Tyson, J.J., Gyorffy, B., and Csikasz-Nagy, A. 2007. Irreversible cell-cycle transitions are due to systems-level feedback. *Nat Cell Biol* 9:724–728.

Pawson, T. 2004. Specificity in signal transduction: From phosphotyrosine-SH2 domain interactions to complex cellular systems. *Cell* 116:191–203.

Pearson, G., Robinson, F., Beers Gibson, T. et al. 2001. Mitogen-activated protein (MAP) kinase pathways: Regulation and physiological functions. *Endocr Rev* 22:153–183.

Raj, A., Rifkin, S.A., Andersen, E., and van Oudenaarden, A. 2010. Variability in gene expression underlies incomplete penetrance. *Nature* 463:913–918.

Raj, A. and van Oudenaarden, A. 2008. Nature, nurture, or chance: Stochastic gene expression and its consequences. *Cell* 135:216–226.

Rehm, M., Dussmann, H., Janicke, R.U. et al. 2002. Single-cell fluorescence resonance energy transfer analysis demonstrates that caspase activation during apoptosis is a rapid process. Role of caspase-3. *J Biol Chem* 277:24506–24514.

Santos, S.D., Verveer, P.J., and Bastiaens, P.I. 2007. Growth factor-induced MAPK network topology shapes Erk response determining PC-12 cell fate. *Nat Cell Biol* 9:324–330.

Sasagawa, S., Ozaki, Y., Fujita, K., and Kuroda, S. 2005. Prediction and validation of the distinct dynamics of transient and sustained ERK activation. *Nat Cell Biol* 7:365–373.

Schulze, A., Nicke, B., Warne, P.H., Tomlinson, S., and Downward, J. 2004. The transcriptional response to Raf activation is almost completely dependent on Mitogen-Activated Protein Kinase Kinase activity and shows a major autocrine component. *Mol Biol Cell* 15:3450–3463.

Seet, B.T., Dikic, I., Zhou, M.M., and Pawson, T. 2006. Reading protein modifications with interaction domains. *Nat Rev Mol Cell Biol* 7:473–483.

Shvartsman, S.Y., Muratov, C.B., and Lauffenburger, D.A. 2002. Modeling and computational analysis of EGF receptor-mediated cell communication in *Drosophila* oogenesis. *Development* 129:2577–2589.

Spencer, S.L., Gaudet, S., Albeck, J.G., Burke, J.M., and Sorger, P.K. 2009. Non-genetic origins of cell-to-cell variability in TRAIL-induced apoptosis. *Nature* 459:428–432.

Srivastava, D. 2006. Making or breaking the heart: From lineage determination to morphogenesis. *Cell* 126:1037–1048.

Strigini, M., and Cohen, S.M. 1999. Formation of morphogen gradients in the *Drosophila* wing. *Semin Cell Dev Biol* 10:335–344.

Takahashi, K. and Yamanaka, S. 2006. Induction of pluripotent stem cells from mouse embryonic and adult fibroblast cultures by defined factors. *Cell* 126:663–676.

Varfolomeev, E.E. and Ashkenazi, A. 2004. Tumor necrosis factor: An apoptosis JuNKie? *Cell* 116:491–497.

Vaudry, D., Stork, P.J., Lazarovici, P., and Eiden, L.E. 2002. Signaling pathways for PC12 cell differentiation: Making the right connections. *Science* 296:1648–1649.

Ventura, J.J., Cogswell, P., Flavell, R.A., Baldwin, A.S., Jr., and Davis, R.J. 2004. JNK potentiates TNF-stimulated necrosis by increasing the production of cytotoxic reactive oxygen species. *Genes Dev* 18:2905–2915.

von Kriegsheim, A., Baiocchi, D., Birtwistle, M. et al. 2009. Cell fate decisions are specified by the dynamic ERK interactome. *Nat Cell Biol* 11:1458–1464.

Weinberger, L.S., Burnett, J.C., Toettcher, J.E., Arkin, A.P., and Schaffer, D.V. 2005. Stochastic gene expression in a lentiviral positive-feedback loop: HIV-1 Tat fluctuations drive phenotypic diversity. *Cell* 122:169–182.

Weissman, I.L. 2000. Stem cells: Units of development, units of regeneration, and units in evolution. *Cell* 100:157–168.

Werlen, G., Hausmann, B., Naeher, D., and Palmer, E. 2003. Signaling life and death in the thymus: Timing is everything. *Science* 299:1859–1863.

Willerth, S.M. and Sakiyama-Elbert, S.E. 2009. Kinetic analysis of neurotrophin-3-mediated differentiation of embryonic stem cells into neurons. *Tissue Eng Part A* 15:307–318.

Yaffe, M.B. 2002. Phosphotyrosine-binding domains in signal transduction. *Nat Rev Mol Cell Biol* 3:177–186.

Yaffe, M.B. and Elia, A.E. 2001. Phosphoserine/threonine-binding domains. *Curr Opin Cell Biol* 13:131–138.

Zelzer, E. and Olsen, B.R. 2003. The genetic basis for skeletal diseases. *Nature* 423:343–348.

4

Biomaterial Scaffolds for Human Embryonic Stem Cell Culture and Differentiation

Stephanie Willerth
University of Victoria
University of British Columbia

David Schaffer
University of California, Berkeley

4.1 Introduction

Owing to their two hallmark properties—the ability to self-renew or expand in an undifferentiated state and the capacity to differentiate into one or more mature cell types—stem cells have enormous therapeutic potential. A wide variety of stem cell types exist, including adult, tissue-specific stem cells, induced pluripotent and embryonic stem cells, and each of these has been investigated for use as potential treatments for a variety of diseases (Ameen et al., 2008, Li and Clevers, 2010, Shi, 2009). The goal of this chapter is to describe the recent progress in investigating the therapeutic potential of a particular stem cell population, human embryonic stem cells (hESCs), in combination with bioactive materials. In particular, Sections 4.2 and 4.3 of this chapter will discuss combining biomaterial scaffolds with hESCs for two distinct applications: the expansion of undifferentiated hESCs and the directed differentiation of these cells into mature phenotypes for various applications. Finally, suggestions for further investigation into combining hESCs and biomaterial scaffolds will be detailed in Section 4.4.

4.1.1 Human Embryonic Stem Cells

Mouse ES cells were first derived in 1981, and hESCs were derived more recently in 1998 (Figure 4.1) (Martin, 1981, Thomson et al., 1998). Human pluripotent stem cells offer considerable therapeutic potential. They are immortal and can be expanded indefinitely, and then be induced to give rise to any cell

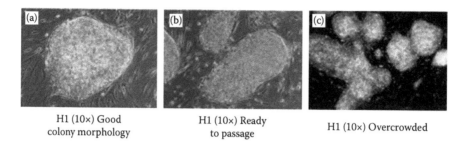

| H1 (10×) Good colony morphology | H1 (10×) Ready to passage | H1 (10×) Overcrowded |

FIGURE 4.1 Morphologies observed in hESC cultures. (a) "Good" cell morphology, (b) hESCs that are ready for passaging, and (c) overcrowded hESCs. (From Kent, L. 2009. *J Vis Exp*, 34, e1427.)

type found in the body. Furthermore, the use of human cell lines reduces the potential for an immune response after transplantation. However, there are considerable challenges associated with controlling these behaviors in culture, in which will be required for the biomedical potential of hESCs to be realized.

Considerable progress has been made in translating knowledge from developmental biology to control analogous fate decisions in hESC cultures and thereby drive cells into particular lineages. As a result, a number of detailed protocols have been developed for differentiating hESCs into a variety of cell types, including neuronal, liver, cartilage, and muscle (Duan et al., 2010, Harkness et al., 2009, Hill et al., 2010, Lindvall and Kokaia, 2009). However, a number of challenges remain, including developing approaches for the uniform differentiation of cells into a specific cell type, as the presence of contaminating cells can pose risks of side effects or tumorigenesis. One promising approach, incorporating the signals inherent in these protocols into "instructive" biomaterial scaffolds, may offer advantages for controlling hESC differentiation in a reproducible, safe, and scaleable manner.

4.1.2 Biomaterial Scaffolds

Biomaterials are defined as materials that comprise either the whole or a part of a living structure or biomedical device that performs, supplements, or replaces a natural function (Williams, 1987). A variety of these materials have been investigated for a range of stem cell and in particular hESC applications. For example, the standard culture methods for undifferentiated hESCS currently utilize a natural biomaterial substrate, Matrigel, and Section 4.2.1 discusses this culture method in depth. Natural materials such as Matrigel and potentially synthetic biomaterials could be used to scale up hESC culture to produce large number of cells for tissue engineering applications. Additional biomaterial scaffolds are under development for controlling hESC cell differentiation and enhancing cell survival upon transplantation. Finally, for regenerative medicine applications, biomaterials are often combined with cells in attempts to engineer replacements for diseased or damaged tissues. Numerous successful examples that integrate biomaterial scaffolds with hESCs are detailed in Section 4.3. One of the considerations in designing such systems is whether to use a natural or synthetic biomaterial, which is the focus of Sections 4.1.2.1 and 4.1.2.2.

4.1.2.1 Natural Biomaterials

As their name suggests, natural biomaterials are derived from natural sources and often consist of proteins, polysaccharides, or combinations of both. Examples of protein-based natural biomaterials include fibrin and collagen, while polysaccharide-based biomaterials often utilize alginate, agarose, hyaluronic acid (HA), or dextran. Some of the advantages of natural biomaterials are that they may contain bioactive motifs which support cell adhesion or may otherwise promote cell proliferation and migration, and they often present excellent biocompatibility when optimally purified. However, it can be challenging to program new properties into natural materials, such as a broad range of mechanical properties, and materials that can suffer from lot-to-lot variability.

4.1.2.2 Synthetic Biomaterials

Synthetic biomaterials have chemically defined compositions and offer some advantages compared to natural biomaterials. They can often be synthesized with a high degree of reproducibility, and certain classes of synthetic materials can be generated with a wide range of structural and mechanical properties. However, synthetic materials intrinsically lack bioactive sites for cell adhesion and often need to be functionalized with ligands, such as peptides, to enable cell engagement. Most of the commonly used synthetic scaffold materials are polymers, including poly(ethylene glycol) and poly(lactic-*co*-glycolic) acid. Polymers can often be combined to further alter the properties of the material.

4.2 Biomaterial Scaffolds for Maintaining hESCs in an Undifferentiated State

This section will address the current standards for culturing undifferentiated hESCs along with the limitations of these methods. It will then discuss studies that have attempted to address these issues through the use of biomaterial scaffolds. Specifically, the use of both heterogeneous and homogeneous naturally derived biomaterial scaffolds will be reviewed, along with synthetic scaffold materials, with an emphasis on strategies that mimic how the extracellular matrix binds to integrins. Furthermore, the utility of high-throughput screening for finding suitable biomaterial substrates maintaining hESCs in their undifferentiated state will be discussed.

4.2.1 Current Standard for Culturing Undifferentiated hESCs

When hESCs were initially derived in 1998, they were cultured upon a layer of mouse embryonic fibroblasts (MEFs), which secrete factors necessary to maintain their pluripotency (Thomson et al., 1998). In 2001, the Thomson lab showed that hESCs could be cultured in a feeder-free system where the cells were grown on a Matrigel substrate in the presence of conditioned media containing factors produced by MEFs (Xu et al., 2001). The substrate Matrigel, which consists of proteins and factors extracted from the basement membrane secreted by Engelbreth–Holm–Swarm (EHS) mouse sarcoma cells (Kleinman and Martin, 2005), contains large amounts of laminin and Type IV collagen along with growth factors, enzymes, and polysaccharides. While effective in preserving the pluripotency of the hESCs, the use of a feeder layer, conditioned media derived from MEFs, or Matrigel in culture systems poses many issues that can potentially prevent hESC-based therapies from reaching clinical relevance. Specifically, the use of animal-derived products and cells could contaminate the hESC culture with immunogenic epitopes that would activate the immune system when transplanted into humans. This issue was confirmed when Martin et al. showed that hESCs grown on feeder layers or in the presence of conditioned media expressed a nonhuman sialic acid, prompting a need to develop defined culture methods for undifferentiated hESCs (Martin et al., 2005).

As a step in this direction, in 2006, Ludwig et al. reported defined conditions for deriving new hESC lines without the use of animal-derived products or feeder layers (Ludwig et al., 2006). They used a chemically defined medium referred to as TeSR1, which contained a basic fibroblast growth factor (FGF-2), lithium chloride, γ-aminobutyric acid, pipecolic acid, and transforming growth factor β. This medium was used to support hESC culture on two different substrates: one that consisted of Matrigel and one with a mixture of human-derived proteins, including collagen IV, fibronectin, laminin, and vitronectin. This important study showed that new hESC lines could be generated in chemically defined liquid medium and without the use of animal-derived products. However, the human-derived proteins can still present challenges with variability, contamination, and economics, particularly when several ECM proteins are involved. Thus, the current standard for the culture of undifferentiated hESCs is the use of a defined liquid medium, such as mTeSR, with Matrigel as the biomaterial substrate, though there are considerable efforts to replace Matrigel to address concerns with this animal-derived substrate.

An alternative approach to these systems involves culturing of hESCs in suspension inside of stirred tank bioreactors, which offers the potential to scale up and produce the large quantities of hESCs necessary for clinical applications (Couture, 2010). Several groups have explored the potential of this approach by examining the necessary media formulations and reactor conditions for achieving continuous passaging of hESCs in suspension culture (Amit et al., 2010, Krawetz et al., 2010, Singh et al., 2010, Steiner et al., 2010).

4.2.2 Natural Biomaterials for Undifferentiated hESC Culture

Various attempts have been made to use naturally derived biomaterial scaffolds in both 2D and 3D settings as a potential method of culturing undifferentiated hESCs. These materials are classified as either heterogeneous or homogeneous, depending on the level of purification conducted prior to scaffold production.

4.2.2.1 Heterogeneous Naturally Derived Biomaterial Scaffolds

As mentioned previously, Matrigel currently serves as the standard biomaterial substrate for the culture of undifferentiated hESCs. One group investigated the use of a 3D microwell culture system that combined Matrigel patterning with physical constraints to maintain hESCs in their undifferentiated state by limiting colony growth (Mohr et al., 2006). In this system, 3D microwells were produced by cross-linking polyurethane prepolymer. The bottoms of the wells were then coated with Matrigel to allow the hESCs to adhere, and the remaining surfaces were coated with a triethylene glycol-terminated alkanethiol self-assembling monolayer (SAM) that resists protein and cell adsorption. The result was that these microwells could then restrict the size of the resulting hESC colonies while preserving their undifferentiated state. Such a strategy would be even more effective if it could be implemented without the use of the Matrigel substrate. Another group investigated the use of microwells seeded with MEFs and found this system could maintain homogeneous colonies of undifferentiated hESCs (Khademhosseini et al., 2006), an intriguing result, though MEFs place some limitations on the long-term applicability of such a system.

More recent work investigated how hESC behavior was altered when Matrigel was adsorbed onto a variety of substrates (Kohen et al., 2009). The following Matrigel-coated surfaces were examined: polystyrene, tissue culture-grade oxygen plasma-treated polystyrene, and glass. While Matrigel formed multilayer networks on all surfaces, these networks had a fibrillar morphology when coated onto the tissue-grade polystyrene or glass, but not on polystyrene. Also, while the network was denser when coated onto glass, they found that the Matrigel-coated tissue culture-grade polystyrene was the most effective for culturing undifferentiated hESCs. The results of this study suggest that modulating both the structure and density of the proteins presented on a 2D surface may be important considerations when trying to engineer a suitable replacement for Matrigel.

Another set of studies has investigated the use of other types of heterogeneous protein mixtures that could serve as potential alternatives to Matrigel. The first examined the use of human serum—which would reduce potential immune-response issues and possible exposure to xenogenic pathogens—as an alternative for coating biomaterial substrates for undifferentiated hESCs culture (Stojkovic et al., 2005). They demonstrated that plates coated with human serum maintained undifferentiated hESCs in their pluripotent state provided that they were cultured in the presence of hESC-conditioned media. They further confirmed that the hESCs had a normal karyotype and maintained the capacity to induce teratoma formation in immunodeficient mice, that is, the formation of all three germ layers of an early embryo that indicates the cells are pluripotent. A more recent study compared a variety of additional extracellular matrix substrates to determine their suitability for culture of undifferentiated hESCs (Hakala et al., 2009), such as fibronectin, a mixture of proteins (collagen IV, vitronectin, fibronectin, and laminin), human serum, fetal bovine serum, and Matrigel. The different substrates were tested in a combination with both the conventional hESC media containing FGF-2 and the hESC-conditioned media. The results indicated that the combination of Matrigel and TeSR1 media was superior to all other media/substrate combinations tested for the capacity to maintain undifferentiated hESCs with continuous passaging. Two other combinations, including using either fetal bovine serum or human serum as a substrate in conjunction with hESC-conditioned media, supported hESC culture for multiple passages (10–14 in total). These studies illustrate the progress and the difficulty of finding a suitable culture substrate replacement for Matrigel.

4.2.2.2 Homogeneous Naturally Derived Biomaterial Scaffolds

The alternative to using a heterogeneous, naturally derived biomaterial is to focus on substrates derived from a single type of biomaterial. Such substrates have a more defined composition compared to Matrigel and serum, and thus the results they yield may experience less variability or reduce the potential for exposing cells to a viral or bacterial contaminant. Interestingly, numerous studies with single-component scaffolds tend to be conducted in 3D, which could offer advantages over the previous approaches detailed in Section 4.2.2.1 that examined biomaterial substrates in 2D. Specifically, the 3D environment may better provide a more biomimetic environment, as well as offer the potential for facile scale-up of cultures.

4.2.2.2.1 Alginate

Multiple research groups have evaluated 3D alginate scaffolds as a potential material for culturing undifferentiated hESCs. Alginate is a negatively charged, linear polysaccharide derived from the cell walls of brown algae, and it can be polymerized to form hydrogels. One study encapsulated hESCs inside alginate hydrogels, in the presence of chemically defined media (Siti-Ismail et al., 2008) (Figure 4.2). This approach allowed hESCs to remain undifferentiated for up to 260 days of culture, likely in part by inhibiting differentiation via restricting the space for cell expansion. After this extended time period, the cells could be released from the hydrogels and differentiated into the three germ layers. This strategy uses chemically defined media while avoiding the use of passaging or animal-derived products, though this alginate encapsulation strategy restricts the capacity for substantial cell expansion. A more recent study cultured hESCs inside scaffolds composed of both alginate and chitosan, another linear polysaccharide derived from chitin, which is found in the shells of crustaceans (Li et al., 2010). The authors showed that the hESCs maintained pluripotency when cultured inside these scaffolds for 21 days. The hESCs expanded exponentially for 12 days and then leveled off into a linear phase. Furthermore, these scaffolds can also be easily decomposed to recover the cells. Both these studies show that 3D alginate scaffolds hold promise for maintaining hESCs in their undifferentiated state without using extracellular matrix proteins.

4.2.2.2.2 Hyaluronic Acid

Hyaluronic acid (HA), a nonsulfated polysaccharide that is highly expressed during embryogenesis, has also been evaluated as a potential 3D biomaterial scaffold material for the culture of undifferentiated hESCs (Gerecht et al., 2007). In this study, hESCs were encapsulated in a 3D HA scaffold and cultured in the presence of mouse embryonic fibroblast conditioned media. This combination of scaffold and media allowed the hESCs to remain viable and undifferentiated. To confirm that these results were due to the

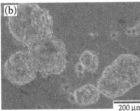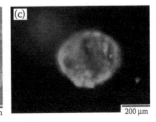

FIGURE 4.2 **(See color insert.)** Morphology and viability of the hESCs encapsulated within alginate hydrogels. (a) hESC aggregates encapsulated for 110 days with no differentiation into the germ layers or cysts being observed, (b) the morphology of the decapsulated at day 110 hESCs cultured in 2D cultures, and (c) the encapsulated hESCs remained viable within the aggregates at day 110. (From Siti-Ismail, N. et al. 2008. *Biomaterials*, 29, 3946–52.)

presence of the HA matrix and not the 3D environment, the same experiments were repeated with a dextran scaffold, which did not maintain hESCs in an undifferentiated state. These results suggest that the HA may provide specific cues necessary for maintaining hESC pluripotency.

4.2.2.2.3 Laminin

A recent study has detailed the use of recombinantly expressed laminin-511 (one of the main components of Matrigel) as a substrate for the long-term culture of hESCs (Rodin et al., 2010). Undifferentiated hESCs were cultured in monolayers on dishes coated with human laminin-511 in the presence of a variant of chemically defined mTeSR media. They were impressively able to passage these cells for 4 months in culture while maintaining pluripotency. This culture shows promise as a way to culture undifferentiated hESCs in an xeno and feeder layer-independent manner, though the use of recombinantly expressed proteins could pose challenges for economics of scale-up.

4.2.2.2.4 Vitronectin

In addition to laminin, the use of recombinantly produced vitronectin has been evaluated as a substrate for hESC culture (Prowse et al., 2010). Vitronectin is found in both the extracellular matrix as well as in serum. It contains multiple binding domains, including those that bind integrins—a desirable property for substrates intended for hESC culture. In this study, they focused on developed a substrate for hESCs that consisted of a recombinantly expressed, truncated version of the vitronectin protein including the somatomedin B domain and an RGD domain. When used in combination with the appropriate defined media, surfaces coated with this protein were able to support hESC passage for over 10 passages, providing an intriguing alternative to the use of Matrigel.

4.2.3 Synthetic Biomaterials for Undifferentiated hESC Culture

As described in the introduction, synthetic biomaterials offer certain advantages over natural biomaterials, including reproducibility and the ability to broadly modulate the structural and mechanical properties of a material, though such materials intrinsically lack bioactive cues for interacting with cells. As one example, when Hakala et al. investigated a wide variety of potential synthetic substrates, they found the following synthetic materials to be unsuitable for supporting the culture of hESCs: titanium, dioxide-coated titanium, zirconium dioxide-coated titanium, poly(desaminotyrosyl-tyrosine-ethyl ester carbonate), and poly-L-D-lactide (Hakala et al., 2009). These results show the difficulty of finding a synthetic material that can support hESC culture. This section will describe two strategies that have proven somewhat effective at maintaining undifferentiated hESCs and that with some refinement could potentially eliminate the need to use Matrigel for hESC culture: (1) designing synthetic materials that mimic how the extracellular matrix interacts with the integrins necessary to maintain hESC culture and (2) high-throughput screens that identify new bioactive ligands or materials for the this purpose.

4.2.3.1 Biomaterials That Mimic the Extracellular Matrix

One strategy for developing synthetic materials that support hESC culture is to use synthetic peptide sequences that mimic the ECM protein motifs that bind cell surface adhesion receptors. In an early work, the Healy group used a synthetic semi-interpenetrating polymer network (sIPN) hydrogel for supporting hESC culture (Li et al., 2006). Both the ligand presentation and mechanical properties of these sIPNs can be tuned, and in this work they functionalized the sIPN with an RGD (arginine–glycine–aspartic acid) peptide, a motif known to bind a subset of integrin receptors. In the presence of MEF-conditioned medium, these sIPNS were able to support hESC culture for up to 5 days, and this study found that higher concentrations of the RGD peptides worked better at maintaining hESCs in their undifferentiated state. A more recent work investigated the specific integrins necessary for hESCs to engage and adhere to a biomaterial surface (Meng et al., 2010). The investigators examined integrin expression on the surface of hESCs by reverse transcription polymerase chain reaction (RT-PCR) and immunostaining, then performed blocking assays using integrin-specific antibodies to identify the receptors necessary for adhesion to Matrigel. Four integrins were involved for this adhesion: $\alpha_V\beta_3$, α_6, β_1, and $\alpha V\beta_3$. The peptides designed to bind these integrins were tested individually and in combination, and a blend of two integrin-binding peptides with a peptide that binds syndecan was sufficient to promote hESC adhesion and short-term maintenance. A more recent study has investigated the use of synthetic peptide-acrylate surfaces as a means of supporting undifferentiated hESC culture (Melkoumian et al., 2010). They found that the combination of peptides derived from bone sialoprotein and vitronectin containing RGD sequences conjugated to acrylate-coated surfaces were able to support undifferentiated hESC culture for 10 passages. They also showed that cells cultured in this manner could also be differentiated into functional cardiomyocytes. The resulting commercially available plates (Corning), however, pose challenges for the economics of scale-up, since bioactivity requires considerable amounts of the peptides on the surfaces.

Another study developed a synthetic polymer coating capable of supporting the passage of undifferentiated hESCs (Villa-Diaz et al., 2010). Coating plates with poly[2-(methacrloyloxy)ethyl dimethyl-(3-sulfopropyl)ammonium hydroxide] to serve as a substrate for hESC adhesion was an effective means for undifferentiated hESC culture in the presence of StemPro media, though the mechanical passaging of individual colonies was apparently involved. It was hypothesized that this surface adsorbed the proteins secreted by the hESCs and thus mimicked the ECM needed to support hESC culture, and future work will likely elucidate the mechanism of this interesting surface as a substrate for undifferentiated hESCs. These studies demonstrate how investigating the mechanisms of cell–matrix engagement can aid in the design of synthetic materials to interact with hESCs and maintain their pluripotency. However, there are still challenges in developing materials that fully emulate the properties of Matrigel.

4.2.3.2 Biomaterials Identified Using High-Throughput Screening

High-throughput screening provides an alternative means to design and identify materials that can support hESC culture. These approaches involve screening a large number of materials or peptide sequences for suitable cell culture substrates. One of the first high-throughput studies focused on screening an array of biomaterials consisting of different combinations of acrylate, diacrylate, dimethacrylate, and triacrylate monomers to determine which would support undifferentiated hESC culture (Anderson et al., 2004). Different combinations of these monomers were present in an array that was then seeded with hESCs. After 6 days, the resulting cultures were assessed for cell survival and differentiation, and this intriguing approach identified several novel materials that could support hESC attachment. A follow-up study examined the effect of hydrophobicity/hydrophilicity and cross-linking density of the various combinations of acrylate monomers on the ability of hESCs to adhere to a surface (Mei et al., 2009). They also assessed the ability of these materials to absorb fibronectin, as this property would correlate to cell adhesion. However, their results showed that materials absorbing similar amounts of fibronectin had differences in their ability to support undifferentiated hESC culture, interestingly suggesting that protein confirmation may play an important role. One caveat with this work is that the specific monomer concentrations that could successfully support hESC culture required presentation on a poly(2-hydroxyethyl methacrylate) pHEMA surface. This high throughput approach to screening synthetic polymer arrays was further extended to identify the first xeno-free, chemically defined synthetic substrate for supporting robust hESC culture (Mei et al., 2010). In this study, various combinations of monomers were arrayed and their roughness, elastic modulus, wettability, and surface chemistry were characterized before plating hESCs on the arrays to determine their ability to support sustained hESC growth. They found a correlation between the acrylate concentration and the ability of the polymer to support hESC culture.

A similar approach was used to identify a synthetic polymer—poly(methyl vinyl ether-alt-maleic anhydride) (PMVE-alt-MA)—that supported the long-term attachment, proliferation and self-renewal of both hESCs and induced pluripotent stem cells (Brafman et al., 2010). This approach screened arrays consisting of 91 polymers at 5 different concentrations for identifying the appropriate substrates for

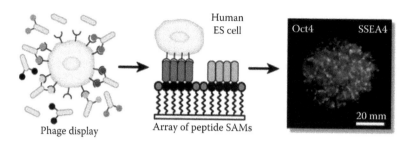

FIGURE 4.3 **(See color insert.)** High-throughput identification of peptides that support hESC attachment and proliferation. (From Derda, R. et al. 2010. *J Am Chem Soc*, 132, 1289–95.)

long term attachment and self renewal of human pluripotent stem cells. The hESCs cultured on PMVE-alt-MA maintained the characteristic colony morphology and expressed OCT4 as well as Nanog.

More recent studies have used other high-throughput methods to address a different problem: the identification of peptide sequences that could potentially support hESC culture. Work from the Kiessling group developed arrays of SAMs that displayed different peptide sequences, which were then screened to determine which peptides supported hESC growth (Derda et al., 2007). They focused on screening peptides derived from the protein laminin, as it is a major component of Matrigel. The work showed that both the sequence and density of the peptide were important for supporting hESC growth, results that were applied to synthesize a 3D scaffold based on an identified peptide sequence (RNIAEIIKDI) that could maintain hESCs in their undifferentiated state for 4 days. Another study from this group selected a phage library presenting random peptide sequences to identify ones that bind to the human embryonic carcinoma cells (Figure 4.3) (Derda et al., 2010). Subsequent testing of 370 clones yielded six with the ability to bind to the embryonic carcinoma cells. Two of these peptide sequences (TVKHRPDALHPQ and LTTAPKLPKVTR) then showed the ability to support hESC culture for 20 days (three passages) when presented via a SAM monolayer in the presence of the TeSR1 media. The most recent study from the Kiessling group applied the same approach to finding peptide substrates to support hESCs and found that the two most effective peptide substrates (GKKQRFRHRNRKG, FHRRIKA, and GWQPPARARI) contained heparin binding sequences (Klim et al., 2010). Collectively, these studies show that high-throughput screening provides a method for identifying materials and peptide sequences that can maintain undifferentiated hESCs without needing prior biological knowledge.

4.3 Biomaterials for Promoting hESC Differentiation into Specific Lineages

In addition to serving as systems for hESC propagation, biomaterials are being developed to direct undifferentiated hESCs into mature lineages, and this section will focus on studies involving differentiation into numerous cell types from each of the three germ layers.

4.3.1 Endoderm

During development, the endoderm gives rise to the tissue that makes up the digestive tract, the respiratory system, bladder, endocrine system, and vasculature. One study showed that when undifferentiated hESCs were encapsulated inside alginate capsules, they tended to primarily differentiate into an endodermal lineage, with markers of other lineages (mesoderm and ectoderm) detected at lower levels (Dean et al., 2006). Another group also investigated the use of alginate microcapsules as a means of directing hESCs to differentiate into an endoderm (Chayosumrit et al., 2010). To promote differentiation, they treated the hESCs in both serum replacement media, as well as in conditioned media obtained from human fetal fibroblasts, in the presence and absence of the small-molecule Rho-associated kinase (ROCK) inhibitor Y-27632. They found that the addition of Y-27632 to either type of medium would direct hESC differentiation into the endoderm as indicated by several cellular markers. In another work that involved spatially confining hESCs, it was shown that micropatterning these cells into colonies 200 μm in diameter promoted differentiation into a definitive endoderm (Lee et al., 2009). Additional studies have focused on further maturing endodermal tissue into specific lineages of therapeutic relevance, including pancreatic, liver, and vascular cells.

4.3.1.1 Pancreas

As part of the endocrine system, the pancreas serves many functions, including secreting digestive enzymes into the small intestine and releasing insulin to regulate glucose uptake throughout the body. Tissue engineering approaches using hESCs have focused on producing β-cells, the insulin-secreting cells found in the pancreatic islets of Langerhans that are lost in type I diabetes due to an autoimmune response. One group successfully developed a five-step protocol for differentiating hESCs into islet-like cells, including β-cells, followed by encapsulation in a biomaterial scaffold for implantation (Mao et al., 2009). To enhance the survival of these cells in an *in vivo* setting, they were encapsulated inside poly(lactic-*co*-glycolic acid) (PLGA) scaffolds prior to transplantation into a mouse model of type I diabetes. The mice subsequently experienced improvements in glucose levels after fasting, suggesting that the implants could reverse the effects of this diabetes model. This study provides an example of how hESCs combined with biomaterial scaffolds could lead to the development of a clinically promising therapy.

4.3.1.2 Liver

Multiple groups have investigated different methods of using biomaterial scaffolds to direct hESCs differentiation into the liver tissue. In 2003, Levenberg et al. investigated the ability of hESCs to give rise to different lineages when seeded inside PLGA scaffolds and found that treating hESCs with insulin-like growth factor and activin-A (ACT) resulted in differentiation into the tissue exhibiting the characteristics of the liver tissue, as indicated by immunostaining (Levenberg et al., 2003) (Figure 4.4). More recently, a study investigated the use of poly(tetrafluoroethylene) (PTFE) coated with poly(amino urethane) (PAU) as a 2D biomaterial substrate for differentiating hESCs into hepatocytes (Soto-Gutierrez et al., 2006). When the hESCs seeded on these scaffolds were treated with FGF-2, a deleted variant of hepatocyte growth factor, and dimethyl sulfoxide (DMSO), the cells differentiated into hepatocytes as indicated by albumin staining and their ability to metabolize ammonia and lidocaine. In another attempt to engineer the liver tissue, one group seeded PLGA scaffolds with hESCs and then transplanted these scaffolds into the liver tissue of immunodeficient mice (Lees et al., 2007). The hESCs differentiated in multiple lineages as indicated by expression of markers for hepatocytes, β-cells, and neurons. Thus, while hepatocytes were successfully generated, with many differentiation systems and materials there can be challenges in generating a high proportion of a specific, desired lineage.

4.3.1.3 Vasculature

A major challenge in engineering tissue is the generation of stable vasculature, the network of vessels that supply blood to a tissue or an organ. This section will detail the numerous attempts at creating vasculature inside biomaterial scaffolds through the differentiation of hESCs into endothelial cells.

One of the first studies in this area examined the ability of hESCs to form vasculature when seeded inside PLGA scaffolds in both *in vitro* and *in vivo* settings (Levenberg et al., 2003). In this work, also discussed earlier in the section on liver, the authors promoted hESC differentiation into a variety of tissues, including neural, liver, and cartilage and interestingly also observed the formation of a 3D network of blood vessels, suggesting the potential of hESCs to form stable vasculature when engineering tissues. Another study examined the ability of embryoid bodies generated from hESCs to form blood vessels inside alginate scaffolds (Gerecht-Nir et al., 2004). The resulting 3D culture system promoted an enhancement in vascular differentiation compared to static and rotary culture of hESCs, suggesting that hESC differentiation can be induced through manipulation of the structural properties and in particular the porosity of a biomaterial scaffold. HA hydrogels have also been investigated as a potential scaffold material for promoting hESC

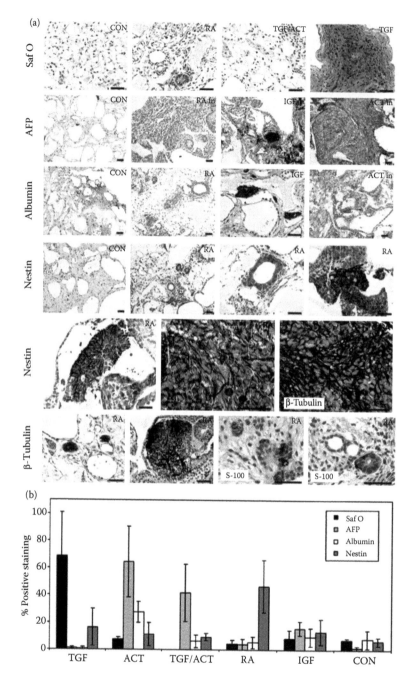

FIGURE 4.4 (See color insert.) Differentiation of hESCs into mesodermal-, ectodermal-, and endodermal-derived cell types and tissue-like structures when seeded inside PLGA scaffolds. (a) Immunostaining of tissue sections taken from hES constructs incubated for 2 weeks with control medium (CON) or medium supplemented with growth factors: TGF-β (TGF), activin-A (ACT), RA, IGF, and a combination of TGF-β and activin-A (TGF/ACT). Samples were stained with Safranin-O (Saf O) or with antibodies against human alpha fetoprotein, albumin, nestin, β_{III}-tubulin, and S-100 (scale bars = 50 μm). (b) Quantitative analysis of antibody staining. The percentage of positive staining corresponds to percentages of the area positively stained with the antibody within the tissue sections. The results shown are mean values (±SD) of sample sections obtained in three different experiments performed in duplicate. (From Levenberg, S. et al. 2003. *Proc Natl Acad Sci USA*, 100, 12741–6).

differentiation into vasculature, based on the finding that some HA degradation products have been shown to induce angiogenesis (Gerecht et al., 2007, Slevin et al., 2007). Gerecht et al. (2007) found that hESCs seeded inside these scaffolds could be encouraged to differentiate into vasculature when treated with vascular endothelial growth factor (VEGF), as indicated by a number of markers including smooth muscle actin.

A different approach to promote vasculature formation from hESCs involved bioactive dextran scaffolds functionalized with RGD peptides and containing microspheres that released VEGF (Ferreira et al., 2007). Embryoid bodies were seeded inside these scaffolds and cultured for 10 days, after which the cells showed a large increase in the expression of the VEGF receptor Flk-1. This expression was not observed under other culture conditions. Ferreira et al. also found that these cells could be further differentiated into more mature endothelial cells by removing them from the scaffolds and culturing them in endothelial growth medium (EGM) supplemented with VEGF.

A more recent study investigated the ability of hESCs to form functional endothelium when seeded inside poly(2-hydroxyl methacrylate) scaffolds *in vivo* (Nourse et al., 2010). The authors pretreated the hESCs with VEGF and used cell sorting to identify CD31 positive endothelial cells before seeding them inside these biomaterial scaffolds, which were then implanted subcutaneously into nude rats. Ten days after implantation, the harvested scaffold constructs contained blood vessels derived from the selected hESCs, showing the potential of poly(2-hydroxyl methacrylate) as a substrate for supporting hESC differentiation.

4.3.2 Mesoderm

The mesoderm is the germ layer that gives rise to muscle, bone and bone marrow, part of the skin, and connective tissue. While many studies have focused on using mesenchymal stem cells (MSCs) to engineer bone and cartilage tissue, primary human MSCs can typically only be cultured and expanded for a limited time. Therefore, hESCs provide a valuable alternative for engineering bone and cartilage. One study showed that patterning hESCs into large colonies (1200 μm in diameter) in the presence of bone morphogenic protein 2 (BMP-2) and activin A resulted in differentiation into mesoderm (Lee et al., 2009), indicating that it is possible to develop approaches to uniformly direct hESCs into a mesodermal lineage, and this section will focus on attempts to further differentiate hESCs into specific cell types produced from the mesoderm.

4.3.2.1 Bone

hESCs offer a potential means for engineering bone tissue that could serve as a replacement for skeletal defects. In 2003, McWeir and colleagues induced hESCs to differentiate into an osteogenic lineage when seeded on gelatin-coated, 2D surfaces in the presence of media supplemented with ascorbic acid phosphate, β-glycerophosphate, and dexamethasone (Sottile et al., 2003). They confirmed osteogenic differentiation by measuring calcium deposition and staining for a variety of markers associated with the bone tissue. A more recent study examined the behavior of hESCs pretreated to differentiate into osteogenic lineages when seeded inside composite scaffolds consisting of PLGA and the mineral hydroxyapatite (HA) for bone tissue engineering applications (Kim et al., 2008). When these scaffolds were implanted *in vivo*, they were able to promote significant mineralization and bone formation when implanted subcutaneously. They also found that the addition of BMP-2 further enhanced hESC differentiation and the resulting formation of bone tissue. Another study focused on the behavior of hESCS on 2D surfaces consisting of fibronectin-coated gold nanoparticles in the presence and absence of electrical stimulation (Woo et al., 2009). They found that the combination of this biomaterial and stimulation promoted an increase in osteogenic markers not observed without electrical stimulation. These studies illustrate how combining biomaterial scaffolds with additional cues can successfully differentiate hESCs toward bone tissue.

4.3.2.2 Cartilage

A number of studies have examined the potential of hESCs to differentiate into cartilage and ligament, which could serve as replacements for the injured and damaged tissue. One study examined hESC differentiation into cartilage when seeded inside PLGA scaffolds and treated with TGF-β *in vitro* and *in vivo* (Levenberg et al., 2003). They assessed cartilage formation through the use of Safranin-O staining and glycosaminoglycan production, and these features were observed in the presence of TGF-β but not other growth factors added to the culture medium.

A different study performed by the Polak group attempted to induce chondrocyte formation from hESCs using a coculture system with primary chondrocytes, then seeded the resulting progenitor cells into PLGA for further *in vivo* testing (Vats et al., 2006). Specifically, after 28 days in the coculture, the resulting cells were characterized based on morphology and immunocytochemistry. The cells not only resembled chondrocytes morphologically after the coculture, but they also expressed collagen II and Sox9, which are associated with cartilage formation. PLGA scaffolds subsequently seeded with these pretreated cells formed cartilage tissue *in vivo* with extensive vascularization when implanted into severe combined immunodeficiency (SCID) mice. Recent work by Elisseeff and colleagues evaluated the ability of hESCs to differentiate into chondrocytes when seeded inside poly(ethylene glycol) diacrylate scaffolds modified to contain RGD sequences (Hwang et al., 2006). The hESCs were first induced to form embryoid bodies, then seeded into the poly (ethylene glycol) diacrylate (PEGDA)-RGD scaffolds and cultured in the presence of TGF-β to further enhance differentiation. hESCs cultured in these scaffolds showed significant increases in Safronin-O staining, glycosaminoglycan secretion, and collagen production, demonstrating the potential of combining hESCs with biomaterial scaffolds for cartilage tissue engineering.

Another study developed a method for differentiating hESCs into cartilage tissue that involved culturing the cells in 2D first followed by 3D culture (Bai et al., 2010). Five-day embryoid bodies containing hESCs were seeded upon gelatin-coated plates, then cultured in a medium that promoted chondrogenic differentiation for 27 days. After this induction process in 2D culture, the cells were encapsulated in alginate and embedded into PLGA scaffolds. These cells differentiated into chondrocytes in both *in vitro* culture and *in vivo* and were able to remain differentiated over extended periods of time compared to the previous work using hESCs to generate a cartilage. This study illustrates the complex processes that can be involved in inducing hESC differentiation in chondrocytes.

Finally, a recent study from the Kaplan lab used bioreactors to culture hESC-derived mesenchymal stem cells (MSCs) seeded on silk scaffolds for engineering cartilage tissue (Tigli et al., 2011). They first induced the hESCs to differentiate into MSCs followed by a selection step to isolate a homogenous MSC population before seeding the cells onto porous silk scaffolds. The cells were then incubated for 5 days on the scaffolds to allow the cells to adhere before being placed into a continuous flow bioreactor. These scaffolds were cultured in the bioreactors for 4 weeks and the resulting constructs analyzed. The cells showed an increase in expression of cartilage associated genes as well as increases in Safronin-O staining and collagen type II production. This study illustrates the complexity and lengthy time course necessary when engineering tissues using hESCs as a starting point.

4.3.2.3 Cardiac Tissue

Microwell culture methods, similar to those described previously in Section 4.2.2.1, have been utilized to direct hESC differentiation in cardiomyocytes (Mohr et al., 2010). Investigators cultured hESCs in microwells with a variety of lateral dimensions, while maintaining a constant well depth of 300 μm. They found that while smaller embryoid bodies were less likely to contract and differentiate into cardiac tissue, those that did contract showed a high percentage of cells differentiating into cardiac tissue. Larger embryoid bodies were more likely to contract, but a small fraction of cells stained positive for the markers, indicating differentiation into cardiomyocytes. This study provides potential criteria for spatially designing hESC differentiation systems for cardiac tissue engineering.

4.3.3 Ectoderm

The nervous system, the top layer of the epidermis, and tooth enamel are all formed from the ectoderm. This section will summarize the studies that have demonstrated the potential of hESCs to generate neural tissue when cultured on and inside biomaterial scaffolds, as there has been little focus to date on generating other regions of the ectoderm.

4.3.3.1 Neural Tissue

Many groups have used a variety of scaffolds and conditions for producing neural tissue from hESCs. One of the first attempts involved seeding hESCs into 3D PLGA scaffolds in the presence of retinoic acid (Levenberg et al., 2003). The cells differentiated into neural tissue, specifically neurons as indicated by β-tubulin III staining after 2 weeks of *in vitro* culture. Similar staining results were observed when, after 2 weeks of *in vitro* culture, the scaffolds were implanted *in vivo* into SCID mice for 2 weeks. A more recent study also examined the behavior of hESCs seeded inside 3D PLGA scaffolds in response to the treatment with the nerve growth factor (NGF) and retinoic acid (Inanc et al., 2008). They found that these conditions upregulated the expression of nestin (a neural progenitor marker) while downregulating the expression of α-fetoprotein (an endoderm marker), confirming the potential of such a scaffold for generating a neural tissue.

Another study used a variety of substrates to determine which extracellular matrix interactions are responsible for regulating the differentiation of hESCs into neural lineages when cells were cultured in a neural induction medium (Ma et al., 2008). They examined the following 2D substrates: poly-D-lysine (PDL), PDL/fibronectin mixture, PDL/laminin mixture, type I collagen, and Matrigel. They found that laminin specifically enhanced the ability of hESCs to differentiate into neural progenitors and eventually neurons by binding to the integrin α6β1. Thus, the use of laminin or laminin motifs may be important when designing biomaterial scaffolds for promoting the generation of a neural tissue from undifferentiated hESCs. Another group examined the ability of hESCs to differentiate into neurons when seeded on 2D thin films made of carbon nanotubes grafted with poly(acrylic acid) (Chao et al., 2009). They found that these thin films enhanced the fraction of hESCs that both adhered to the surface and differentiated into neurons, as indicated by β-tubulin III staining, compared to the culture on 2D surfaces coated with poly-L-ornithine, illustrating the potential of carbon films for directing hESC differentiation into neurons.

4.4 Conclusions

The work on materials for hESC culture and differentiation illustrates the utility, variety, and versatility of biomaterial scaffolds. Synthetic materials offer particular advantages in reproducibility and design flexibility; however, there are challenges in functionalizing and endowing them with the complex signaling cues inherent in natural materials. Therefore, high-throughput screening and selection tools offer considerable potential for identifying and designing synthetic systems with the activity of their natural counterparts. Furthermore, while most studies till date have been conducted in 2D, an emerging number of results indicate the value of 3D studies as a better means to emulate natural tissue and thereby potentially better direct cell fates.

Numerous examples in Section 4.3 illustrate how a scaffold material can influence hESC differentiation and survival both *in vitro* and *in vivo*. One of the next major challenges presented by developing clinically relevant hESC-based therapies is how to control the desired, and prevent unwanted, differentiation, as the presence of contaminating cell types may lead to undesired side effects and possible tumorigenicity. The design of biomaterials to provide potent and uniform differentiation cues may aid in addressing this challenge. Furthermore, there are often challenges with the viability of the resulting cells upon implantation, and the use of scaffolds also has the potential to address this problem. Therefore, the research progress described in this review will have a considerable promise in increasingly being translated toward hESC-based therapies.

References

Ameen, C., Strehl, R., Bjorquist, P., Lindahl, A., Hyllner, J., and Sartipy, P. 2008. Human embryonic stem cells: Current technologies and emerging industrial applications. *Crit Rev Oncol Hematol*, 65, 54–80.

Amit, M., Chebath, J., Margulets, V., Laevsky, I., Miropolsky, Y., Shariki, K., Peri, M. et al. 2010. Suspension culture of undifferentiated human embryonic and induced pluripotent stem cells. *Stem Cell Rev*, 6, 248–59.

Anderson, D. G., Levenberg, S., and Langer, R. 2004. Nanoliter-scale synthesis of arrayed biomaterials and application to human embryonic stem cells. *Nat Biotechnol*, 22, 863–6.

Bai, H. Y., Chen, G. A., Mao, G. H., Song, T. R., and Wang, Y. X. 2010. Three step derivation of cartilage like tissue from human embryonic stem cells by 2D-3D sequential culture *in vitro* and further implantation *in vivo* on alginate/PLGA scaffolds. *J Biomed Mater Res A*, 94(2), 539–46.

Brafman, D. A., Chang, C. W., Fernandez, A., Willert, K., Varghese, S., and Chien, S. 2010. Long-term human pluripotent stem cell self-renewal on synthetic polymer surfaces. *Biomaterials*, 31, 9135–44.

Chao, T. I., Xiang, S., Chen, C. S., Chin, W. C., Nelson, A. J., Wang, C., and Lu, J. 2009. Carbon nanotubes promote neuron differentiation from human embryonic stem cells. *Biochem Biophys Res Commun*, 384, 426–30.

Chayosumrit, M., Tuch, B., and Sidhu, K. 2010. Alginate microcapsule for propagation and directed differentiation of hESCs to definitive endoderm. *Biomaterials*, 31, 505–14.

Couture, L. A. 2010. Scalable pluripotent stem cell culture. *Nat Biotechnol*, 28, 562–3.

Dean, S. K., Yulyana, Y., Williams, G., Sidhu, K. S., and Tuch, B. E. 2006. Differentiation of encapsulated embryonic stem cells after transplantation. *Transplantation*, 82, 1175–84.

Derda, R., Li, L., Orner, B. P., Lewis, R. L., Thomson, J. A., and Kiessling, L. L. 2007. Defined substrates for human embryonic stem cell growth identified from surface arrays. *ACS Chem Biol*, 2, 347–55.

Derda, R., Musah, S., Orner, B. P., Klim, J. R., Li, L., and Kiessling, L. L. 2010. High-throughput discovery of synthetic surfaces that support proliferation of pluripotent cells. *J Am Chem Soc*, 132, 1289–95.

Duan, Y., MA, X., Zou, W., Wang, C., Behbahan, I. S., Ahuja, T. P., Tolstikov, V., and Zern, M. A. 2010. Differentiation and characterization of metabolically functioning hepatocytes from human embryonic stem cells. *Stem Cells* 28(4), 674–86.

Ferreira, L. S., Gerecht, S., Fuller, J., Shieh, H. F., Vunjak-Novakovic, G., and Langer, R. 2007. Bioactive hydrogel scaffolds for controllable vascular differentiation of human embryonic stem cells. *Biomaterials*, 28, 2706–17.

Gerecht, S., Burdick, J. A., Ferreira, L. S., Townsend, S. A., Langer, R., and Vunjak-Novakovic, G. 2007. Hyaluronic acid hydrogel for controlled self-renewal and differentiation of human embryonic stem cells. *Proc Natl Acad Sci USA*, 104, 11298–303.

Gerecht-Nir, S., Cohen, S., Ziskind, A., and Itskovitz-Eldor, J. 2004. Three-dimensional porous alginate scaffolds provide a conducive environment for generation of well-vascularized embryoid bodies from human embryonic stem cells. *Biotechnol Bioeng*, 88, 313–20.

Hakala, H., Rajala, K., Ojala, M., Panula, S., Areva, S., Kellomaki, M., Suuronen, R., and Skottman, H. 2009. Comparison of biomaterials and extracellular matrices as a culture platform for multiple, independently derived human embryonic stem cell lines. *Tissue Eng Part A*, 15, 1775–85.

Harkness, L., Taipaleenmaki, H., Mahmood, A., Frandsen, U., Saamanen, A. M., Kassem, M., and Abdallah, B. M. 2009. Isolation and differentiation of chondrocytic cells derived from human embryonic stem cells using dlk1/Fa1 as a novel surface marker. *Stem Cell Rev*, 5, 353–68.

Hill, K. L., Obrtlikova, P., Alvarez, D. F., King, J. A., Keirstead, S. A., Allred, J. R., and Kaufman, D. S. 2010. Human embryonic stem cell-derived vascular progenitor cells capable of endothelial and smooth muscle cell function. *Exp Hematol*, 38, 246–257.

Hwang, N. S., Varghese, S., Zhang, Z., and Elisseeff, J. 2006. Chondrogenic differentiation of human embryonic stem cell-derived cells in arginine-glycine-aspartate-modified hydrogels. *Tissue Eng*, 12, 2695–706.

Inanc, B., Elcin, A. E., and Elcin, Y. M. 2008. Human embryonic stem cell differentiation on tissue engineering scaffolds: Effects of NGF and retinoic acid induction. *Tissue Eng Part A*, 14(6), 955–64.

Kent, L. 2009. Culture and maintenance of human embryonic stem cells. *J Vis Exp*, 34, e1427.

Khademhosseini, A., Ferreira, L., Blumling, J., Yeh, J., Karp, J. M., Fukuda, J., and Langer, R. 2006. Co-culture of human embryonic stem cells with murine embryonic fibroblasts on microwell-patterned substrates. *Biomaterials*, 27, 5968–77.

Kim, S., Kim, S. S., Lee, S. H., Eun, A. H. N. S., Gwak, S. J., Song, J. H., Kim, B. S., and Chung, H. M. 2008. *In vivo* bone formation from human embryonic stem cell-derived osteogenic cells in poly(d,l-lactic-*co*-glycolic acid)/hydroxyapatite composite scaffolds. *Biomaterials*, 29, 1043–53.

Kleinman, H. K., and Martin, G. R. 2005. Matrigel: Basement membrane matrix with biological activity. *Semin Cancer Biol*, 15, 378–86.

Klim, J. R., Li, L., Wrighton, P. J., Piekarczyk, M. S., and Kiessling, L. L. 2010. A defined glycosaminoglycan-binding substratum for human pluripotent stem cells. *Nat Methods*, 7, 989–94.

Kohen, N. T., Little, L. E., and Healy, K. E. 2009. Characterization of Matrigel interfaces during defined human embryonic stem cell culture. *Biointerphases*, 4, 69–79.

Krawetz, R., Taiani, J. T., Liu, S., Meng, G., Li, X., Kallos, M. S., and Rancourt, D. E. 2010. Large-scale expansion of pluripotent human embryonic stem cells in stirred-suspension bioreactors. *Tissue Eng Part C Methods*, 16, 573–82.

Lee, L. H., Peerani, R., Ungrin, M., Joshi, C., Kumacheva, E., and Zandstra, P. 2009. Micropatterning of human embryonic stem cells dissects the mesoderm and endoderm lineages. *Stem Cell Res*, 2, 155–62.

Lees, J. G., Lim, S. A., Croll, T., Williams, G., Lui, S., Cooper-White, J., Mcquade, L. R., Mathiyalagan, B., and Tuch, B. E. 2007. Transplantation of 3D scaffolds seeded with human embryonic stem cells: Biological features of surrogate tissue and teratoma-forming potential. *Regen Med*, 2, 289–300.

Levenberg, S., Huang, N. F., Lavik, E., Rogers, A. B., Itskovitz-Eldor, J., and Langer, R. 2003. Differentiation of human embryonic stem cells on three-dimensional polymer scaffolds. *Proc Natl Acad Sci USA*, 100, 12741–6.

Li, L. and Clevers, H. 2010. Coexistence of quiescent and active adult stem cells in mammals. *Science*, 327, 542–5.

Li, Y. J., Chung, E. H., Rodriguez, R. T., Firpo, M. T., and Healy, K. E. 2006. Hydrogels as artificial matrices for human embryonic stem cell self-renewal. *J Biomed Mater Res A*, 79, 1–5.

Li, Z., Leung, M., Hopper, R., Ellenbogen, R., and Zhang, M. 2010. Feeder-free self-renewal of human embryonic stem cells in 3D porous natural polymer scaffolds. *Biomaterials*, 31, 404–12.

Lindvall, O. and Kokaia, Z. 2009. Prospects of stem cell therapy for replacing dopamine neurons in Parkinson's disease. *Trends Pharmacol Sci*, 30, 260–7.

Ludwig, T. E., Levenstein, M. E., Jones, J. M., Berggren, W. T., Mitchen, E. R., Frane, J. L., Crandall, L. J. et al. 2006. Derivation of human embryonic stem cells in defined conditions. *Nat Biotechnol*, 24, 185–7.

Ma, W., Tavakoli, T., Derby, E., Serebryakova, Y., Rao, M. S., and Mattson, M. P. 2008. Cell-extracellular matrix interactions regulate neural differentiation of human embryonic stem cells. *BMC Dev Biol*, 8, 90.

Mao, G. H., Chen, G. A., Bai, H. Y., Song, T. R., and Wang, Y. X. 2009. The reversal of hyperglycaemia in diabetic mice using PLGA scaffolds seeded with islet-like cells derived from human embryonic stem cells. *Biomaterials*, 30, 1706–14.

Martin, G. R. 1981. Isolation of a pluripotent cell line from early mouse embryos cultured in medium conditioned by teratocarcinoma stem cells. *Proc Natl Acad Sci USA*, 78, 7634–8.

Martin, M. J., Muotri, A., Gage, F., and Varki, A. 2005. Human embryonic stem cells express an immunogenic nonhuman sialic acid. *Nat Med*, 11, 228–32.

Mei, Y., Gerecht, S., Taylor, M., Urquhart, A., Bogatyrev, S., Cho, S., Davies, M., Alexander, M., Langer, R. S., and Anderson, D. G. 2009. Mapping the interactions among biomaterials, adsorbed proteins, and human embryonic stem cells. *Adv Mater*, 21, 2781–2786.

Mei, Y., Saha, K., Bogatyrev, S. R., Yang, J., Hook, A. L., Kalcioglu, Z. I., Cho, S. W. et al. 2010. Combinatorial development of biomaterials for clonal growth of human pluripotent stem cells. *Nat Mater*, 9, 768–78.

Melkoumian, Z., Weber, J. L., Weber, D. M., Fadeev, A. G., Zhou, Y., Dolley-Sonneville, P., Yang, J. et al. 2010. Synthetic peptide-acrylate surfaces for long-term self-renewal and cardiomyocyte differentiation of human embryonic stem cells. *Nat Biotechnol*, 28, 606–10.

Meng, Y., Eshghi, S., Li, Y. J., Schmidt, R., Schaffer, D. V., and Healy, K. E. 2010. Characterization of integrin engagement during defined human embryonic stem cell culture. *FASEB J*, 24(4), 1056–65.

Mohr, J. C., De Pablo, J. J., and Palecek, S. P. 2006. 3-D microwell culture of human embryonic stem cells. *Biomaterials*, 27, 6032–42.

Mohr, J. C., Zhang, J., Azarin, S. M., Soerens, A. G., De Pablo, J. J., Thomson, J. A., Lyons, G. E., Palecek, S. P., and Kamp, T. J. 2010. The microwell control of embryoid body size in order to regulate cardiac differentiation of human embryonic stem cells. *Biomaterials*, 31, 1885–93.

Nourse, M. B., Halpin, D. E., Scatena, M., Mortisen, D. J., Tulloch, N. L., Hauch, K. D., Torok-Storb, B., Ratner, B. D., Pabon, L., and Murry, C. E. 2010. VEGF induces differentiation of functional endothelium from human embryonic stem cells: Implications for tissue engineering. *Arterioscler Thromb Vasc Biol*, 30, 80–9.

Prowse, A. B., Doran, M. R., Cooper-White, J. J., Chong, F., Munro, T. P., Fitzpatrick, J., Chung, T. L., Haylock, D. N., Gray, P. P., and Wolvetang, E. J. 2010. Long term culture of human embryonic stem cells on recombinant vitronectin in ascorbate free media. *Biomaterials*, 31, 8281–8.

Rodin, S., Domogatskaya, A., Strom, S., Hansson, E. M., Chien, K. R., Inzunza, J., Hovatta, O., and Tryggvason, K. 2010. Long-term self-renewal of human pluripotent stem cells on human recombinant laminin-511. *Nat Biotechnol*, 28, 611–5.

Shi, Y. 2009. Induced pluripotent stem cells, new tools for drug discovery and new hope for stem cell therapies. *Curr Mol Pharmacol*, 2, 15–8.

Singh, H., Mok, P., Balakrishnan, T., Rahmat, S. N., and Zweigerdt, R. 2010. Up-scaling single cell-inoculated suspension culture of human embryonic stem cells. *Stem Cell Res*, 4, 165–79.

Siti-Ismail, N., Bishop, A. E., Polak, J. M., and Mantalaris, A. 2008. The benefit of human embryonic stem cell encapsulation for prolonged feeder-free maintenance. *Biomaterials*, 29, 3946–52.

Slevin, M., Krupinski, J., Gaffney, J., Matou, S., West, D., Delisser, H., Savani, R. C., and Kumar, S. 2007. Hyaluronan-mediated angiogenesis in vascular disease: Uncovering Rhamm and Cd44 receptor signaling pathways. *Matrix Biol*, 26, 58–68.

Soto-Gutierrez, A., Navarro-Alvarez, N., Rivas-Carrillo, J. D., Chen, Y., Yamatsuji, T., Tanaka, N., and Kobayashi, N. 2006. Differentiation of human embryonic stem cells to hepatocytes using deleted variant of Hgf and poly-amino-urethane-coated nonwoven polytetrafluoroethylene fabric. *Cell Transplant*, 15, 335–41.

Sottile, V., Thomson, A., and Mcwhir, J. 2003. *In vitro* osteogenic differentiation of human ES cells. *Cloning Stem Cells*, 5, 149–55.

Steiner, D., Khaner, H., Cohen, M., Even-Ram, S., Gil, Y., Itsykson, P., Turetsky, T. et al. 2010. Derivation, propagation and controlled differentiation of human embryonic stem cells in suspension. *Nat Biotechnol*, 28, 361–4.

Stojkovic, P., Lako, M., Przyborski, S., Stewart, R., Armstrong, L., Evans, J., Zhang, X., and Stojkovic, M. 2005. Human-serum matrix supports undifferentiated growth of human embryonic stem cells. *Stem Cells*, 23, 895–902.

Thomson, J. A., Itskovitz-Eldor, J., Shapiro, S. S., Waknitz, M. A., Swiergiel, J. J., Marshall, V. S., and Jones, J. M. 1998. Embryonic stem cell lines derived from human blastocysts. *Science*, 282, 1145–7.

Tigli, R. S., Cannizaro, C., Gumusderelioglu, M., and Kaplan, D. L. 2011. Chondrogenesis in perfusion bioreactors using porous silk scaffolds and hESC-derived MSCs. *J Biomed Mater Res A*, 96, 21–8.

Vats, A., Bielby, R. C., Tolley, N., Dickinson, S. C., Boccaccini, A. R., Hollander, A. P., Bishop, A. E., and Polak, J. M. 2006. Chondrogenic differentiation of human embryonic stem cells: The effect of the microenvironment. *Tissue Eng*, 12, 1687–97.

Villa-Diaz, L. G., Nandivada, H., Ding, J., Nogueira-DE-Souza, N. C., Krebsbach, P. H., O'shea, K. S., Lahann, J., and Smith, G. D. 2010. Synthetic polymer coatings for long-term growth of human embryonic stem cells. *Nat Biotechnol*, 28, 581–3.

Williams, D. F. 1987. *Definitions in Biomaterials: Proceedings of a Consensus Conference of the European Society of Biomaterials, Chester, England, March 3–5,* Amsterdam, Elsevier.

Woo, D. G., Shim, M. S., Park, J. S., Yang, H. N., Lee, D. R., and Park, K. H. 2009. The effect of electrical stimulation on the differentiation of hESCs adhered onto fibronectin-coated gold nanoparticles. *Biomaterials*, 30, 5631–8.

Xu, C., Inokuma, M. S., Denham, J., Golds, K., Kundu, P., Gold, J. D., and Carpenter, M. K. 2001. Feeder-free growth of undifferentiated human embryonic stem cells. *Nat Biotechnol*, 19, 971–4.

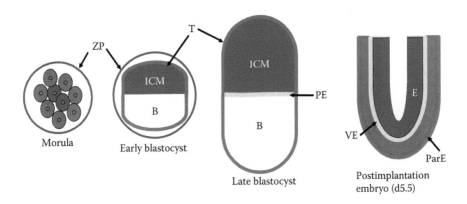

FIGURE 1.1 Development of the early mouse embryo. In the morula, the inner (blue) cells will form ICM and the outer (pink) cells will form trophoblast. In the early blastocyst, a cavity (the blastocoele, B) forms between the ICM and the trophoblast; the embryo is still enclosed in the zona pellucida (ZP). By the late blastocyst stage, the ICM cells in contact with the blastocoele differentiate into the PE, which later forms visceral endoderm on the epiblast side and parietal endoderm on the trophoblast side. At implantation the proamniotic cavity begins to form within the ICM. Cells of the ICM differentiate into an epithelial layer, the epiblast (E). (Adapted from O'Shea, K. S. 2004. *Biol Reprod* **71**(6): 1755–65.)

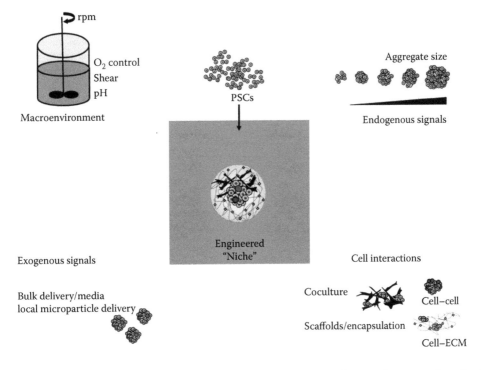

FIGURE 1.4 The PSC differentiation niche can be manipulated in a variety of ways to direct mesoderm development. The macroenvironment can be controlled to manipulate medium oxygen tension and pH, and fluid shear stress. The level of endogenous signaling can be modulated by manipulating PSC aggregate size, thereby varying local cell density. Coculture systems, or ECM strategies, can be used to manipulate cell interactions that promote mesoderm induction. The most commonly employed method for directed mesoderm differentiation is the addition of exogenous signals that are known to promote mesoderm development. This is usually carried out by direct addition of cytokines to the bulk medium, but can be more precisely controlled by using local microparticle delivery.

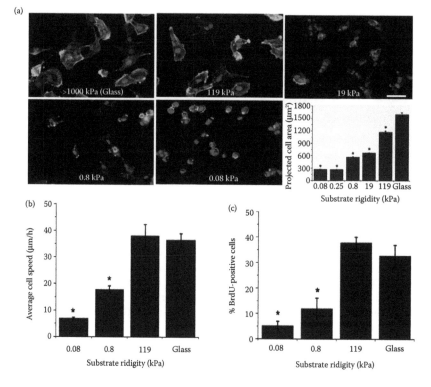

FIGURE 2.3 Mechanobiological control of glioma cell behavior. The effect of mechanics on the morphology, motility, and proliferation of U373 MG glioblastoma multiforme tumor cells was assessed by plating cells on variable-stiffness polyacrylamide substrates coated with fibronectin. (a) Effect on cell morphology and adhesion. Cell morphology shows a steep dependence on substrate stiffness, with cells spreading extensively and forming well-defined focal adhesions and stress fibers on glass or stiff substrates, but not on softer substrates. Immunofluorescence images depict nuclear DNA (blue), F-actin (green), and the proliferation marker Ki67 (red). (b) Effect on motility. Increasing substrate stiffness increases the speed of random cell migration. (c) Effect on proliferation. Substrate stiffness also influences proliferation, with a greater fraction of BrdU-positive cells seen on stiffer substrates. (Reproduced with permission from Ulrich, T.A. et al. *Cancer Research* 2009, 69(10), 4167–4174.)

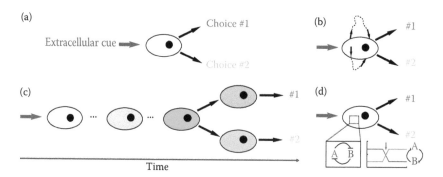

FIGURE 3.1 Recurring systems-level themes in cell-fate decisions. (a) Cell signaling connects extracellular cues with cellular decision-making. Cues are transduced through the cell by networks of signaling pathways (not shown), which together coordinate cell-fate choices. (b) Autocrine circuits are a component of cell-signal processing. Regulated release of intrinsic autocrine cues provides microenvironment-dependent feedback to reinforce cell fate. (c) Signaling dynamics allow time-dependent evolution of cell state toward key decision points. The rate and trajectory of these signaling events ultimately determine the cell-fate choice. (d) Modeling the role of network topology in cell decisions. Certain signaling architectures, such as double-negative feedback (shown), can give rise to bistable networks that can flip in response to a cue (red).

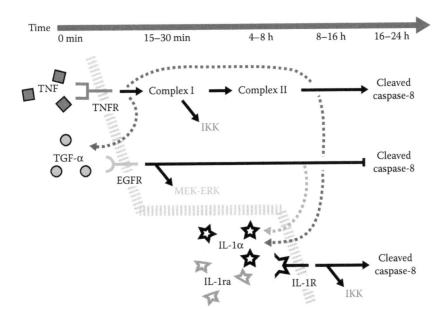

FIGURE 3.2 A time-dependent autocrine cascade induced by TNF. TNF drives the early release of TGF-α, which cooperates with TNF to induce IL-1α at intermediate times. Later, TNF induces the release of IL-1ra to inhibit IL-1α signaling. TNF induces early IKK signaling through the TNF receptor Complex I (Micheau and Tschopp 2003), whereas TGF-α induces early MEK–ERK signaling and IL-1α induces late IKK signaling. TNF and IL-1α promote cleavage and activation of the initiator apoptotic enzyme, caspase-8, whereas TGF-α inhibits caspase-8 cleavage. The plasma membrane (gray) is staggered to show the sequence of induced autocrine factors.

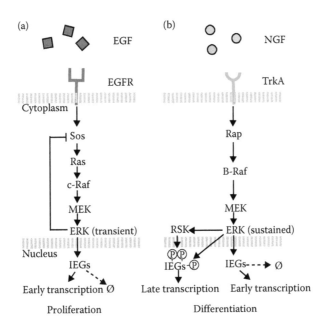

FIGURE 3.3 Changes in ERK activity kinetics and gene expression induced by two different extracellular cues. (a) EGF induces transient ERK activity and proliferation in PC12 cells. ERK deactivation is achieved by negative feedback of SOS. Activated ERK induces IEGs that are rapidly degraded without a sustained ERK signal. (b) NGF induces sustained ERK activity and differentiation in PC12 cells. NGF-induced ERK activity does not require SOS and thus is not subject to ERK-mediated negative feedback. Prolonged ERK activation causes hyperphosphorylation of IEGs through ERK and RSK, which stabilize IEGs and lead to prolonged gene expression.

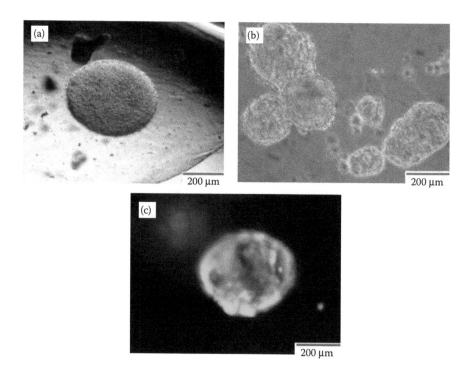

FIGURE 4.2 Morphology and viability of the hESCs encapsulated within alginate hydrogels. (a) hESC aggregates encapsulated for 110 days with no differentiation into the germ layers or cysts being observed, (b) the morphology of the decapsulated at day 110 hESCs cultured in 2D cultures, and (c) the encapsulated hESCs remained viable within the aggregates at day 110. (From Siti-Ismail, N. et al. 2008. *Biomaterials*, 29, 3946–52.)

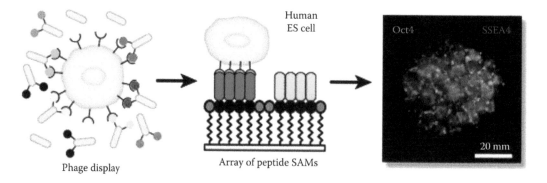

FIGURE 4.3 High-throughput identification of peptides that support hESC attachment and proliferation. (From Derda, R. et al. 2010. *J Am Chem Soc*, 132, 1289–95.)

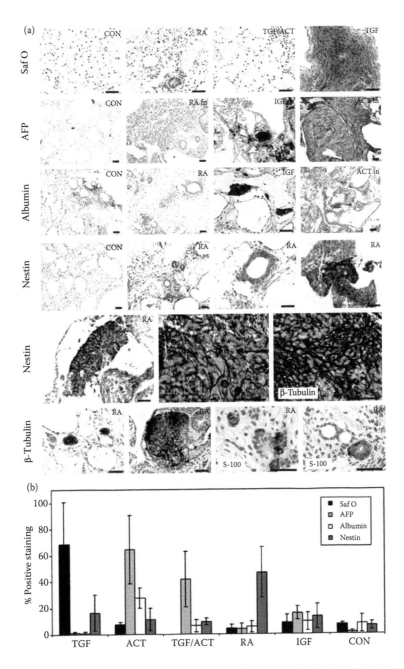

FIGURE 4.4 Differentiation of hESCs into mesodermal-, ectodermal-, and endodermal-derived cell types and tissue-like structures when seeded inside PLGA scaffolds. (a) Immunostaining of tissue sections taken from hES constructs incubated for 2 weeks with control medium (CON) or medium supplemented with growth factors: TGF-β (TGF), activin-A (ACT), RA, IGF, and a combination of TGF-β and activin-A (TGF/ACT). Samples were stained with Safranin-O (Saf O) or with antibodies against human alpha fetoprotein, albumin, nestin, β_{III}-tubulin, and S-100 (scale bars = 50 μm). (b) Quantitative analysis of antibody staining. The percentage of positive staining corresponds to percentages of the area positively stained with the antibody within the tissue sections. The results shown are mean values (±SD) of sample sections obtained in three different experiments performed in duplicate. (From Levenberg, S. et al. 2003. *Proc Natl Acad Sci USA*, 100, 12741–6).

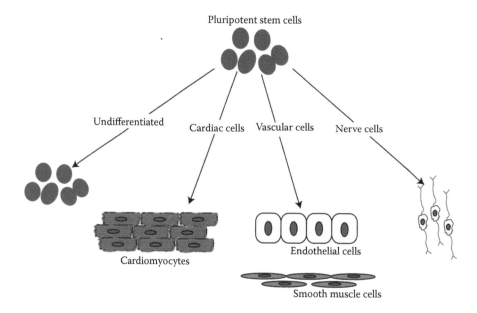

FIGURE 6.3 Pluripotent embryonic stem cells can either remain undifferentiated, or differentiate into cells found within the myocardium. Differentiated cells include cardiac cells (myocytes), vascular cells (endothelial, smooth muscle), and peripheral nerve cells.

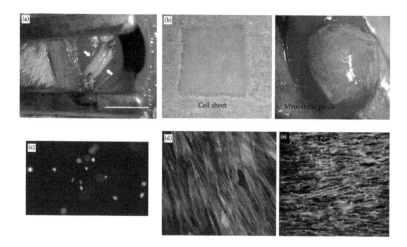

FIGURE 6.4 Stem cells have been assembled in a number of scaffolds. (a) A thick 3D collagen scaffold constructed *in vitro* was sutured onto an infarct. (Adapted from Zimmermann, W. H., Melnychenko, I., and Eschenhagen, T. 2004. *Biomaterials*, 25, 1639–1647.) (b) Cell sheets have been assembled and layers of sheets were sutured onto the infarct region. (Adapted from Shimizu, T. et al. 2009. *Curr. Pharm. Des.*, 15, 2807–2814.) (c) Stem cells have been delivered within a synthetic hydrogel. (Adapted from Wall, S. et al. 2010. *J. Biomed. Mater. Res.: Part A*, 95(4), 1055–1066.) (d) Cells have been anistropically aligned *in vitro* by culturing on aligned fiber scaffolds produced by electrospinning. (Adapted from Dang, J. M. and Leong, K. W. 2007. *Adv. Mater.*, 19, 2775–2779; Kenar, H., Kose, G. T., and Hasirci, V. 2010. *J. Mater. Sci.-Mater. Med.*, 21, 989–997.)

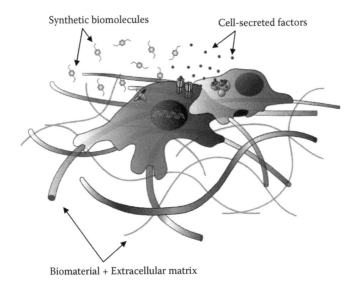

FIGURE 8.1 Reciprocal interactions between cells and the extracellular microenvironment, comprising the extracellular matrix, cell-secreted factors, and other cells. (From Phadke, A., Chang, C.-W., and Varghese, S. 2010a. *Functional biomaterials for controlling stem cell differentiation.* In Roy, K. (Ed.) *Biomaterials as Stem Cell Niche,* Berlin/Heidelberg, Springer-Verlag, pp. 19–44.)

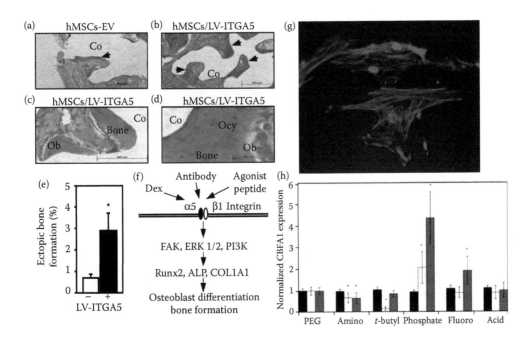

FIGURE 8.2 (a–f) Coral/hydroxyapatite (labeled Co) scaffolds seeded with hMSCs transduced with either empty vector (EV) or LV-ITGA5 (overexpression of integrin α5), implanted into ectopic sites in nude mice. (From Hamidouche, Z. 2009. *Proc Natl Acad Sci,* 106, 18587–18591.) (a) Control cells (EV), (b–d) cells transduced with LV-ITGA5, showing forced overexpression of integrin α5. Ectopic bone is labeled pink with plump osteoblasts (Ob) and osteocytes (Ocy). (e) Quantification of ectopic bone formation of LV-ITGA5 hMSCs. Error bars represent standard deviation. (f) Proposed mechanism by which activity of α5β1 promotes osteogenic differentiation. (g, h) Osteogenic differentiation of hMSCs on phosphate-functionalized surfaces. (From Benoit, D. S. W. et al. 2008. *Nat Mater,* 7, 816–823.) (g) hMSCs cultured on phosphate functionalized hydrogels (red: actin filaments, blue: nuclei) (h) Expression of osteogenic marker CBFA1 at day 0 (black), day 4 (white), and day 5 (gray) on gels functionalized with various groups. Error bars represent standard deviation. * Indicates statistical significance as compared with PEG.

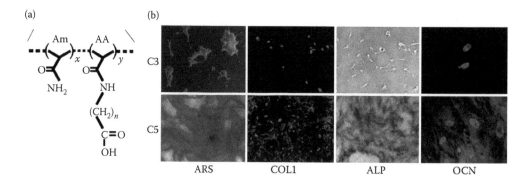

FIGURE 8.3 Effect of tunable matrix hydrophobicity on adhesion, proliferation, and osteogenic differentiation of hMSCs. (a) Schematic representation of polymeric surfaces used. To obtain hydrogels with varying matrix hydrophobicity but identical functionality, n was varied from 1 to 7. (b) Histological staining hMSCs cultured on C3 and C5 gels for alkaline phosphatase (ALP) and matrix mineralization as evidence by alizarin red-S (ARS), along with immunohistochemical staining for COL1 and OCN. (From Ayala, R. et al. 2011. *Biomaterials*, 32, 3700–3711.)

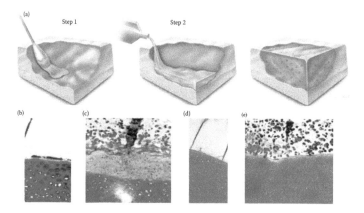

FIGURE 8.4 (a–e) Functionalized chondroitin sulfate (CS) used as a bioadhesive for integration between polymeric cartilage constructs and native cartilage tissue. (a) Schematic diagram, representing (step 1) application of chondroitin sulfate adhesive and (step 2) application of polymeric precursor with suspended cells, ultimately resulting in the construct, integrated with native tissue. (b) Acellular hydrogel, attached to cartilage explant. (c) Cartilage explants attached to cell–polymer construct with formation of neocartilage. (Red: Safranin-O staining, indicating proteoglycans). (d) Acellular hydrogel attached to cartilage defect in athymic mice. (e) Cell-seeded hydrogel attached to native cartilage *in vivo*. (From Wang, D.-A. et al. 2007. *Nat Mater*, 6, 385–392.)

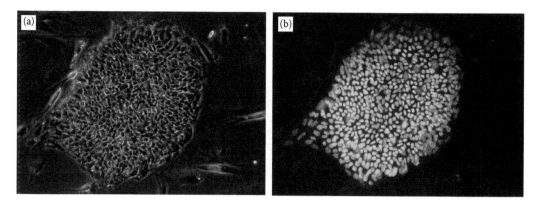

FIGURE 9.1 Phase contrast image (a) and OCT4 immunofluorescence image (b) of an H1 hESC colony cocultured with MEF feeders.

5

Stem Cells and Regenerative Medicine in the Nervous System

Shelly
Sakiyama-Elbert
*Washington University in
St. Louis*

Stem cells play a critical role in the regeneration of many tissues. The discovery of stem cells in the adult mammalian brain in two small niches revolutionized the view of the adult central nervous system (Eriksson et al. 1998, Gage et al. 1998). Within these niches, adult neural stem cells (NSCs) continue to proliferate and differentiate well into adulthood. However, this discovery led to many new questions, such as how are these niches so different from the rest of the adult brain, and why, despite the presence of endogenous stem cells, is regeneration so limited in the adult mammalian nervous system. Better understanding of the internal and external cues governing proliferation and differentiation of mammalian NSCs will help to answer these questions and direct the development of cell-based therapies to treat nerve injuries and neurodegenerative disorders. This chapter will explore potential sources of stem cells for treatment of neurological injuries/disorders using regenerative medicine and also potential applications for use. There are other potential cell-based therapies for the treatment of neurological injuries and disorders; however, they will not be discussed in this chapter.

5.1 Stem Cell Sources

There are many criteria to examine when selecting a cell source for regenerative medicine. The cells should be readily available, easily differentiated into the cell type of interest, easily expanded for scale-up to the cell number needed for human therapies, have the potential for clinical translation (human cells from either an allogenic or autologous source), and have a low risk of unwanted differentiation (into nondesirable cell types) or proliferation after transplantation (after purification). However, no cell type currently available meets all of these criteria, so for now the risks and benefits of each cell type must be weighed for the desired application or research study. In this section, sources of stem cells for neurological injury and disorder will be discussed.

5.1.1 Embryonic Stem Cells

Embryonic stem cells (ESCs) are derived from the preimplantation blastocysts of mouse and human embryos. They are pluripotent and can give rise to cells from all three germ layers (endoderm, mesoderm, and ectoderm), including neural cells. They can be expanded in culture and genetic engineering can be readily performed to generate cell lines that express reporter proteins or selection markers. Mouse and human ESC lines are commercially available for research use. However, because they can generate many cell types, induction and purification of the desired cell type is important when using ESCs. To generate neural progenitor cells, similar to NSCs (or other neural cell types) from ESCs, researchers use induction protocols that seek to recapitulate the signals found in the stem cell niche to promote neural induction and ultimately differentiation.

Many researchers have utilized embryonic stem cell-derived neural progenitor cells (ESNPCs) for research and animal studies. Others have also developed more specific induction protocols to obtain more restricted lineages of progenitor cells, such as progenitor motor neurons (pMNs) or oligodendrocyte progenitor cells (OPCs) to have better control over the fate of differentiation. Overall, these cells have the advantage of being derived from mouse or human ESCs, and thus can be generated from well-characterized cell lines in relatively abundant quantities. Restrictions on the use of human ESCs in the United States prior to 2009 limited research with human cells, but these restrictions have recently been relaxed, allowing more broad usage of U.S. government funds for research.

One disadvantage of using ESCs is that they can by definition form teratomas upon transplantation (Wobus et al. 1984), so one must be sure to purify out any remaining undifferentiated ESCs or confirm that no undifferentiated cells remain prior to transplantation into patients (Park et al. 2011, Riess et al. 2007). Another challenge is that much of the research in this area was originally performed with mouse ESCs, and these protocols must be adapted for human cells, which often require longer culture times (typically double the length) to attain the same differentiation states.

ESNPCs are used by many labs for research and are most commonly derived using retinoic acid (RA) addition to embryoid body (EB) suspension cultures in an 8-day protocol (4−/4+) developed by Bain et al. (1995). These cells can then be dissociated or left as EBs (Willerth et al. 2006) and used in *in vitro* studies or transplanted (either as whole EBs or single cells, after dissociation) (Johnson et al. 2010b, McDonald et al. 1999). They are generally amenable to dissociation when used for two-dimensional (2D) cell culture, but have low viability *in vitro* or *in vivo* in three-dimensional (3D) scaffolds/culture systems. Another disadvantage is that there are residual undifferentiated ESCs (10–30%) at the end of this protocol that need to be purified prior to transplantation to prevent tumor formation (Johnson et al. 2010a).

Other researchers have developed protocols for alternative methods of induction from ESCs to generate more restricted progenitor cells or avoid the use of EB cultures. Induction of oligodendrocyte progenitors from ESCs can be performed using a protocol developed by Brustle et al. (1999) and adapted for human cells by Nistor et al. (2005). These cells can then be transplanted to promote remyelination after spinal cord injury (SCI), and could also be used as a potential treatment for multiple sclerosis. Currently, human clinical trials using these cells are underway using cells from Geron (Alper 2009).

Other more restricted neural progenitors can also be used, for example, pMNs have the potential to differentiate into motor neurons and oligodendrocytes (Miles et al. 2004, Mukouyama et al. 2006, Wichterle et al. 2002) making them attractive for a number of applications (e.g., SCI or amyotrophic lateral sclerosis [ALS]) where astrocytes are not a desired cell type and motor neurons/oligodendrocytes are needed. Retinal pigment epithelium cells and retinal progenitors have been derived from human embryonic stem cells (hESCs) for generation of tissue engineering retinal layers (Klimanskaya et al. 2004, Lund et al. 2006, Nistor et al. 2010, Seiler et al. 2010). ESC differentiation protocols for dopaminergic (DA) neurons, Purkinjie cells, and other types of neurons have also been developed for mouse and in some cases human ESCs for treatment of Parkinson disease (PD) and cerebellar ataxia. The use of ESCs for the treatment of neurological injuries and disorders is discussed in the applications section.

The questions that needs to be answered with ES-derived cell therapies are in general whether they have been purified sufficiently to reduce the risk of tumor formation in humans, whether immune suppression will be required (as existing human ESC lines have limited ethnic and immunological diversity (Laurent et al. 2010)), and whether they can be generated in sufficient number for human therapies. Some of these issues, such as immunological compatibility can be addressed through the use of inducible pluripotent stem cells (iPSCs), which are described below.

5.1.2 Neural Stem Cells

NSCs isolated from embryonic and adult brain are another well-studied source of cells for regenerative medicine (Okano 2010). They can give rise to all three cell types of the central nervous system, neurons, oligodendrocytes, and astrocytes. They can also be propagated in cell culture as neurospheres. These cells are similar in many ways to the ESNPCs; however, they are not contaminated with undifferentiated ESCs, and do not require induction (such as the 4–/4+ RA protocol). They can be propagated as neural spheres in suspension culture (similar to the EB cultures) and can be expanded at the tissue-specific stem cell stage, which is not easy to do with ESNPCs. The disadvantage is that there are very limited sources of human NSCs for clinical research. Fetal NSCs are available in limited supply from aborted fetal tissue; however, ethical concerns may limit the use of this tissue. Adult human NSCs are not available except in rare circumstances. Both cell types are restricted to allogenic tissue, which will likely require immunosuppression for patients receiving therapies based on these cells. However, another advantage is that NSCs will only differentiate into neural lineage cells (neurons, oligodendrocytes, and astrocytes); thus, the concern for overproliferation is reduced (to neural types) and cues directing differentiation do not have to be as specific as those required for ESCs. The use of NSCs for different nerve injuries/degenerative disorders is discussed below in the applications section.

5.1.3 Inducible Pluripotent Stem Cells

The development of methods to reprogram somatic (adult and fetal) cells into ESC-like iPSCs has revolutionized the field of stem cell biology. iPSCs are considered to be similar to ESCs, in that they are pluripotent and can be expanded in culture. However, it remains to be determined whether they are truly identical to ESCs. The methods for iPSC derivation were first developed for mouse cells by Yamanaka's lab (Takahashi and Yamanaka 2006) and then rapidly translated to human cells (Park et al. 2008b, Takahashi et al. 2007, Yu et al. 2007). These cells hold the potential of generating patient-specific stem cells for any tissue of interest. The current standard method of reprogramming is to transfect the cells of interest with four retroviral vectors to force expression of the genes required for reprogramming (*Oct4, Sox2, c-myc, Klf4*). Ongoing research is focused on developing small molecule or nonretroviral methods for reprogramming and increasing the efficiency in nongenetically engineered cells (Carey et al. 2009, Stadtfeld et al. 2008, Wernig et al. 2008a). Cells derived by this method have differentiation potential similar to that of ESCs, and thus this approach holds great promise for regenerative medicine. Additionally, cells derived by reprogramming could also be used to establish cell culture model of neurons with neurodegenerative disorders (e.g., PD, ALS, Huntington disease, Alzheimer disease [AD]) that could be valuable for screening potential therapeutics (Dimos et al. 2008).

Currently, low reprogramming efficiency and safety risks surrounding random genomic insertion and reactivation of the reprogramming genes (especially the oncogene *c-myc*) limit utility for clinical studies. These cells also face similar concerns regarding the need for purification as ESCs, in that undifferentiated iPSCs must be removed to limit the risk of tumor formation. Recent research has found that iPSC derived from adult mouse tissues retain residual DNA methylation signatures that are characteristic of their somatic tissue of origin, resulting in a bias toward lineages related to that of the donor tissue and restricting alternative cell fates (Kim et al. 2010b). These findings may limit the source of donor tissue for iPSCs for neural applications or rendering the utility of iPSCs to be more limited than originally anticipated; however, other studies demonstrate that additional passaging of the cells reduces these effects (Polo et al. 2010).

5.1.4 Direct Induction of Neurons from Somatic Cells

Another approach for reprogramming that may limit the safety concerns surrounding iPSCs involves reprogramming somatic cells directly to neurons (or other neural cell types). Directly reprogrammed cells are similar to neurons harvested from primary cultures in many respects, and thus cannot be expanded in culture. Recent work has shown that fibroblasts can be reprogrammed directly to neurons using a similar approach to that used to generate iPSCs (*Ascl1, Brn2* [also called *Pou3f2*] and *Myt1l*) (Vierbuchen et al. 2010). Direct reprogramming would allow generation of a specific cell type of interest directly from a patient without requiring an induction/differentiation period. However, this study was performed with cells from a genetically modified mouse (Tau-green fluorescent protein) to allow purification of successfully reprogrammed cells and embryonic tissue was used, which may be more plastic than adult tissue. Additionally, these neurons cannot be expanded once they are generated in culture, so an additional expansion step prior to reprogramming may be required to generate the number of cells needed for clinical studies from a small skin biopsy. However, this work represents an exciting alternative source for neural cells for both cell culture testing and regenerative medicine.

5.1.5 Skin-Derived Precursors

Skin-derived precursors (SKPs) cells represent an interesting neural crest-derived lineage that could be used to generate many cell types of interest for regenerative medicine. They were identified by the Miller lab, which found that SKPs can give rise to neurons, glia, smooth muscle cells, and adipocytes (Nagoshi et al. 2008, Toma et al. 2001, 2005) and can be propagated in culture. They could provide a clinically accessible source of patient-derived cells for treatment of nerve injury. SKPs have been shown to differentiate into Schwann cells (SCs), which are of great benefit for the treatment of peripheral nerve injury and can also myelinate neurons when transplanted into the central nervous system (Biernaskie et al. 2007, McKenzie et al. 2006). Combinations of SKP-derived SCs with conduits (e.g., Neurogen/Integra) or decellularized nerve grafts (Avance/Axogen) could provide an alternative to allografts that would not require immunosuppression for patients lacking sufficient autograft source material.

5.1.6 Bone Marrow Stromal Cells

Many researchers have evaluated the potential of nonneural tissue stem cells, such as those derived from bone marrow stromal cells (BMSCs) for the treatment of neural injury or neurodegenerative disorders. These cells are a highly desirable cell type for clinical use because autologous cells are readily available from patients, and they are frequently used without expansion cultures that can be time consuming. While these cells can be beneficial by providing trophic support, extracellular matrix (ECM) synthesis, and anti-inflammatory signals, they cannot differentiate in substantial numbers into nonhematopoietic tissues (Wagers et al. 2002). Rat and human BMSCs can differentiate into cells that express markers for neuronal cells (Deng et al. 2001, Sanchez-Ramos et al. 2000, Woodbury et al. 2000), but later studies indicate that these cells do not express voltage-gated ion channels required for functional integration (Hofstetter et al. 2002). Additionally, studies have shown that old patients have fewer stem cells resident in their bone marrow, so there may not be sufficient stem cells for autologous therapies in older patients. These concerns aside, transplantation of MSCs can still provide a more hospitable environment for nerve regeneration through trophic support, permissive ECM synthesis, and anti-inflammatory effects, and thus may still prove beneficial for the treatment of nerve injury.

5.2 Applications

5.2.1 Spinal Cord Injury

SCI is a significant clinical problem that currently lacks a treatment to enhance functional recovery. The standard of care is administration of steroids (methylprednisolone), and waiting to see whether the paralysis is temporary due to inflammation and initially trauma or permanent. While the population affected by SCI is modest (12,000 new patients per year in the United States), most patients are 16–32 years of age and the impact on their long-term quality life and productivity is enormous (NSCISC 2012). After SCI, an astroglial scar forms to isolate the spared neural tissue from the lesion site, which also serve to inhibit regeneration into the lesion. Secondary injury follows, resulting in the death of both neurons and oligodendrocytes, inflicting addition functional damage beyond the initial injury. Currently, research is focusing on two main approaches: (1) rewiring local neural circuitry around the lesion site, and (2) remyelinating axons that have lost myelination (often due to secondary injury) to improve function of spared tissue. Previously, approaches to promote long axon tract regeneration were studied extensively; however, there has been little success to date with such research (Webber et al. 2007).

One of the early studies that used mouse ESCs for treatment of SCI utilized a multistep differentiation protocol to obtain oligodendrocytes and astrocytes for transplantation (Brustle et al. 1999). Once transplanted into the spinal cord, these cells demonstrated the ability to restore myelination in myelin-deficient *shiverer* rats, indicating the potential of mouse ESCs to treat SCI. This work was followed by two studies performed by the McDonald lab to evaluate the ability of mouse ESCs to promote recovery in SCI models (Liu et al. 2000, McDonald et al. 1999). In the first study, the cells were induced using the 4−/4+ RA protocol (Bain et al. 1995), partially dissociated and injected 9 days after contusion injury (McDonald et al. 1999). These cells differentiated into primarily oligodendrocytes and astrocytes and promoted a modest increase in functional recovery (as assessed by the Basso, Beattie, and Bresnahan [BBB] score). The BBB is a locomotor rating scale that is commonly used in the SCI community to assess functional recovery in rats, named after the three authors that developed the scale (Basso et al. 1995). The second study focused on producing oligodendrocytes from mouse ESCs for remyelination after SCI (Liu et al. 2000). These oligodendrocytes were found to remyelinate axons both *in vitro* and in a chemically induced demyelination in the spinal cord. More recently, whole EBs containing ESNPCs have been transplanted with fibrin-based scaffold delivering neurotrophin-3 (NT-3) and platelet-derived growth factor AA in a subacute rat dorsal hemisection. In the presence of the scaffolds, the cells differentiated into predominately neurons and an increase in functional recovery was observed (Johnson et al. 2010a,b).

The focus on remyelination therapy has been studied by many researchers (Brustle et al. 1999, Cao et al. 2010, Li et al. 2010, Marques et al. 2010, Tsuji et al. 2010). The Keirstead lab has adapted this approach for use with human ESCs (Nistor et al. 2005) and developed a protocol to obtain OPCs. They transplanted these cells into a rat T10 contusion SCI model 7 days and 10 months after injury. In both cases, they found that the cells survived and differentiated into oligodendrocytes; however, only in the case of transplantation after 7 days did the OPCs improve functional recovery (BBB score) (Cloutier et al. 2006, Faulkner and Keirstead 2005, Keirstead et al. 2005). More recently, OPCs in combination with motoneuron progenitors from human ESCs were found to improve functional recovery with acute transplantation in a rat complete transection injury model (Erceg et al. 2010).

Other researchers have explored the transplantation of rat embryonic NSCs. Cao et al. (2001) investigated implanting NSCs into both normal and lesioned (T8 contusion) spinal cord. They found that the majority of the implanted cells differentiated into astrocytes, suggesting that these cells may need to be predifferentiated before implantation (Cao et al. 2002). Additionally, they observed that factors present after SCI restrict differentiation of these cells. To try and overcome some of these signals, specifically bone morphogenetic protein (BMP), NSCs were engineered to express noggin, a BMP agonist, to attempt to obtain better differentiation into neurons and oligodendrocytes (Enzmann et al. 2005). This strategy was unsuccessful at preventing differentiation and these cells actually produced an increase in lesion size. However, a different study successfully implemented this strategy and achieved differentiation of NSCs expressing noggin into neurons and oligodendrocytes after being implanted into a T8 contusion injury as well as promoted an increase in BBB scores 3 weeks after injury (Setoguchi et al. 2004). These studies illustrate the importance of controlling protein expression levels when performing genetic manipulation. Recently the Nakashima lab demonstrated that valproic acid (VPA) could be used to enhance neuronal differentiation of NSCs in a mouse contusion SCI model (Abematsu et al. 2010). They found that both donor and host neurons contributed to the routing of anterograde signals through the lesion site resulting in enhanced functional recovery with both VPA and NSCs compared to either treatment alone. The Anderson lab has shown that human fetal NSCs can moderately enhance functional recovery in a NOD-SCID mouse SCI model as late as 30 days after injury, helping to define a window for intervention with cell transplantation in rodents (Gelain et al. 2010).

Other research has focused the use of scaffolds in combination with cell transplantation to enhance cell survival and/or to control cell differentiation after SCI. The Langer lab developed polylactic glycolic acid (PLGA) scaffolds for transplantation of mouse NSCs (Teng et al. 2002). These scaffolds were then implanted into the lesion site resulting from a T9/10 lateral hemisection. This approach produced improved functional recovery (BBB scores) after 3 weeks compared to animals receiving only cells and the lesion-only (untreated) animals. This increase in recovery was observed throughout the rest of the 10-week study even though the transplanted cells did not stain positive for mature cell markers, indicating that they remained undifferentiated, suggesting that the effect of the cells was likely due to neuroprotection (decreased secondary injury) rather than cell replacement. The authors also hypothesize that a reduction in glial scar formation due to the scaffold and/or cells may also have contributed to the improved outcomes.

The Lavik lab has further expanded the work with these scaffolds to assess the effect of transplanting NSCs in combination with endothelial cells, to explore the effects of vascularization on recovery after SCI (Rauch et al. 2009). They found that transplantation of both cell types within a polyethylene glycol/poly-L-lysine hydrogel inside the PLGA scaffolds increased functional blood vessel formation over controls; however, there were fewer NSCs in the cotransplant groups compared to those with NSCs alone. No functional recovery was observed in this study; however, this may be an interesting approach to evaluate recapitulation of a neurovascular stem cell niche.

Other researchers have explored the effect of genetic modification of NSCs to express exogenous growth factors. Tuszynski and colleagues developed mouse NSC lines that constitutively secrete a variety of growth factors, including nerve growth factor (NGF), brain-derived neurotrophic factor (BDNF), and glial-derived neurotrophic factor (GDNF), *in vitro* and *in vivo* (Lu et al. 2003). *In vivo*, the transplanted NSCs promoted axonal sprouting in a C3 hemisection injury. They then genetically modified these NSCs to produce NT-3, which when tested in the same *in vivo* model, enhanced the axonal sprouting that was previously observed. Another approach combined NSC transplantation with antibodies that neutralize the effects of ciliary neurotrophic factor (CNTF) (Ishii et al. 2006). This study showed that neutralizing CNTF reduced the amount of NSCs that differentiated into astrocytes while promoting regeneration of the corticospinal tract. No behavioral analysis was reported for this study. A different study by the Schwartz lab looked at transplanting adult NSCs 7 days postinjury in combination with myelin-specific T cells to determine the effect on SCI in a mouse model (T12 contusion injury) (Ziv et al. 2006). The combination of cells was able to produce functional recovery as evidenced by an increase in the Basso mouse scale (Engesser-Cesar et al. 2005). However, generally the challenge with growth factor-secreting NSCs is that they tend to produce a "candy store" effect, where axons grow into the region of high growth factor expression (e.g., cell transplantation site) and then are trapped, unable to grow down the gradient of trophic factor on the opposite side and therefore failing to innervate their intended target (Tannemaat et al. 2008).

The Shoichet lab developed examined implantation of extramedullary chitosan channels seeded with NSCs from either brain or spinal cord (Nomura et al. 2008). At 14 weeks, they found that tissue bridges in scaffolds seeded with both types of cells and long-term NSC survival was observed; however, the bridges were thicker with the brain-derived cells. The NSCs differentiated primarily into astrocytes and oligodendrocyte, and no functional recovery was observed (Zahir et al. 2008).

Hofstetter et al. transplanted BMSCs into T7 contusion injury model of SCI immediately following injury and with a 7 days delay (Hofstetter et al. 2002). The cells transplanted 7 days postinjury showed improved rates of survival and formed bundles that bridged the lesion. Implantation of BMSCs also led to an increase in BBB scores compared to controls. A long-term study by the Vaquero lab looked at the long-term effects of transplanting BMSCs 3 months after a crush injury (Zurita and Vaquero 2006). This study showed that functional recovery steadily increased over the course of a year as indicated by BBB scores. Finally, human BMSCs transplanted into a T9 contusion model of SCI also promoted functional recovery as indicated by BBB scores, suggesting that this therapy has the potential to work in humans (Himes et al. 2006).

Additional work has been done to clarify the mechanisms by which BMSCs promote functional recovery. One study demonstrated that BMSCs help guide regenerating axons across the injury site when implanted 2 days after a T8 contusion injury and can help promote recovery by restoring the stepping control circuitry (Ankeny et al. 2004). A more recent study showed that BMSCs express the gamma amino butyric acid receptor (Yano et al. 2006). BMSCs stimulate phosphoinositide-3-kinase and mitogen-activated protein kinase signaling in neurons, which promotes their survival (Isele et al. 2007). A recent paper from the Silver lab demonstrated that multi-potent adult progenitor cells (MAPCs) derived from bone marrow can prevent axonal dieback by decreasing the release of matrix metalloproteinase 9 from macrophages and by altering the macrophage phenotype from the "classically activated" M1 to the "anti-inflammatory M2 state" (Busch et al. 2011). All these mechanisms contribute to the success of BMSC transplantation as a treatment for SCI.

A pair of studies from the Tuszynski lab investigated the use of BMSCs that were genetically modified to express growth factors to treat a dorsal column transaction injury at C3 (Lu et al. 2005, 2007b). The first study looked at the effects of transplanting BMSCs that secreted BDNF immediately following injury. The BDNF-expressing cells induced more robust axonal growth into the lesion site compared to normal BMSCs (Lu et al. 2005). Functional recovery, judged by a tape removal task and rope walking, was not observed. The second study examined the ability of BMSCs modified to express NT-3 to induce axonal growth through chronic glial scars (Lu et al. 2007b). These cells were implanted 6 weeks postinjury at the C3 level and the scar was not resected. These cells were able to promote regeneration of axons through the scars and into the lesion site, demonstrating that the glial scar can be penetrated. These studies illustrate the additional benefits of genetic modification when used in combination with BMSCs.

Recently, the Fischer lab has demonstrated that a mixed population of neuonal and glial progenitor cells derived from the embryonic spinal cord can be grafted into a dorsal column sensory lesion acutely and guided to their target, the dorsal column nuclei (DCN), using a gradient of BDNF (Bonner et al. 2011). Active synapses between graft neurons and host DCN were observed by electrophysiology. This provides the first report of neural progenitor cell derived neurons providing an active relay in the injured spinal cord.

The use of iPSC derived neural progenitors has also been studied for transplantation following spinal cord injury. Okano's lab evaluated "safe" (non-tumor forming) neurosphere clones, found they differentiated into all three neural lineages, and promoted functional recovery. In contrast, "unsafe" clones formed tumors and resulted in functional loss late after injury and initial recovery (Tsuji et al. 2010). This work as well as recent studies on the potential immune response to iPSCs even from mice of the same genetic background suggest that iPSCs need to be evaluated carefully for safety concerns after differentiation in a given injury model (Zhao et al. 2011).

5.2.2 Peripheral Nerve

Severe peripheral nerve injuries are common and result in significant long-term functional morbidity. In the United States alone, 360,000 people suffer from upper extremity nerve injuries annually, resulting in over 8.5 million restricted activity days and almost 5 million bed/disability days (Barton 1998). Despite recent advances in the understanding of the neurobiology related to nerve regeneration and refinement in surgical techniques, complete functional recovery after repair of a damaged nerve is rare (Bunge 1994). Strategies that enhance nerve regeneration and thus improve functional recovery following these injuries would have an important clinical impact. The major barriers to functional recovery following nerve injury are long regeneration times and appropriate reinnervation of end organ targets. The human nerve regenerates at a pace of ~1 mm a day which can result in years of regeneration before end organ contact. SCs, the glial cells of the peripheral nervous system, play an important role during regeneration of peripheral nerves. They provide trophic support and secrete ECM molecules during nerve regeneration, as well as myelinating intact and regenerated axons. With extended regeneration times, SCs of the distal nerve stump become less supportive of regenerating axons (Gordon 2010, Sulaiman and Gordon 2009). Additionally, the denervated muscle begins to atrophy, resulting in profound, irreversible muscle damage and fibrosis (Fu and Gordon 1995, 1997, Kobayashi et al. 1997, Mackinnon and Dellon 1988). The combined impact is fewer regenerating axons innervating atrophied muscle resulting in poor functional recovery.

Regenerative medicine strategies in the peripheral nerve are largely focused on methods to obtain SCs from either ESCs or neural crest stem cells. Cui et al. examined the effects of transplanting ESNPCs into an epineurium-based conduit in a 10 mm rat sciatic nerve gap injury (Cui et al. 2008). One to three months after injury, the ESNPCs expressed S100 and appeared to differentiate into SCs. Retrograde labeling and functional assessment suggests that the cells improved recovery and the overall nerve diameter was similar to that of the uninjured control.

The use of SKPs to provide a source of SCs for peripheral nerve injury was mentioned above. The Miller lab demonstrated that they were able to use SKPs as a source of cells to myelinate the injured peripheral nervous system, as well as the dysmyelinated neonatal brain of *shiverer* mice (McKenzie et al. 2006). Midha's lab continued this work and demonstrated that the regenerative capacity of an acellular (freeze/thawed) nerve graft (with reduced antigenicity) could be enhanced by the addition of autologous SKP-derived SCs (Walsh et al. 2009). The SKPs show comparable regeneration to mature SCs and autografts as assessed by fiber number and cross-sectional area of myelinated axons at 8 weeks in the distal nerve. Electrophysiology showed similar results to autografts. Their lab also demonstrated the SKPs could improve the regenerative capacity of chronically denervated nerve with higher counts of regenerated motor neurons and histological recovery similar to that of immediate nerve repair (Walsh et al. 2010). These results correlated with superior muscle reinnervation, as measured by compound muscle action potentials and wet muscle weights. These results suggest that SKPs hold great potential as a source of SCs and as a potential treatment for peripheral nerve injury.

5.2.3 Traumatic Brain Injury

Traumatic brain injury (TBI) is a major health problem in the United States with over 1.4 million cases reported annually, and a high incidence among veterans from the Iraq and Afghanistan wars (Langlois et al. 2003). Currently, there is no clinical treatment to repair damaged neural tissue in the brain. The primary type of cell studied for treatment of TBI is adult mesenchymal stromal cells (MSCs), which can be derived from bone marrow. Several studies have shown that transplantation of MSCs can improve cognitive function after TBI in preclinical models when transplanted with collagen cylinders as scaffolds (Lu et al. 2007a, Qu et al. 2009, Xiong et al. 2009). These effects are assumed to be due to secretion of growth factors and chemokines that are neuroprotective and stimulate host remodeling. Kim et al. observed a similar effect when human MSCs were administered intravenously in a rat TBI model. Expression of NGF, BDNF, and NT-3 were all increased in the cell treatment group and caspase expression was decreased, suggesting an antiapoptotic effect as well (Kim et al. 2010a).

Other researchers have explored using NSCs as a treatment for TBI. Riess et al. transplanted mouse NSC line (C17.2) into a mouse TBI model by injecting into the hippocampus 3 days after injury (Riess et al. 2002). Improvements in motor function were observed at 12 weeks, but cognitive improvements were not observed at either 3 or 12 weeks. Differentiation of the cells into both neurons and glia was observed. Conversely, Gao et al. "preprimed" fetal human NSCs (cultured with 10 ng/mL basic fibroblast growth factor, 2.5 μg/mL heparin, and 1 μg/mL laminin for 5 days prior to harvest) and then transplanted them into the hippocampus 1 day after TBI in a rat model. The cells differentiated primarily into neurons. They observed an increase in cognitive function and increased expression of GDNF in the cell-treated group (Gao et al. 2006).

Researchers have also studied the transplantation of ESNPCs into TBI models. Shindo et al. (2006) transplanted mouse ESNPCs into both mild and severe mouse TBI models. They observed better survival and differentiation of NPCs in the mild injury model compared to the severe injury model. Additional differences in synaptic formation and growth factor expression were also observed between the mild and severe injury models.

The effect of biomaterial scaffolds on NSCs has been investigated following TBI. Tate et al. (2002) transplanted mouse NSCs with or without fibronectin/collagen I (Fn) scaffolds 1 week after TBI in a mouse model. Cells transplanted with the Fn scaffolds showed increased survival and migration to the hippocampus compared to cells alone. They also examined using laminin scaffolds in a second study that both scaffolds improved cell survival at 8 weeks and the laminin scaffolds showed improved cognitive function at 5 weeks (Tate et al. 2009). These studies demonstrate that stem cell transplantation may hold a potential as a treatment for TBI and that biomaterials scaffolds may improve cell survival and functional outcomes.

5.2.4 Neurodegenerative Disorders

Neurodegenerative disorders are another potential application for stem cell therapies in the nervous system. They can be grouped into two major categories, those that are characterized by a loss of a specific cell population, such as DA neurons in PD, and those where more widespread degeneration occurs, resulting in the loss of major types of neurons, such AD over a period of years. Similar to the treatment for acute injury conditions, there are two main strategies for therapy: (1) cell replacement of neurons or glia that are lost, and (2) neuroprotective or immunomodulatory indirect therapies that focus on stimulating endogenous cells to regenerate or protecting those that are left from insult.

The loss of DA neurons in the nigrostriatal pathway is the main pathology of PD; therefore, cell replacement and possibly neuroprotection (in early disease phases) of DA neurons is considered the primary target for cell-based therapies. PD is a good target for cell replacement therapies because there is one specific cell population that is targeted in a specific location within the brain. Clinical trials with fetal ventral mesencephalon tissue have shown some benefits with improved function/slowed disease progression; however, limited tissue availability and ethical concerns have prevented large-scale application (along with variable clinical outcomes) (Kopyov et al. 1997, Piccini et al. 1999).

Researchers have tested the effect of ESC-derived DA neurons in a rat model of PD and found that the cells improved behavioral recovery (Kim et al. 2002). They developed a five-stage protocol for differentiation of DA neurons from nuclear receptor related-1 (Nurr1, a transcription factor that has a role in the differentiation of midbrain precursors into dopamine neurons) expressing ESCs (Lee et al. 2000, Okabe et al. 1996) and demonstrated that the TH+ neurons integrate into the transplant site by electrophysiology. More recently, Shim et al. have shown that using Nurr1 overexpression they can generate DA neurons from NSCs that release DA in response to depolarizing stimuli (Shim et al. 2007). They found these cells differentiated into TH+ neurons in the adult rat brain. Other researchers have explored the potential of iPSCs for treatment of PD. Werning et al. (2008b) differentiated iPSCs into DA neurons, and transplantation of these cells was shown to improve behavior in a rat model of PD.

ALS is another good target for cell replacement therapy because it is characterized by the loss of a specific cell population, motor neurons. However, it is a more challenging target because the motor neuron loss occurs more widely throughout the spinal cord. Thus, cell transplants must be more widely distributed throughout the spinal cord and integrate into spinal circuitry, as well as innervating long-range targets. Neuroprotection from microglia may also be required.

A great deal of research has been done on generating motor neurons from ESCs, NSCs, and iPSCs (Bohl et al. 2008, Dimos et al. 2008, Karumbayaram et al. 2009, Lee et al. 2007, Li et al. 2005, 2008, Singh Roy et al. 2005, Wichterle et al. 2002). These are generally developed using RA and Sonic hedgehog induction, as developed in the Jessell lab. Also, Neural Stem Inc. has developed a cell line from human fetal spinal cord NSI-566RSC for spinal cord transplant therapy (Cizkova et al. 2007, Riley et al. 2009). This line was prepared from the cervical-upper thoracic cord of a single 8-week human fetus after elective abortion and tissue donation (Xu et al. 2006). When transplanted into a rat superoxide dismutase (SOD) model of ALS, ~70% of the human NSCs differentiated into neurons (positive for beta tubulin) and many synapsed onto rat motor neurons, but they do not appear to replace degenerated neuromuscular units. Rats with live NSCs showed fewer functional deficits than control animals, perhaps due to increased levels of GDNF and BDNF in the presence of the NSCs. NeuralStem is currently conducting phase I clinical trials for ALS with the cell line at Emory University (2010). Because of the rapid progress of ALS and the prognosis for patients, the risks of cell therapy that are acceptable patients may be higher than for less severe injuries or more slowly progressing diseases.

AD is another potential application for stem cell therapies. AD is characterized by memory impairment, cognitive decline, and dementia due to a widespread, progressive axonal pathology. Neuronal and synaptic loss and deposition of plaques occurs in the amygdale, hippocampus, and basal forebrain cholinergic system, as well as in the cortical areas. Because of the widespread pathology, AD is a difficult target for neuronal replacement. Currently, approaches have focused on either neuroprotection or replacement of forebrain cholinergic neurons, based on the temporary improvement in some AD patients when acetylcholinesterase inhibitors are administered (Manabe et al. 2007). There may also be a loss of function of endogenous NSCs in the subgranular zone of the dentate gyrus resulting in increased proliferation early in the disease and depletion of the NSCs in the later phase (Zhao et al. 2008). Clearance of amyloid plaques by immunotherapy or microglia is also a potential target of neuroprotection approaches such as BMSC transplantation (Lee et al. 2010). Thus, while stem cells hold potential as therapeutics for neurodegenerative disorders, much remains to be understood about their mechanism of action in the treatment of such diseases.

5.2.5 *In Vitro* Testing Platforms

Stem cells also provide a unique source of cells to develop *in vitro* culture models of neurons and glia that have not been available previously. This is particularly due to patient-derived iPSCs (Dimos et al. 2008, Park et al. 2008a). These cells provide a source for large quantities of neurons and glial from patients with familial and sporadic neurodegenerative disease. It will allow testing to determine what is different in these cells at both a genetic and biochemical level to assess the potential of therapeutics for neuroprotection and regeneration. For sporadic and genetically complex diseases such as PD and AD, this will be a huge benefit since animal models do not have the same pathology as humans. These *in vitro* testing platforms may be one of the earliest realizations of the potential of stem cells for regenerative medicine in the nervous system.

Currently, models have been developed for several neurological disorders, including ALS (Dimos et al. 2008), PD (Wernig et al. 2008b), familial dysautonomia (Lee et al. 2009), spinal muscular atrophy (Ebert et al. 2009), and Rett syndrome (Ballas et al. 2009). Some of the challenges for *in vitro* model design include protocol development for the differentiation of both neurons (that die) and other cell types, such as glia that may contribute to cell death. However, these models hold great potential as tools for new drug discovery and testing.

5.2.6 Caveats Regarding the Use of Stem Cells for Treatment of Neurological Disorders/Injuries

While stem cells hold great potential as treatments for neurological disorders and injury, there remain many challenges to be overcome before their full potential is realized. In this section, I outline some caveats that are critical to keep in mind when considering translation of stem cell therapies from pre-clinical animal models to human patients: (1) animal models may not fully mimic human pathology and are often induced by a drug or genetic mutation that does not full recapitulate the disease in humans; (2) the behavior of stem cells in animal models may only partially reflect their behavior in human disease models, due to differences in cell type, immune response, or disease model; and (3) the longevity of studies is often limited in animal models. Therefore, it is important to weigh the risk of the treatment with the potential benefit to the patient's quality of life and make sure that the patient is fully aware of the potential risks from any potential treatment. Finally, it is important to determine the biological mechanisms of the effect in the animal model (e.g., neuroprotective, neuron replacement, improved local circuitry, remyelination, immunomodulation) in order to better understand the expected outcomes in patients.

5.3 Conclusions

Recent advances in the field of stem cell biology, including the development of iPSCs, make it an exciting time to be working in the field of neural tissue engineering. Previous research has demonstrated the stem cells from multiple sources can be transplanted into models of neurological injury and disease to improve functional outcomes in preclinical models. Currently, the challenge in many of these models lies in determining the mechanism of action of the stems cells (e.g., neuroprotection neuroreplacement) and verifying that it is the same in humans. Future work also needs to focus on long-term safety of therapies that will be translated to the clinic and developing better *in vitro*/preclinical models of human disease/injury.

References

Abematsu, M., Tsujimura, K., Yamano, M., Saito, M., Kohno, K., Kohyama, J., Namihira, M., Komiya, S., and Nakashima, K. 2010. Neurons derived from transplanted neural stem cells restore disrupted neuronal circuitry in a mouse model of spinal cord injury. *Journal of Clinical Investigation,* 120: 3255–66.

Alper, J. 2009. Geron gets green light for human trial of ES cell-derived product. *Nat Biotechnol,* 27: 213–4.

Ankeny, D. P., Mctigue, D. M., and Jakeman, L. B. 2004. Bone marrow transplants provide tissue protection and directional guidance for axons after contusive spinal cord injury in rats. *Exp Neurol,* 190: 17–31.

Bain, G., Kitchens, D., Yao, M., Huettner, J. E., and Gottlieb, D. I. 1995. Embryonic stem cells express neuronal properties in vitro. *Dev Biol,* 168: 342–57.

Ballas, N., Lioy, D. T., Grunseich, C., and Mandel, G. 2009. Non-cell autonomous influence of MeCP2-deficient glia on neuronal dendritic morphology. *Nat Neurosci,* 12: 311–7.

Barton, N. 1998. Upper Extremity Disorders: Frequency, Impact and Cost. *J Hand Surg [Br]* 23: 255.

Basso, D. M., Beattie, M. S., and Bresnahan, J. C. 1995. A sensitive and reliable locomotor rating scale for open field testing in rats. *J Neurotrauma,* 12: 1–21.

Biernaskie, J., Sparling, J. S., Liu, J., Shannon, C. P., Plemel, J. R., Xie, Y., Miller, F. D., and Tetzlaff, W. 2007. Skin-derived precursors generate myelinating Schwann cells that promote remyelination and functional recovery after contusion spinal cord injury. *J Neurosci,* 27: 9545–59.

Bohl, D., Liu, S., Blanchard, S., Hocquemiller, M., Haase, G., and Heard, J. M. 2008. Directed evolution of motor neurons from genetically engineered neural precursors. *Stem Cells,* 26: 2564–75.

Bonner, J. F., Connors, T. M., Silverman, W. F., Kowalski, D. P., Lemay, M. A., and Fischer, I. 2011. Grafted neural progenitors integrate and restore synaptic connectivity across the injured spinal cord. *Journal of Neuroscience,* 31: 4675–86.

Brustle, O., Jones, K. N., Learish, R. D., Karram, K., Choudhary, K., Wiestler, O. D., Duncan, I. D., and Mckay, R. D. 1999. Embryonic stem cell-derived glial precursors: A source of myelinating transplants. *Science,* 285: 754–6.

Bunge, R. P. 1994. The role of the Schwann cell in trophic support and regeneration. *J Neurol,* 242: S19–21.

Busch, S. A., Hamilton, J. A., Horn, K. P., Cuascut, F. X., Cutrone, R., Lehman, N., Deans, R. J., Ting, A. E., Mays, R. W., and Silver, J. 2011. Multipotent adult progenitor cells prevent macrophage-mediated axonal dieback and promote regrowth after spinal cord injury. *Journal of Neuroscience,* 31: 944–53.

Cao, Q., He, Q., Wang, Y., Cheng, X., Howard, R. M., Zhang, Y., Devries, W. H., Shields, C. B., Magnuson, D. S., Xu, X. M., Kim, D. H., and Whittemore, S. R. 2010. Transplantation of ciliary neurotrophic factor-expressing adult oligodendrocyte precursor cells promotes remyelination and functional recovery after spinal cord injury. *J Neurosci,* 30: 2989–3001.

Cao, Q. L., Howard, R. M., Dennison, J. B., and Whittemore, S. R. 2002. Differentiation of engrafted neuronal-restricted precursor cells is inhibited in the traumatically injured spinal cord. *Exp Neurol,* 177: 349–59.

Cao, Q. L., Zhang, Y. P., Howard, R. M., Walters, W. M., Tsoulfas, P., and Whittemore, S. R. 2001. Pluripotent stem cells engrafted into the normal or lesioned adult rat spinal cord are restricted to a glial lineage. *Exp Neurol,* 167: 48–58.

Carey, B. W., Markoulaki, S., Hanna, J., Saha, K., Gao, Q., Mitalipova, M., and Jaenisch, R. 2009. Reprogramming of murine and human somatic cells using a single polycistronic vector. *Proc Natl Acad Sci USA,* 106: 157–62.

Cizkova, D., Kakinohana, O., Kucharova, K., Marsala, S., Johe, K., Hazel, T., Hefferan, M. P., and Marsala, M. 2007. Functional recovery in rats with ischemic paraplegia after spinal grafting of human spinal stem cells. *Neuroscience,* 147: 546–60.

Cloutier, F., Siegenthaler, M. M., Nistor, G., and Keirstead, H. S. 2006. Transplantation of human embryonic stem cell-derived oligodendrocyte progenitors into rat spinal cord injuries does not cause harm. *Regen Med,* 1: 469–79.

Cui, L., Jiang, J., Wei, L., Zhou, X., Fraser, J. L., Snider, B. J., and Yu, S. P. 2008. Transplantation of embryonic stem cells improves nerve repair and functional recovery after severe sciatic nerve axotomy in rats. *Stem Cells,* 26: 1356–65.

Deng, W., Obrocka, M., Fischer, I., and Prockop, D. J. 2001. *In vitro* differentiation of human marrow stromal cells into early progenitors of neural cells by conditions that increase intracellular cyclic AMP. *Biochem Biophys Res Commun,* 282: 148–52.

Dimos, J. T., Rodolfa, K. T., Niakan, K. K., Weisenthal, L. M., Mitsumoto, H., Chung, W., Croft, G. F., Saphier, G., Leibel, R., Goland, R., Wichterle, H., Henderson, C. E., and Eggan, K. 2008. Induced pluripotent stem cells generated from patients with ALS can be differentiated into motor neurons. *Science,* 321: 1218–21.

Ebert, A. D., Yu, J., Rose, F. F., Jr., Mattis, V. B., Lorson, C. L., Thomson, J. A., and Svendsen, C. N. 2009. Induced pluripotent stem cells from a spinal muscular atrophy patient. *Nature,* 457: 277–80.

Engesser-Cesar, C., Anderson, A. J., Basso, D. M., Edgerton, V. R., and Cotman, C. W. 2005. Voluntary wheel running improves recovery from a moderate spinal cord injury. *J Neurotrauma,* 22: 157–71.

Enzmann, G. U., Benton, R. L., Woock, J. P., Howard, R. M., Tsoulfas, P., and Whittemore, S. R. 2005. Consequences of noggin expression by neural stem, glial, and neuronal precursor cells engrafted into the injured spinal cord. *Exp Neurol,* 195: 293–304.

Erceg, S., Ronaghi, M., Oria, M., García Roselló, M., Aragó, M. A. P., Lopez, M. G., Radojevic, I. et al. 2010. Transplanted oligodendrocytes and motoneuron progenitors generated from human embryonic stem cells promote locomotor recovery after spinal cord transection. *Stem Cells,* 28: 1541–49.

Eriksson, P. S., Perfilieva, E., Bjork-Eriksson, T., Alborn, A. M., Nordborg, C., Peterson, D. A., and Gage, F. H. 1998. Neurogenesis in the adult human hippocampus. *Nat Med,* 4: 1313–7.

Faulkner, J. and Keirstead, H. S. 2005. Human embryonic stem cell-derived oligodendrocyte progenitors for the treatment of spinal cord injury. *Transpl Immunol,* 15: 131–42.

Fu, S. Y. and Gordon, T. 1995. Contributing factors to poor functional recovery after delayed nerve repair: Prolonged axotomy. *J Neurosci,* 15: 3876–85.

Fu, S. Y. and Gordon, T. 1997. The cellular and molecular basis of peripheral nerve regeneration. *Mol Neurobiol,* 14: 67–116.

Gage, F. H., Kempermann, G., Palmer, T. D., Peterson, D. A., and Ray, J. 1998. Multipotent progenitor cells in the adult dentate gyrus. *J Neurobiol,* 36: 249–66.

Gao, J., Prough, D. S., Mcadoo, D. J., Grady, J. J., Parsley, M. O., Ma, L., Tarensenko, Y. I., and Wu, P. 2006. Transplantation of primed human fetal neural stem cells improves cognitive function in rats after traumatic brain injury. *Exp Neurol,* 201: 281–92.

Gelain, F., Salazar, D. L., Uchida, N., Hamers, F. P. T., Cummings, B. J., and Anderson, A. J. 2010. Human neural stem cells differentiate and promote locomotor recovery in an early chronic spinal coRd Injury NOD-scid Mouse Model. *PLoS ONE,* 5: e12272.

Gordon, T. 2010. The physiology of neural injury and regeneration: The role of neurotrophic factors. *J Commun Disord,* 43: 265–73.

Himes, B. T., Neuhuber, B., Coleman, C., Kushner, R., Swanger, S. A., Kopen, G. C., Wagner, J., Shumsky, J. S., and Fischer, I. 2006. Recovery of function following grafting of human bone marrow-derived stromal cells into the injured spinal cord. *Neurorehabil Neural Repair,* 20: 278–96.

Hofstetter, C. P., Schwarz, E. J., Hess, D., Widenfalk, J., El Manira, A., Prockop, D. J., and Olson, L. 2002. Marrow stromal cells form guiding strands in the injured spinal cord and promote recovery. *Proc Natl Acad Sci USA,* 99: 2199–204.

Isele, N. B., Lee, H. S., Landshamer, S., Straube, A., Padovan, C. S., Plesnila, N., and Culmsee, C. 2007. Bone marrow stromal cells mediate protection through stimulation of PI3-K/Akt and MAPK signaling in neurons. *Neurochem Int,* 50: 243–50.

Ishii, K., Nakamura, M., Dai, H., Finn, T. P., Okano, H., Toyama, Y., and Bregman, B. S. 2006. Neutralization of ciliary neurotrophic factor reduces astrocyte production from transplanted neural stem cells and promotes regeneration of corticospinal tract fibers in spinal cord injury. *J Neurosci Res,* 84: 1669–81.

Johnson, P., Tatara, A., Mccreedy, D., Shiu, A., and Sakiyama-Elbert, S. 2010a. Tissue engineered fibrin scaffolds containing neural progenitors enhance functional recovery in a subacute model of SCI. *Soft Matter,* 6: 5127–37.

Johnson, P. J., Tatara, A., Shiu, A., and Sakiyama-Elbert, S. E. 2010b. Controlled release of neurotrophin-3 and platelet-derived growth factor from fibrin scaffolds containing neural progenitor cells enhances survival and differentiation into neurons in a subacute model of SCI. *Cell Transplant,* 19: 89–101.

Karumbayaram, S., Kelly, T. K., Paucar, A. A., Roe, A. J., Umbach, J. A., Charles, A., Goldman, S. A., Kornblum, H. I., and Wiedau-Pazos, M. 2009. Human embryonic stem cell-derived motor neurons expressing SOD1 mutants exhibit typical signs of motor neuron degeneration linked to ALS. *Dis Model Mech,* 2: 189–95.

Keirstead, H. S., Nistor, G., Bernal, G., Totoiu, M., Cloutier, F., Sharp, K., and Steward, O. 2005. Human embryonic stem cell-derived oligodendrocyte progenitor cell transplants remyelinate and restore locomotion after spinal cord injury. *J Neurosci,* 25: 4694–705.

Kim, H. J., Lee, J. H., and Kim, S. H. 2010a. Therapeutic effects of human mesenchymal stem cells on traumatic brain injury in rats: Secretion of neurotrophic factors and inhibition of apoptosis. *J Neurotrauma,* 27: 131–8.

Kim, J. H., Auerbach, J. M., Rodriguez-Gomez, J. A., Velasco, I., Gavin, D., Lumelsky, N., Lee, S. H., Nguyen, J., Sanchez-Pernaute, R., Bankiewicz, K., and Mckay, R. 2002. Dopamine neurons derived from embryonic stem cells function in an animal model of Parkinson's disease. *Nature,* 418: 50–6.

Kim, K., Doi, A., Wen, B., Ng, K., Zhao, R., Cahan, P., Kim, J. et al. 2010b. Epigenetic memory in induced pluripotent stem cells. *Nature,* 467: 285–90

Klimanskaya, I., Hipp, J., Rezai, K. A., West, M., Atala, A., and Lanza, R. 2004. Derivation and comparative assessment of retinal pigment epithelium from human embryonic stem cells using transcriptomics. *Cloning Stem Cells,* 6: 217–45.

Kobayashi, J., Mackinnon, S. E., Watanabe, O., Ball, D. J., Gu, X. M., Hunter, D. A., and Kuzon, W. M., Jr. 1997. The effect of duration of muscle denervation on functional recovery in the rat model. *Muscle Nerve,* 20: 858–66.

Kopyov, O. V., Jacques, D. S., Lieberman, A., Duma, C. M., and Rogers, R. L. 1997. Outcome following intrastriatal fetal mesencephalic grafts for Parkinson's patients is directly related to the volume of grafted tissue. *Exp Neurol*, 146: 536–45.

Langlois, J.A., Kegler, S.R., Butler, J.A., Gotsch, K.E., Johnson, R. L., and Reichard, A.A. Traumatic Brain Injury-related hospital discharges. results from a 14-State Surveillance System 1997, M. S. S., pp. 1–120. 2003. Traumatic brain injury-related hospital discharges. Results from a 14-State Surveillance System *MMWR Surveill Summ*, 52: 1–20.

Laurent, L. C., Nievergelt, C. M., Lynch, C., Fakunle, E., Harness, J. V., Schmidt, U., Galat, V. et al. 2010. Restricted ethnic diversity in human embryonic stem cell lines. *Nat Methods*, 7: 6–7.

Lee, G., Papapetrou, E. P., Kim, H., Chambers, S. M., Tomishima, M. J., Fasano, C. A., Ganat, Y. M. et al. 2009. Modelling pathogenesis and treatment of familial dysautonomia using patient-specific iPSCs. *Nature*, 461: 402–6.

Lee, H., Al Shamy, G., Elkabetz, Y., Schoefield, C. M., Harrsion, N. L., Panagiotakos, G., Socci, N. D., Tabar, V., and Studer, L. 2007. Directed differentiation and transplantation of human embryonic stem cell derived motoneurons. *Stem Cells*, 25: 1931–39.

Lee, J. K., Jin, H. K., Endo, S., Schuchman, E. H., Carter, J. E., and Bae, J. S. 2010. Intracerebral transplantation of bone marrow-derived mesenchymal stem cells reduces amyloid-beta deposition and rescues memory deficits in Alzheimer's disease mice by modulation of immune responses. *Stem Cells*, 28: 329–43.

Lee, S. H., Lumelsky, N., Studer, L., Auerbach, J. M., and Mckay, R. D. 2000. Efficient generation of midbrain and hindbrain neurons from mouse embryonic stem cells. *Nat Biotechnol*, 18: 675–9.

Li, Q., Brus-Ramer, M., Martin, J. H., and Mcdonald, J. W. 2010. Electrical stimulation of the medullary pyramid promotes proliferation and differentiation of oligodendrocyte progenitor cells in the corticospinal tract of the adult rat. *Neurosci Lett*, 479: 128–33.

Li, X. J., Du, Z. W., Zarnowska, E. D., Pankratz, M., Hansen, L. O., Pearce, R. A., and Zhang, S. C. 2005. Specification of motoneurons from human embryonic stem cells. *Nat Biotechnol*, 23: 215–21.

Li, X. J., Hu, B. Y., Jones, S. A., Zhang, Y. S., Lavaute, T., Du, Z. W., and Zhang, S. C. 2008. Directed differentiation of ventral spinal progenitors and motor neurons from human embryonic stem cells by small molecules. *Stem Cells*, 26: 886–93.

Liu, S., Qu, Y., Stewart, T. J., Howard, M. J., Chakrabortty, S., Holekamp, T. F., and Mcdonald, J. W. 2000. Embryonic stem cells differentiate into oligodendrocytes and myelinate in culture and after spinal cord transplantation. *Proc Natl Acad Sci USA*, 97: 6126–31.

Lu, D., Mahmood, A., Qu, C., Hong, X., Kaplan, D., and Chopp, M. 2007a. Collagen scaffolds populated with human marrow stromal cells reduce lesion volume and improve functional outcome after traumatic brain injury. *Neurosurgery*, 61: 596–602; discussion 602–3.

Lu, P., Jones, L. L., Snyder, E. Y., and Tuszynski, M. H. 2003. Neural stem cells constitutively secrete neurotrophic factors and promote extensive host axonal growth after spinal cord injury. *Exp Neurol*, 181: 115–29.

Lu, P., Jones, L. L. and Tuszynski, M. H. 2005. BDNF-expressing marrow stromal cells support extensive axonal growth at sites of spinal cord injury. *Exp Neurol*, 191: 344–60.

Lu, P., Jones, L. L. and Tuszynski, M. H., 2007b. Axon regeneration through scars and into sites of chronic spinal cord injury. *Exp Neurol*, 203: 8–21.

Lund, R. D., Wang, S., Klimanskaya, I., Holmes, T., Ramos-Kelsey, R., Lu, B., Girman, S., Bischoff, N., Sauve, Y., and Lanza, R. 2006. Human embryonic stem cell-derived cells rescue visual function in dystrophic RCS rats. *Cloning Stem Cells*, 8: 189–99.

Mackinnon, S. E. and Dellon, A. L., 1988. *Surgery of the peripheral nerve*, New York, Thieme Medical Publishers.

Manabe, T., Tatsumi, K., Inoue, M., Makinodan, M., Yamauchi, T., Makinodan, E., Yokoyama, S., Sakumura, R., and Wanaka, A. 2007. L3/Lhx8 is a pivotal factor for cholinergic differentiation of murine embryonic stem cells. *Cell Death Differ*, 14: 1080–5.

Marques, S. A., Almeida, F. M., Fernandes, A. M., Dos Santos Souza, C., Cadilhe, D. V., Rehen, S. K., and Martinez, A. M. 2010. Predifferentiated embryonic stem cells promote functional recovery after spinal cord compressive injury. *Brain Res*, 1349: 115–28.

McDonald, J. W., Liu, X. Z., Qu, Y., Liu, S., Mickey, S. K., Turetsky, D., Gottlieb, D. I., and Choi, D. W. 1999. Transplanted embryonic stem cells survive, differentiate and promote recovery in injured rat spinal cord. *Nat Med,* 5: 1410–2.

McKenzie, I. A., Biernaskie, J., Toma, J. G., Midha, R., and Miller, F. D. 2006. Skin-derived precursors generate myelinating Schwann cells for the injured and dysmyelinated nervous system. *J Neurosci,* 26: 6651–60.

Miles, G. B., Yohn, D. C., Wichterle, H., Jessell, T. M., Rafuse, V. F., and Brownstone, R. M. 2004. Functional properties of motoneurons derived from mouse embryonic stem cells. *J Neurosci,* 24: 7848–58.

Mukouyama, Y. S., Deneen, B., Lukaszewicz, A., Novitch, B. G., Wichterle, H., Jessell, T. M., and Anderson, D. J. 2006. Olig2+ neuroepithelial motoneuron progenitors are not multipotent stem cells in vivo. *Proc Natl Acad Sci USA,* 103: 1551–6.

Nagoshi, N., Shibata, S., Kubota, Y., Nakamura, M., Nagai, Y., Satoh, E., Morikawa, S. et al. 2008. Ontogeny and multipotency of neural crest-derived stem cells in mouse bone marrow, dorsal root ganglia, and whisker pad. *Cell Stem Cell,* 2: 392–403.

Nistor, G., Seiler, M. J., Yan, F., Ferguson, D., and Keirstead, H. S. 2010. Three-dimensional early retinal progenitor 3D tissue constructs derived from human embryonic stem cells. *J Neurosci Methods,* 190: 63–70.

Nistor, G. I., Totoiu, M. O., Haque, N., Carpenter, M. K., and Keirstead, H. S. 2005. Human embryonic stem cells differentiate into oligodendrocytes in high purity and myelinate after spinal cord transplantation. *Glia,* 49: 385–96.

Nomura, H., Zahir, T., Kim, H., Katayama, Y., Kulbatski, I., Morshead, C. M., Shoichet, M. S., and Tator, C. H. 2008. Extramedullary chitosan channels promote survival of transplanted neural stem and progenitor cells and create a tissue bridge after complete spinal cord transection. *Tissue Eng Part A,* 14: 649–65.

NSCISC, 2012. National Spinal Cord Injury Statistical Center, Birmingham, Alabama. Available at: https://www.nscisc.uab.edu/PublicDocuments/fact_figures_docs/Facts%202012%20Feb%20 Final.pdf

Okabe, S., Forsberg-Nilsson, K., Spiro, A. C., Segal, M., and Mckay, R. D. 1996. Development of neuronal precursor cells and functional postmitotic neurons from embryonic stem cells in vitro. *Mech Dev,* 59: 89–102.

Okano, H. 2010. Neural stem cells and strategies for the regeneration of the central nervous system. *Proc Jpn Acad Ser B Phys Biol Sci,* 86: 438–50.

Park, I. H., Arora, N., Huo, H., Maherali, N., Ahfeldt, T., Shimamura, A., Lensch, M. W., Cowan, C., Hochedlinger, K., and Daley, G. Q. 2008a. Disease-specific induced pluripotent stem cells. *Cell,* 134: 877–86.

Park, I. H., Zhao, R., West, J. A., Yabuuchi, A., Huo, H., Ince, T. A., Lerou, P. H., Lensch, M. W., and Daley, G. Q. 2008b. Reprogramming of human somatic cells to pluripotency with defined factors. *Nature,* 451: 141–6.

Park, K. D., Seong, S. K., Park, Y. M., Choi, Y., Park, J. H., Lee, S. H., Baek, D. H. et al. 2011. Telomerase reverse transcriptase (TERT) Related with Telomerase Activity Regulates Tumorigenic Potential of Mouse Embryonic Stem Cells. *Stem Cells Dev,* 20: 149–57.

Piccini, P., Brooks, D. J., Bjorklund, A., Gunn, R. N., Grasby, P. M., Rimoldi, O., Brundin, P. et al. 1999. Dopamine release from nigral transplants visualized in vivo in a Parkinson's patient. *Nat Neurosci,* 2: 1137–40.

Polo, J. M., Liu, S., Figueroa, M. E., Kulalert, W., Eminli, S., Tan, K. Y., Apostolou, E. et al. 2010. Cell type of origin influences the molecular and functional properties of mouse induced pluripotent stem cells. *Nat Biotechnol,* 28: 848–55.

Qu, C., Xiong, Y., Mahmood, A., Kaplan, D. L., Goussev, A., Ning, R., and Chopp, M. 2009. Treatment of traumatic brain injury in mice with bone marrow stromal cell-impregnated collagen scaffolds. *J Neurosurg,* 111: 658–65.

Rauch, M. F., Hynes, S. R., Bertram, J., Redmond, A., Robinson, R., Williams, C., Xu, H., Madri, J. A., and Lavik, E. B. 2009. Engineering angiogenesis following spinal cord injury: A coculture of neural progenitor and endothelial cells in a degradable polymer implant leads to an increase in vessel density and formation of the blood-spinal cord barrier. *Eur J Neurosci,* 29: 132–45.

Riess, P., Molcanyi, M., Bentz, K., Maegele, M., Simanski, C., Carlitscheck, C., Schneider, A. et al. 2007. Embryonic stem cell transplantation after experimental traumatic brain injury dramatically improves neurological outcome, but may cause tumors. *J Neurotrauma,* 24: 216–25.

Riess, P., Zhang, C., Saatman, K. E., Laurer, H. L., Longhi, L. G., Raghupathi, R., Lenzlinger, P. M. et al. 2002. Transplanted neural stem cells survive, differentiate, and improve neurological motor function after experimental traumatic brain injury. *Neurosurgery,* 51: 1043–52; discussion 1052–4.

Riley, J., Federici, T., Park, J., Suzuki, M., Franz, C. K., Tork, C., Mchugh, J., Teng, Q., Svendsen, C., and Boulis, N. M. 2009. Cervical spinal cord therapeutics delivery: Preclinical safety validation of a stabilized microinjection platform. *Neurosurgery,* 65: 754–61; discussion 761–2.

Sanchez-Ramos, J., Song, S., Cardozo-Pelaez, F., Hazzi, C., Stedeford, T., Willing, A., Freeman, T. B. et al. 2000. Adult bone marrow stromal cells differentiate into neural cells in vitro. *Exp Neurol,* 164: 247–56.

Seiler, M. J., Aramant, R. B., Thomas, B. B., Peng, Q., Sadda, S. R., and Keirstead, H. S. 2010. Visual restoration and transplant connectivity in degenerate rats implanted with retinal progenitor sheets. *Eur J Neurosci,* 31: 508–20.

Setoguchi, T., Nakashima, K., Takizawa, T., Yanagisawa, M., Ochiai, W., Okabe, M., Yone, K., Komiya, S., and Taga, T. 2004. Treatment of spinal cord injury by transplantation of fetal neural precursor cells engineered to express BMP inhibitor. *Exp Neurol,* 189: 33–44.

Shim, J. W., Park, C. H., Bae, Y. C., Bae, J. Y., Chung, S., Chang, M. Y., Koh, H. C. et al. 2007. Generation of functional dopamine neurons from neural precursor cells isolated from the subventricular zone and white matter of the adult rat brain using Nurr1 overexpression. *Stem Cells,* 25: 1252–62.

Shindo, T., Matsumoto, Y., Wang, Q., Kawai, N., Tamiya, T., and Nagao, S. 2006. Differences in the neuronal stem cells survival, neuronal differentiation and neurological improvement after transplantation of neural stem cells between mild and severe experimental traumatic brain injury. *J Med Invest,* 53: 42–51.

Singh Roy, N., Nakano, T., Xuing, L., Kang, J., Nedergaard, M., and Goldman, S. A. 2005. Enhancer-specified GFP-based FACS purification of human spinal motor neurons from embryonic stem cells. *Exp Neurol,* 196: 224–34.

Stadtfeld, M., Nagaya, M., Utikal, J., Weir, G., and Hochedlinger, K. 2008. Induced pluripotent stem cells generated without viral integration. *Science,* 322: 945–9.

Sulaiman, O. A. and Gordon, T. 2009. Role of chronic Schwann cell denervation in poor functional recovery after nerve injuries and experimental strategies to combat it. *Neurosurgery,* 65: A105–14.

Takahashi, K., Tanabe, K., Ohnuki, M., Narita, M., Ichisaka, T., Tomoda, K., and Yamanaka, S. 2007. Induction of pluripotent stem cells from adult human fibroblasts by defined factors. *Cell,* 131: 861–72.

Takahashi, K. and Yamanaka, S. 2006. Induction of pluripotent stem cells from mouse embryonic and adult fibroblast cultures by defined factors. *Cell,* 126: 663–76.

Tannemaat, M. R., Eggers, R., Hendriks, W. T., De Ruiter, G. C., Van Heerikhuize, J. J., Pool, C. W., Malessy, M. J., Boer, G. J., and Verhaagen, J. 2008. Differential effects of lentiviral vector-mediated overexpression of nerve growth factor and glial cell line-derived neurotrophic factor on regenerating sensory and motor axons in the transected peripheral nerve. *Eur J Neurosci,* 28: 1467–79.

Tate, C. C., Shear, D. A., Tate, M. C., Archer, D. R., Stein, D. G., and Laplaca, M. C. 2009. Laminin and fibronectin scaffolds enhance neural stem cell transplantation into the injured brain. *J Tissue Eng Regen Med,* 3: 208–17.

Tate, M. C., Shear, D. A., Hoffman, S. W., Stein, D. G., Archer, D. R., and Laplaca, M. C. 2002. Fibronectin promotes survival and migration of primary neural stem cells transplanted into the traumatically injured mouse brain. *Cell Transplant,* 11: 283–95.

Teng, Y. D., Lavik, E. B., Qu, X., Park, K. I., Ourednik, J., Zurakowski, D., Langer, R., and Snyder, E. Y. 2002. Functional recovery following traumatic spinal cord injury mediated by a unique polymer scaffold seeded with neural stem cells. *Proc Natl Acad Sci USA*, 99: 3024–9.

Toma, J. G., Akhavan, M., Fernandes, K. J., Barnabe-Heider, F., Sadikot, A., Kaplan, D. R., and Miller, F. D. 2001. Isolation of multipotent adult stem cells from the dermis of mammalian skin. *Nat Cell Biol*, 3: 778–84.

Toma, J. G., Mckenzie, I. A., Bagli, D., and Miller, F. D. 2005. Isolation and characterization of multipotent skin-derived precursors from human skin. *Stem Cells*, 23: 727–37.

Tsuji, O., Miura, K., Okada, Y., Fujiyoshi, K., Mukaino, M., Nagoshi, N., Kitamura, K. et al. 2010. Therapeutic potential of appropriately evaluated safe-induced pluripotent stem cells for spinal cord injury. *Proc Natl Acad Sci USA*, 107: 12704–9.

Vierbuchen, T., Ostermeier, A., Pang, Z. P., Kokubu, Y., Sudhof, T. C., and Wernig, M. 2010. Direct conversion of fibroblasts to functional neurons by defined factors. *Nature*, 463: 1035–41.

Wagers, A. J., Sherwood, R. I., Christensen, J. L., and Weissman, I. L. 2002. Little evidence for developmental plasticity of adult hematopoietic stem cells. *Science*, 297: 2256–9.

Walsh, S., Biernaskie, J., Kemp, S. W., and Midha, R. 2009. Supplementation of acellular nerve grafts with skin derived precursor cells promotes peripheral nerve regeneration. *Neuroscience*, 164: 1097–107.

Walsh, S. K., Gordon, T., Addas, B. M., Kemp, S. W., and Midha, R. 2010. Skin-derived precursor cells enhance peripheral nerve regeneration following chronic denervation. *Exp Neurol*, 223: 221–8.

Webber, D. J., Bradbury, E. J., Mcmahon, S. B., and Minger, S. L. 2007. Transplanted neural progenitor cells survive and differentiate but achieve limited functional recovery in the lesioned adult rat spinal cord. *Regen Med*, 2: 929–45.

Wernig, M., Lengner, C. J., Hanna, J., Lodato, M. A., Steine, E., Foreman, R., Staerk, J., Markoulaki, S., and Jaenisch, R. 2008a. A drug-inducible transgenic system for direct reprogramming of multiple somatic cell types. *Nat Biotechnol*, 26: 916–24.

Wernig, M., Zhao, J. P., Pruszak, J., Hedlund, E., Fu, D., Soldner, F., Broccoli, V., Constantine-Paton, M., Isacson, O., and Jaenisch, R. 2008b. Neurons derived from reprogrammed fibroblasts functionally integrate into the fetal brain and improve symptoms of rats with Parkinson's disease. *Proc Natl Acad Sci USA*, 105: 5856–61.

Wichterle, H., Lieberam, I., Porter, J. A., and Jessell, T. M. 2002. Directed differentiation of embryonic stem cells into motor neurons. *Cell*, 110: 385–97.

Willerth, S. M., Arendas, K. J., Gottlieb, D. I., and Sakiyama-Elbert, S. E. 2006. Optimization of fibrin scaffolds for differentiation of murine embryonic stem cells into neural lineage cells. *Biomaterials*, 27: 5990–6003.

Wobus, A. M., Holzhausen, H., Jakel, P., and Schoneich, J. 1984. Characterization of a pluripotent stem cell line derived from a mouse embryo. *Exp Cell Res*, 152: 212–9.

Woodbury, D., Schwarz, E. J., Prockop, D. J., and Black, I. B. 2000. Adult rat and human bone marrow stromal cells differentiate into neurons. *J Neurosci Res*, 61: 364–70.

Xiong, Y., Qu, C., Mahmood, A., Liu, Z., Ning, R., Li, Y., Kaplan, D. L., Schallert, T., and Chopp, M. 2009. Delayed transplantation of human marrow stromal cell-seeded scaffolds increases transcallosal neural fiber length, angiogenesis, and hippocampal neuronal survival and improves functional outcome after traumatic brain injury in rats. *Brain Res*, 1263: 183–91.

Xu, L., Yan, J., Chen, D., Welsh, A. M., Hazel, T., Johe, K., Hatfield, G., and Koliatsos, V. E. 2006. Human neural stem cell grafts ameliorate motor neuron disease in SOD-1 transgenic rats. *Transplantation*, 82: 865–75.

Yano, S., Kuroda, S., Shichinohe, H., Seki, T., Ohnishi, T., Tamagami, H., Hida, K., and Iwasaki, Y. 2006. Bone marrow stromal cell transplantation preserves gammaaminobutyric acid receptor function in the injured spinal cord. *J Neurotrauma*, 23: 1682–92.

Yu, J., Vodyanik, M. A., Smuga-Otto, K., Antosiewicz-Bourget, J., Frane, J. L., Tian, S., Nie, J. et al. 2007. Induced pluripotent stem cell lines derived from human somatic cells. *Science*, 318: 1917–20.

Zahir, T., Nomura, H., Guo, X. D., Kim, H., Tator, C., Morshead, C., and Shoichet, M. 2008. Bioengineering neural stem/progenitor cell-coated tubes for spinal cord injury repair. *Cell Transplant,* 17: 245–54.

Zhao, C., Deng, W., and Gage, F. H. 2008. Mechanisms and functional implications of adult neurogenesis. *Cell,* 132: 645–60.

Zhao, T., Zhang, Z.-N., Rong, Z., and Xu, Y. 2011. Immunogenicity of induced pluripotent stem cells. *Nature,* 474, 212–15.

Ziv, Y., Avidan, H., Pluchino, S., Martino, G., and Schwartz, M. 2006. Synergy between immune cells and adult neural stem/progenitor cells promotes functional recovery from spinal cord injury. *Proc Natl Acad Sci USA,* 103: 13174–9.

Zurita, M. and Vaquero, J. 2006. Bone marrow stromal cells can achieve cure of chronic paraplegic rats: Functional and morphological outcome one year after transplantation. *Neurosci Lett,* 402: 51–6.

6

Stem Cells and Regenerative Medicine for Treating Damaged Myocardium

Rohini Gupta
University of California, Berkeley

Kunal Mehtani
Kaiser Permanente

Kimberly R. Kam
University of California, Berkeley

Kevin E. Healy
University of California, Berkeley

6.1 Introduction

Coronary heart disease (CHD) is a major health problem which causes significant morbidity and mortality worldwide. According to recent statistics, CHD is responsible for one-third of all deaths in individuals over the age of 35, making it the leading cause of death in the United States (Lloyd-Jones et al., 2010). According to current estimates, nearly one-half of middle-aged men and one-third of middle-aged women will develop CHD (Lloyd-Jones et al., 1999). CHD is characterized as a failure of coronary circulation to provide adequate blood supply to the cardiac muscle. The narrowing of coronary arteries due to atherosclerosis can lead to exertional angina and even to myocardial infarction (MI) (i.e., a heart attack). A complete lack of blood supply to the myocardium quickly leads to death of cardiac myocytes. Advances in early treatment for patients suffering from a MI have led to a decrease in early mortality; however, there is a higher incidence of heart failure (HF) among survivors (Velagaleti et al., 2008).

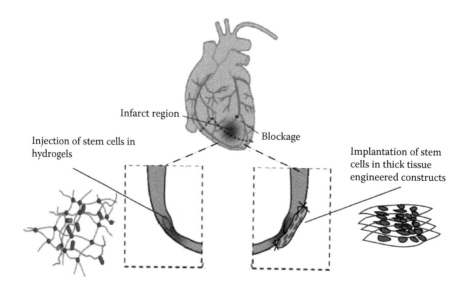

Infarct region

Blockage

Injection of stem cells in
hydrogels

Implantation of stem
cells in thick tissue
engineered constructs

FIGURE 6.1 Schematic of a myocardial infarct. Blockage of coronary vessels can lead to hypoxic and cardiac tissue death (infarct region). Therapies to regenerate cardiac tissue include injection of stem cells in combination with biomaterials. Stem cells are either injected within biomaterials directly into the infarct region or cultured within 3D scaffolds and the tissue engineered constructs are sutured onto the infarct region.

Estimates show that nearly 6 million patients suffer from HF in the United States and about 23 million people are afflicted by HF worldwide (Lloyd-Jones et al., 2010, McMurray et al., 1998). Although these estimates include all causes of HF, ischemic heart disease is still the most common cause of HF. Several medical therapies improve hemodynamics in systolic HF and mortality. However, none of these medical therapies address the loss of cardiac myocytes. Due to a lack of availability of donor hearts, cardiac transplantation is usually not a viable option for most patients.

To truly reverse the trend of increasing morbidity and mortality due to CHD in our society, research and clinical scientists must develop new strategies for directly altering the progressive course of worsening myocardial function in the face of known etiologies for HF. Stem cell transplantation is the only current intervention that deals with cardiac cell death and may represent the first realistic strategy for actually reversing the deleterious effects of what has, until now, been considered irreversible damage to the heart resulting from MI or dilated cardiomyopathy. This novel therapy focuses on regeneration of portions of the infarcted myocardium and improvement of overall systolic function. Typically, stem cells are either transplanted directly, within a supportive "prosurvival" matrix, or as part of a tissue engineered "myocardial patch" (Figure 6.1). Recent studies suggest that various types of stem cells and progenitor cells have the potential to regenerate the myocardium and the interstitial that supports the myocardium. A great challenge facing researchers is to define the elements of experimental cell transplantation schemes that bear the most relevance to human clinical success, and to optimize their application toward a practical therapy. For example, the optimal cell type, number of cells, and the timing of their delivery are not known. It may seem obvious that such a process of rational evaluation and design requires a firm foundation in the scientific and clinical principles underlying a novel therapeutic strategy such as stem cell transplantation; however, despite an enormous increase in reports published in the scientific literature regarding the potential application of this approach to the treatment of human hearts, surprisingly little is known regarding the mechanisms of the modest benefits observed in experimental animal models and some clinical trials. Ultimately, the goals of stem cell transplantation into the diseased myocardial tissue are to provide mortality benefit, improve the quality to life, symptom control, and improve cardiovascular hemodynamics.

6.2 Natural Cells of the Myocardium

The myocardium is an elegant structured tissue of cardiac myocytes, fibroblasts, neural and vascular cells in a highly organized extracellular matrix (ECM) structure. Cardiac myocytes (or cardiomyocytes) are the major cellular component of the myocardium. These specialized muscle cells are 10–15 µm in diameter and can span up to 100 µm in length. Their contractile protein unit, the sarcomere, consists of thin (actin) filaments bound by transverse Z-discs and thick (myosin) filaments held together by M-band that slide past each other during contraction. Elastic protein complexes, titin, connected Z-disc to the M-band and the Z-disc span across to neighboring cells and connect cardiac myocytes (Agarkova and Perriard, 2005). Cell contraction is regulated through intracellular Ca^{2+} channels. Intercellular connection between myocytes is orchestrated through intercalated discs which enable mechanical coupling and rapid electrical propagation via adheren and gap junctions, respectively. Mechanical coupling is provided through adheren junctions that link the actin cytoskeleton and hold myocytes tightly together during expansion and contraction. Gap junctions mediate cellular communication via passive diffusion of ions, water, and other metabolites. The electrical propagation is mediated through a set of connexin proteins which are shared by the two adjoining cells (Noorman et al., 2009); the most abundant of which in the ventricle is Connexin 43.

The second most abundant cell type in the myocardium is the cardiac fibroblast which surrounds the myocardial tissue layers. They produce interstitial collagen and play an active role in maintaining the ECM structure by producing and degrading ECM components as needed. In time of injury, fibroblasts play a key role in remodeling the ECM structure, and in fact excessive fibroblast proliferation and collagen deposition is a hallmark of cardiac dysfunction. Recently, it has been suggested that cardiac fibroblasts may also play a role in modulating electrophysiology (Camelliti et al., 2005).

Additionally, both the sympathetic and autonomic nerve fiber networks are found in the myocardium. While nerve cells do not directly stimulate cardiac myocytes, they do modulate myocardial function (Chen et al., 2001). Finally, the heart consists of a rich supply of blood vessels, the coronary vessels, which supply the large metabolic demands. The intricate blood flow adapts quickly to the variable oxygen tension during rest and exercise. A dense capillary network runs parallel to the myocardial bundles to meet these demands (Spaan et al., 2008). Ideally, transplanted stem cells or a tissue engineered myocardial patch would recapitulate this complex cardiac multicellular milieu. Clearly, sourcing all the various cardiac cells and spatially arranging them in an appropriate structure is a large unmet challenge.

6.3 The ECM of the Myocardium

The myocardial ECM consists mainly of type I and IV collagens arranged in three layers: the endomysium, perimysium, and epimysium (Figure 6.2). The basic building block of the endomysium is a thin sheath of collagen fibers that wraps around individual cardiac myocytes. Collagen struts (crosslinked fibers) extend out from the sheaths and connect cardiac myocytes to other myocytes and capillaries; during systole, the collagen sheath bears the majority of the force ensuring that the myocytes stay aligned. Groups of cardiac myocytes within this sheath are further bundled to form the perimysium, large coiled fibers ranging from 0.5 to 2.5 µm in diameter that provide tensile stiffness to the myocardium. Overall the pericardial fibers have a longitudinal arrangement. Finally, the epimysium is a large connective tissue that surrounds the entire myocardium. The arrangement and mechanics of the ECM enable the critical contractions of the heart, and meet the variable pumping demands of the body (Anderson et al., 2005, Braunwald et al., 1967). More than just providing structural support, the ECM also modulates myocyte function through force and mechanotransduction. Various integrins presented within the ECM bind to integrin receptors and mediate myocyte mechanical and electrical conduction. Thus, for stem cell transplantation and cardiac tissue engineering, it is critical to design biomaterials that can ultimately stimulate regeneration such that the structure and function of the myocardial ECM is preserved.

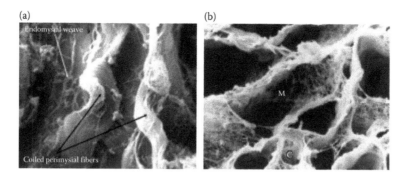

FIGURE 6.2 SEM images of the myocardial ECM. (a) Collagen fibers surround individual myocytes in a thin endomysial sheet and groups of myocytes are further bundled to form the large coiled perimysial fibers. (Adapted from Anderson, R. H. et al. 2005. *Eur. J. Cardiothorac. Surg.*, 28, 517–525.) (b) Detailed view of the endomysial sheet surrounding individual myocytes (M) and supporting vascular cells (C). (Adapted from Macchiarelli, G. et al. 2002. *Histol. Histopathol.*, 17, 699–706.)

6.4 Types of Stem Cells

Stem cells from a number of sources have been explored for regenerating the myocardium. Both multipotent adult stem cells and pluripotent embryonic stem (ES) cells have been used to generate vascular cells and cardiac myocytes for therapy (Figure 6.3). A number of clinical studies have explored adult stem cells for their therapeutic potential in treating ischemic heart disease. While ES cells have only been investigated to date in preclinical models, a new clinical study is expected to assess safety of ES cell transplantation (Alper, 2009). In addition to stem cells, adult skeletal muscle cells have also been explored for clinical cardiac cell therapy. Each cell type is briefly reviewed in the context of myocardium regeneration.

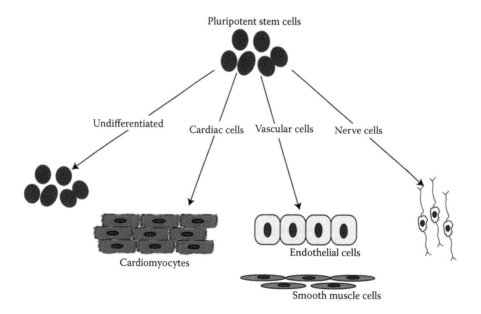

FIGURE 6.3 (**See color insert.**) Pluripotent embryonic stem cells can either remain undifferentiated, or differentiate into cells found within the myocardium. Differentiated cells include cardiac cells (myocytes), vascular cells (endothelial, smooth muscle), and peripheral nerve cells.

6.4.1 Adult Bone Marrow Stem Cells

The bone marrow is a rich source of multipotent stem cells. For cardiac repair, many investigators use unfractionated bone marrow cells (BMC) which include hematopoietic stem cells (HSC), endothelial progenitor cells (EPC), and mesenchymal stem cells (MSC). This strategy provides ease in cell accessibility and does not require extensive manipulation *in vitro*. HSC are located in the bone marrow and are multipotent stem cells that give rise to several types of cells including red blood cells, white blood cells, and platelets. Interestingly, it has been noted that HSC transplanted into murine myocardium may differentiate into cardiomyocytes and blood vessels (Orlic et al., 2003). EPC can be mobilized using cytokines and can be isolated from peripheral blood or bone marrow. Several reports have demonstrated the ability of EPC to revascularize myocardium after an MI (Kocher et al., 2001).

The stromal component of the bone marrow produces MSC that can give rise to a number of vascular cells, including cardiomyocytes. In culture, they exhibit a fibroblast like morphology and can be characterized by a number of markers such as CD34−, CD45−, CD73+, CD90+, and CD105+ (Pittenger and Martin, 2004). One advantage MSC have over other adult BMC is that they appear to possess immunosuppressive properties that promote cell survival after transplantation in an allogeneic environment (Le Blanc and Ringden, 2005). Although a number of studies have shown that they improve cardiac function, the current thought is that rather than contributing directly to functional cardiomyocytes, MSC improve vascularization through paracrine signaling mechanisms (Psaltis et al., 2008). In spite of this observation, the fact that they can be harvested rather easily and are a allotolerant source makes them an attractive candidate for stem cell therapy (Pittenger and Martin, 2004, Psaltis et al., 2008).

6.4.2 Resident Cardiac Stem Cells

There has been a good deal of excitement about the potential of the heart to regenerate itself via resident stem cells. These stem cells have been identified in the human myocardium after an infarct suggesting that they may participate in the repair mechanism. Resident cardiac stem cells (CSC) are generally identified as being Sca1+ or c-kit+, are clonogeneic, multipotent, and can differentiate into myocytes and vascular cells both *in vitro* and *in vivo* (Bearzi et al., 2007). Their origin remains unclear; they may "home" to the heart from the bone marrow after injury or may reside there throughout development. Overall, CSC reside in very low numbers in the adult human heart, estimates range from 2% to 10% of the total cell population (Torella et al., 2006). Therefore, current challenges with CSC include techniques of expanding these cells in large numbers to exploit their regenerative potential.

6.4.3 Embryonic and Induced Pluripotent Stem Cells

Perhaps the most exciting and controversial source of regenerative cells are ES cells. A wealth of research has identified means of successfully inducing differentiation of human embryonic stem cells to form both vascular cells and cardiomyocytes (hES-CM). Differentiation efficiencies range from 2% to 30%, and some challenges remain in isolating and purifying cardiomyocytes from a mixed population (Murry and Keller, 2008). Improvements with directed differentiation protocols are expected to produce human cardiomyocytes from hES cells reliably in the near future (Irion et al., 2008). From these successful differentiation protocols, hES cells can form beating cardiomyocytes that last several months in culture and when transplanted into infarct regions can improve myocardial function in small animal models (Caspi et al., 2007, Laflamme et al., 2007, Leor et al., 2007). Electrophysiological characteristics of the ES-CM suggest that in general, the cardiomyocytes resemble immature fetal cardiomyocytes (Murry and Keller, 2008). Current research efforts are exploring techniques of maturing these cardiomyocytes *in vitro* to produce more adult-like physiology. A number of challenges persist with hES-CM including the need to fully differentiate cells into cardiomyocytes before transplantation to avoid teratoma formation, separation of nondifferentiated cells, and immunological rejection of ES derived CM. Mouse models demonstrate that ES derived CMs almost always give rise to a teratoma upon implantation. In order to prevent the formation of teratoma, isolating highly purified preparation of ES derived CM that are free of undifferentiated stem cells is of great importance (Laflamme and Murry, 2005).

Induced pluripotent stem (iPS) cells represent another source of human cardiomyocytes. Adult mouse and human fibroblasts can be reprogrammed to an embryonic state by overexpression of specific transcription factors. The reprogrammed cells resemble hES cells and can further be differentiated into a number of functional cells including cardiomyocytes (Yoshida and Yamanaka, 2010, Zhang et al., 2009). Recently, an exciting study successfully showed that instead of reprogramming to ES cells, mouse fibroblasts can be directly reprogrammed into cardiomyocytes (Ieda et al., 2010). Similar observations with human cells will accelerate the potential of iPS cells toward clinical myocardial therapy.

6.5 Direct Stem Cell Transplantation in the Clinic

There is currently a large body of preclinical studies showing transplantation of adult stem cells modestly improves contractile function after a MI (Wollert and Drexler, 2005). These positive results noted in animal models have justified clinical trials into stem cell transplantation for regenerating the damaged myocardium. Several randomized clinical trials have evaluated the benefit of stem cell transplantation in CHD as summarized in Table 6.1. Most of these studies have utilized BMC.

The BOOST trial was one of the first randomized clinical trials assessing outcomes of BMC transplantation in post MI patients. Sixty patients were randomly assigned to transplantation of nucleated BMC or standard therapy with no cell transplant. After 6 months, an improved left ventricular ejection fraction (EF)—a measure of overall systolic function—was noted. However, no significant improvement in EF was noted at 18 and 61 months (Wollert et al., 2004). However, in posthoc analysis, patients who sustained transmural or larger infarcts appeared to benefit from BMC transplant as evident by sustained improvement in left ventricular ejection fraction (LVEF) after 18 and 61 months (Meyer et al., 2009, Wollert et al., 2004). The HEBE trial randomly assigned 189 patients with an MI to three groups: patients received either mononucleated BMC, mononucleated cells from peripheral blood, or standard therapy. The results showed that infusion of BMC or cells from peripheral blood did not improve the overall systolic function at 4 months (van der Laan et al., 2008). Similarly, the ASTAMI trail randomized 97 patients to receive mononucleated BMC or standard therapy. The results showed no significant difference in the EF, left

TABLE 6.1 Summary of Randomized Clinical Trials Evaluating Cell Therapy for Acute Myocardial Infarction or Ischemic HF

Trial Name	Type of Cell	Number of Patients	Results
BOOST	BMC	60	Improvement in LVEF short term, Meyer et al. (2009)
REPAIR-AMI	BMC	187	Improved LVEF for 12 months, Schachinger et al. (2006c)
HEBE	BMC	200	No change in systolic function at 4 months, van der Laan et al. (2008)
ASTAMI	BMC	97	No difference after 6 or 12 months, Lunde and Aakhus (2008)
FINCELL	BMC	77	Improvement in LVEF, Miettinen et al. (2010)
REGENT	BMC	200	Improvement in LVEF only in severe patients, Tendera et al. (2009)
Leuven-AMI	BMC	67	No change in LVEF, reduction in myocardial infarct size, Janssens et al. (2006)
MAGIC	Skeletal muscle cells	97	No change in LVEF; reduction in LV end dystolic and systolic volume, Menasche et al. (2008)
TOPCARE-CHD	BMC versus blood derived CPC	58	Improvement in LVEF with BMC, not with CPC, Schachinger et al. (2006a)

ventricular (LV) volumes, or infarct size after 6 and 12 months (Lunde and Aakhus, 2008). At least three other clinical randomized trials have attempted to evaluate the efficacy of BMC transplantation in the post-MI setting (Janssens et al., 2006, Miettinen et al., 2010, Tendera et al., 2009).

The REPAIR-AMI trial randomly assigned 204 patients to receive mononucleated BMC or placebo after percutaneous intervention (PCI) for an MI. The BMC-treated patients showed increased LVEF of 2.5% after 4 months as compared to the control group. A group of 54 patients were followed with serial MRIs. The patients who received mononucleated BMC infusion were noted to have a 2.8% effect on LVEF at 12 months. Interestingly, this multicenter randomized control trial also found reduced mortality in the patients treated with BMC. One year after intervention, six deaths were reported in the placebo group and two deaths in the BMC group. In addition, six patients in the placebo group had a recurring MI, and none were reported in the BMC-treated group at 1 year (Dill et al., 2009, Schachinger et al., 2006b). Although these differences were statistically significant, the limited number of outcomes reported in the study demands larger trials. The FINCELL trial randomized 80 patients to receive either mononucleated BMC infusion or placebo post-PCI. Six months after cell transfer, patients who were provided cell therapy had improved LVEF recovery compared to controls (Huikuri et al., 2008). Similarly, the REGENT trial also showed an improved LVEF 3 months after therapy with unselected mononuclear BMC compared to controls. However, these benefits were not sustained after 6 months of therapy (Tendera et al., 2009).

Similarly, stem cell therapy has also been assessed in HF patients. Skeletal myoblasts have been transplanted into dyskinetic myocardium or infarct scar. It has been previously noted that autologous skeletal myoblasts differentiate into myotubes (Wollert and Drexler, 2010). In the MAGIC trial 97 patients were randomized to receive transepicardial injections of myoblasts versus placebo during coronary artery bypass grafting surgery. These autologous skeletal myoblasts were delivered in and around the dyskinetic myocardium. After 6 months, no improvement was noted in regional or global left ventricular function. But, interestingly a significant reduction in LV end systolic and diastolic volumes was noted after 6 months (Menasche et al., 2008). TOPCARE-CHD trial tested the difference between mononucleated BMC, blood-derived progenitor cells or no cells in the setting of ischemic HF. Three months after therapy, LV EF was assessed by angiography. A significantly greater LVEF was noted in the mononucleated BMC group as compared to the blood-derived progenitor cells or no cell infusion (Assmus et al., 2006).

Although these clinical studies of stem cell regeneration of human myocardial tissue have demonstrated adequate safety, they have had contradictory performance results. The HEBE and ASTAMI trials showed no benefit with infusion of BMC. The BOOST trial showed a short-term benefit from stem cell transplantation, and the REPAIR-AMI trial indicated a sustained improvement in LVEF with infusion of BMC. There may be many reasons for the differences in outcomes. Notably, there was no standardized protocol for collection or infusion of stem cells in these studies. Even the types of stem cells used for transplantation were not consistent among trials. The number of cells transplanted in each study varied greatly and thus could drastically change outcomes. Finally, the numbers of patients in these trials were small. Therefore, larger trials that use an established protocol for stem cell transplantation may provide more consistent and reproducible results. Currently, there are numerous ongoing trials evaluating the safety and efficacy of cell infusion (Wollert and Drexler, 2010).

6.6 Biomaterials for Transplantation of CSC

Regardless of the cell type used, many barriers must be overcome for the field of CSC transplantation therapy to move forward and become clinically applicable. For example, it is not yet known how to keep a majority of the transplanted cells alive for more than a few days (Zhang et al., 2010), and therefore, for any cell-based therapy to work effectively, a prosurvival strategy should be developed, since in the harsh hypoxic environment of the infarcted heart one expects the level of cell death and fibrosis to be significant. In addition, the impact of parameters such as the optimal number and timing of stem cell transplantation post-MI are currently unknown. The success of cell transplantation for cardiac tissue

TABLE 6.2 Natural Biomaterials Used for Cardiac Cell Delivery and Tissue Engineering

Material	Applications
Collagen	Cell delivery into infarct region, Kofidis et al. (2005a), Kofidis et al. (2005b), Kutschka et al. (2006), Xiang et al. (2006)
	3D thick EHT, Eschenhagen et al. (1997), Guo et al. (2006), Hosseinkhani et al. (2010), Radisic et al. (2004)
Alginate	Cell delivery into infarct region, Leor et al. (2000, 2007)
Fibrin	Cell delivery into infarct region, Christman et al. (2004b), Ryu et al. (2005), Simpson et al. (2007)
Chitosan	Thermoreversible, cell delivery into infarct region, Lu et al. (2009); electrospun fibrous scaffold, Dang and Leong (2007)

regeneration therefore hinges upon enhanced cell survival, subsequent promotion of their functional integration into existing tissue, and temporal expansion in transplant size. Another challenge is to find an optimal way of delivering the cells. After injection of stem cells into the myocardium, it is estimated that ~90% of cells are removed by the circulating blood (Orlic et al., 2002, Wu et al., 2009) and of the cells that remain, very few are viable and contribute to functional replacement. Ways of improving stem cell retention and survival include the delivery of prosurvival growth factors (Laflamme et al., 2007) and incorporating cells within biomaterials (Christman et al., 2004b, Leor et al., 2007, Wall et al., 2010). Biomaterials have been explored as a tool to both improve cell retention and cell survival within the myocardium. While there have been a number of studies that have explored the application of biomaterials to improve myocardial function without the use of stem cells, the focus of this chapter will be on biomaterials designed for stem cell therapy. Typically, cells are injected directly into an infarct region along with the biomaterials that provide structural integrity and retain the cells in place. Materials that are suitable for CSC transplantation range from naturally derived (Table 6.2) to synthetic biomaterials (Table 6.3). The following describes the structure and mechanical properties of the biomaterials most relevant for CSC applications.

6.6.1 Naturally Derived Materials as Scaffolds

Naturally derived materials have frequently been used as scaffolds for cardiac tissue engineering. These materials have advantages over their synthetic counterparts because they are components of or have similar biological properties to the natural ECM. Natural materials engage cell surface receptors and also provide the physiological environment to regulate cell function. However, use of natural materials for tissue engineering applications also has disadvantages such as lot-to-lot variability, immune rejection due to xenogeneic protein components, and high contamination potential (Drury and Mooney, 2003).

Collagen is a major component of connective tissue, and makes up ~30% of all protein in the human body. It is widespread and found in the heart, skin, bone, fascia, cartilage, and in most areas requiring strength and flexibility. There are 28 different types of collagen of which collagen type I is the most prevalent (Gordon and Hahn, 2010). All types of collagen are typically composed of three repeating peptide subunits that are each ~1050 amino acid residues long and show a strong sequence homology (Gordon and Hahn, 2010). These chains coil to form a triple helix that is crosslinked together through covalent and hydrogen bonds. The resulting collagen fibrils offer opportunities for specific cell adhesion events, since collagen contains integrin binding domains such as arginine–glycine–aspartic acid (RGD) and GFOGER (glycine–phenylalanine–hydroxyproline–glycine–glutamic acid–arginine) to aid in cell attachment (Reyes and Garcia, 2004). Additionally, the free ε-amines on the lysine residues can be used for chemical modification with bioactive molecules, such as peptides. Moreover, collagen is broken down by various collagenases and serine proteases which allows for localized biodegradation when cells are present. Type I collagen is most commonly used in tissue engineering applications including vascular grafts (Wallace and Rosenblatt, 2003). Type I collagen can be isolated and solubilized through proteolytic enzymes and maintained in acidic conditions at low temperatures until

TABLE 6.3   Synthetic Biomaterials Used for Cardiac Cell Delivery and Tissue Engineering

Material	Repeating Monomer Unit	Applications
Poly(lactide) (Isomers: D-lactide, L-lactide, D, L lactide)		Cell delivery into infarct region, Jin et al. (2009); Electrospun fibrous scaffold, Zong et al. (2005)
Poly(glycolide)		Electrospun fibrous scaffold, Hosseinkhani et al. (2010), Li et al. (2006)
Poly-ε-caprolactone		Scaffold for ventricular repair, Miyagi et al. (2010); Electrospun fibrous stacked scaffolds, Heydarkhan-Hagvall et al. (2008)
Poly-N-isopropylacrylamide		Thermoresponsive hydrogel for stem cell delivery into infarct region, Wall et al. (2010); 3D cell sheet engineering, Haraguchi et al. (2006)
Polyurethane	(chain extender)$_n$-(soft segment)$_n$-(hard segment)$_{2n}$ Example polyurethane: PCL1250/Phe where chain extender (L-phenylalanine) = Soft segment (polycaprolactone diol) = Hard segment (diisocynanato methylcaproate) =	Electrospun fibrous scaffold, Fromstein et al. (2008)

use (Gelse et al., 2003). Gelatin, a denatured form of collagen, is also widely used as biomaterial and is derived through acid or basic isolation typically from bovine or porcine skin (Olsen et al., 2003). Yet another type of popular collagen-based material is Matrigel', secreted from Engelbroth–Holm Swarm mouse sarcoma cells, which is in soluble form at low temperatures (4°C) and gels at 37°C. It is composed of many basement membrane proteins including collagen (Type I and IV), fibronectin, laminin, and growth factors; however, the exact composition of Matrigel is largely unknown (Hughes et al., 2010, Kleinman and Martin, 2005).

Alginate is a naturally derived polysaccharide harvested from brown algae that is composed of β-D-mannuronic acid and α-L-guluronic acid (Rowley et al., 1999). In the presence of divalent cations like Ca^{2+}, adjacent alginate chains cooperatively bind to form ionic interchain bridges (Rowley et al., 1999). Therefore, alginate is an attractive material for injectable scaffolds since it gels in the presence of Ca^{2+}. As alginate is derived from algae, it has little biological interaction with mammalian cells, but can be chemically modified with specific peptides via the carboxylic acid functional groups to promote cell anchorage and cell–material interaction (Rowley et al., 1999). Therefore, this biopolymer is comparable to synthetic polymers in that it elicits minimal biological response and can be decorated with integrin-engaging peptides with a high signal-to-noise biological response.

Fibrin, also referred to as fibrin-glue, is a plasma-derived biopolymer that is used as a biodegradable tissue sealant for numerous surgical applications (Radosevich et al., 1997). During the last step of the coagulation cascade, thrombin enzymatically cleaves fibrinogen to polymerize a semirigid fibrin clot. This biopolymer acts as a tissue sealant by binding to biological tissue via covalent, hydrogen, or electrostatic bonds. Mechanical interlocking also plays a large role in the anchoring of the fibrin clot to the tissue (Radosevich et al., 1997). Furthermore, many studies have reported that fibrin has angiogenic properties (Christman et al., 2004a,b) as its degradation products stimulate the migration of monocytes and subsequent macrophages to the clot which remove the degraded fibrin by-products via phagocytosis. Fibroblasts bind and migrate into the clot network and secrete plasminogen activators that lyse fibrin, favoring neovascularization (Radosevich et al., 1997). Fibrin also serves as an attractive cardiac tissue scaffold material since it is biocompatible, FDA approved, does not exhibit extensive fibrosis or tissue necrosis, is biodegradable, and promotes angiogenesis.

Chitosan is a linear polysaccharide of (1–4)-glycosidic bonds derived from the exoskeletons of animals such as crustaceans, mollusks, and insects (Alves and Mano, 2008, Lu et al., 2009). It can also be extracted from the fungal fermentation processes. It is a biocompatible and biodegradable cationic biopolymer obtained from deacetylation of chitin in an alkaline environment (Alves and Mano, 2008), but is known to cause slight inflammation in mammals. The biodegradation rate of chitosan is determined by the residual acetyl content, a parameter that can be easily tuned. The major pathway for the biodegradation *in vivo* is through lysozyme which depolymerizes the polysaccharide. It has been used for many medical applications due to its low-toxicity and acceptable biocompatibility. Recently, temperature-sensitive variations of chitosan have been developed.

6.6.2 Synthetic Biomaterials as Scaffolds

Synthetic biomaterials offer advantages over naturally derived materials as scaffolds since their chemistry, structure, and mechanical properties can be well controlled and systematically optimized with batch-to-batch reproducibility. They also offer a range of chemistries unavailable to natural materials which have the potential for various applications such as time-controlled release of bioactive compounds from the matrix and tunability of mechanical properties (Saha et al., 2007). Furthermore, biomaterials can be redesigned and modified in an iterative manner for improved performance which is a benefit over naturally occurring materials, which typically have little latitude in modification. Finally, synthetic biomaterials do not have the problems associated with injecting naturally derived materials *in vivo* such as disease transmission. However, a downside to synthetic materials is that degradation products can often induce inflammatory responses. For example, the acidic degradation products from poly(lactide)

(PL) and poly(glycolide) (PG) can lower the pH of the microenvironment and cause serious physiological effects such as a chronic inflammatory response.

The polyesters belonging to the poly (α-hydroxy acid ester) family such as PG, PL, and poly(ε-caprolactone) (PCL) are the most widely used synthetic biodegradable polymers in medicine. PL is typically synthesized by ring opening polymerization of the lactide monomer, a chiral molecule with three forms (i.e., D, L, and *meso*), that can produce poly(D-lactide) (PDL), poly(L-lactide) (PLL), or poly(DL-lactide) (PDLL) (Amass et al., 1998). [*Note:* PL is also referred to as poly(lactic acid), PLA, and PG as poly(glycolic acid), PGA, based on earlier synthesis methods exploiting condensation polymerization of lactic or glycolic acid.] While PL is a hydrophobic polymer, due to its extra methyl group that retards hydrolytic degradation, PG is relatively more hydrophilic and therefore degrades more rapidly. To improve mechanics and vary degradation rates, lactide monomers are generally copolymerized with glycolide monomers to make a wide range of copolymers (Amass et al., 1998, Griffith, 2002). Poly(lactide-co-glycolide) (PLG) copolymers exhibit a highly tunable degradation rate over the individual polymers and are commonly used as controlled-released carriers for various exogenous agents (Wang et al., 2009). PL, PG, and PLG are FDA approved in devices due to their proven biocompatibility. PCL is polymerized from a cyclic lactone monomer to produce a semicrystalline polymer which degrades more slowly *in vivo*. Copolymers of PCL with other polyesters such as PLL are commonly produced for tissue engineering applications (Dong et al., 2009, Ye et al., 1997). Polyesters are also used to generate another set of biodegradable polymers, polyurethanes. Polyesters and other polyols are used as intermediates in combination with isocyanates to form urethane linkages for polyurethanes. By changing the intermediate polymers, a great deal of flexibility can be introduced into the polyurethanes including their biocompatibility and degradability (Guelcher, 2008, Santerre et al., 2005).

Thermoresponsive materials are attractive candidates for noninvasive cardiac therapies because they preclude the need for aggressive open heart chest surgeries. One such polymer used for this application is poly-*N*-isopropylacrylamide P(NIPAAm), an amphiphilic thermoresponsive polymer that swells many times its weight in mass to form a hydrated 3D hydrogel network. When P(NIPAAm) is heated it undergoes a coil to globule transformation at its lower critical solution temperature (LCST), which occurs at ~34°C. The polymer exists as a free-flowing viscous liquid at room temperature and a viscoelastic solid above the LCST. Thermoresponsive materials like P(NIPAAm) behave this way due to secondary bonding such as hydrogen bond formation between the polymer and the solvent, and since the hydrogen bonds are thermally labile, an increase in temperature results in a decrease in hydrogen bonding which leads to phase separation.

6.7 Experimental Observation of Biomaterials for Stem Cell Transplantation into the Heart

Clinically, stem cell transplantation into the infarct region occurs either through an intracoronary transcutaneous (nonsurgical) intervention or a surgical procedure (usually in combination with a coronary artery bypass graft). Stem cell delivery in preclinical studies, both with and without biomaterials, is typically performed via a surgical procedure in small animal models, and with an intracoronary transcutaneous (nonsurgical) intervention in appropriately sized animals. Typically, stem cells within a matrix are injected either directly into the infarct or at the edge of the infarct region (i.e., the border zone) using a small gauge needle. In the case of delivering a tissue engineered patch (usually beating), the patch is directly sutured onto the injured myocardium via surgical procedure (Figure 6.4).

6.7.1 Natural Biomaterials for Cardiac Tissue Engineering

The most common biomaterial used for cardiac tissue engineering is collagen and its denatured derivative, gelatin. One of the few clinical trials that evaluated biomaterials for stem cell therapy implanted clinical grade Type I collagen scaffolds with autologous mononuclear BMC into patients

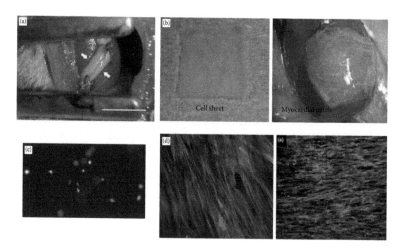

FIGURE 6.4 **(See color insert.)** Stem cells have been assembled in a number of scaffolds. (a) A thick 3D collagen scaffold constructed *in vitro* was sutured onto an infarct. (Adapted from Zimmermann, W. H., Melnychenko, I., and Eschenhagen, T. 2004. *Biomaterials*, 25, 1639–1647.) (b) Cell sheets have been assembled and layers of sheets were sutured onto the infarct region. (Adapted from Shimizu, T. et al. 2009. *Curr. Pharm. Des.*, 15, 2807–2814.) (c) Stem cells have been delivered within a synthetic hydrogel. (Adapted from Wall, S. et al. 2010. *J. Biomed. Mater. Res.: Part A*, 95(4), 1055–1066.) (d) Cells have been anistropically aligned *in vitro* by culturing on aligned fiber scaffolds produced by electrospinning. (Adapted from Dang, J. M. and Leong, K. W. 2007. *Adv. Mater.*, 19, 2775–2779; Kenar, H., Kose, G. T., and Hasirci, V. 2010. *J. Mater. Sci.-Mater. Med.*, 21, 989–997.)

with a myocardial infarct. During a coronary artery bypass graft, either BMC alone or BMC within a collagen scaffold were implanted directly onto the scar. Results strongly suggested that the addition of scaffold limited ventricular remodeling and improved diastolic function (Chachques et al., 2007). While larger randomized control trials are required to substantiate these results, this study provides strong support for application of biomaterials to assist in stem cell transplantation. Complementing these results, several animal studies have shown that cell retention is improved when delivered within collagen matrices. Dai et al. have shown that injecting collagen matrix (Zyderm®, a purified bovine Type 1 and III collagen mix) considerably improves the localization of transplanted rat MSC in the infarcted myocardium preventing cell loss to the noninfarcted myocardium and improving cardiac output. Interestingly they found that while collagen alone or MSC alone improved left ventricular EF, collagen with MSC did not improve function (Dai et al., 2009). This indicates that cell retention may not be sufficient and biomaterials must be designed to improve both cell interaction and overall biological function. Modified collagen including collagen–glycosaminoglycan scaffolds were also successful for delivery of MSC (Xiang et al., 2006). Kutschka et al. also investigated the application of gelatin foam (derived from porcine skin) in combination with Matrigel and other growth factors for delivering rat cardiomyocytes (H9c-2 cells) into an intramyocardial pouch in a heterotopic ischemic heart model. While there was significant donor cell loss immediately after transplantation, cell retention was improved within the Gelatin/Matrigel scaffolds which led to improved angiogenesis and LV function (Kutschka et al., 2006). It is likely that growth factors within Matrigel improved cardiomyocyte survival and function. In fact, they have previously shown that injection of mouse ESC within Matrigel into an infarct region improves cardiac function suggesting that Matrigel promotes stem cell survival (Kofidis et al., 2005b). However, Matrigel cannot be used for clinical implantation due to its xenogeneic source and uncontrolled composition containing various growth factors (Hughes et al., 2010). Recently, the same group transplanted undifferentiated mouse ES cells in collagen scaffolds (no growth factors) in a heterotopic ischemic heart model. After 2 weeks, transplanted ES cells within the scaffolds expressed connexin 43 (albeit they did not assume myocyte phenotype) and improved overall tissue contractions (Kofidis et al., 2005a); longer term studies in large animals in "working" heart model will better elucidate the clinical applicability of this technique. Overall, these studies suggest that while collagen and its derivatives safely enable stem cell engraftment into the heart infarct region, other growth factors and biological ligands may be required to improve eventual cell viability and regeneration potential.

Naturally occurring alginate and fibrin have shown to attenuate the negative remodeling process in myocardial infarcted hearts. In fact, one of the longest preclinical studies from Leor et al. report that fetal cardiac cells injected with alginate into the infarcted region of rat hearts resulted in significant vascularization and attenuation of LV dilatation until 9 weeks (Leor et al., 2007). Recently, the same group injected hES cell embryoid bodies and cardiomyocytes either alone or in an alginate scaffold (pre-implanted into the MI) in nude rats. Interestingly, they report that while hES derived CM attenuate left ventricular function and scar thinning, transplanted cells did not differentiate into functional vascular or cardiac myocytes in the tissue and the addition of gelatin did not improve this cell differentiation (Leor et al., 2007). This study highlights not only the need to carefully modulate the differentiation of ES cells into mature tissue but also the fact that in some cases there is no additional therapeutic benefit with biomaterial-based stem cell therapy. MSC and skeletal myoblasts show improved regeneration capability including improved blood flow when transplanted in a fibrin matrix as compared to saline control to the ischemic region (Christman et al., 2004b, Ryu et al., 2005). Recently the addition of hepatocyte growth factor within a fibrin matrix significantly improved MSC survival and left ventricular function in a mouse model of MI (Zhang et al., 2008). Thermoresponsive biomaterials are excellent candidate for injecting stem cells since the cells can be well dispersed in their liquid state (below their LSCT) and they will form a stiffer hydrogel upon injection. As such, temperature sensitive chitosan has been explored to deliver mouse ES cells into a rat infarct model. ESC injected in chitosan had improved retention, cardiac function, and exhibited significantly higher microvessel densities than the phosphate buffered saline (PBS) group at 4 weeks (Lu et al., 2009).

6.7.2 Synthetic Scaffolds for Cardiac Tissue Engineering

Synthetic porous scaffolds, or matrices, have also been used for injecting stem cells. PG matrices have been seeded with mouse ES cells and injected into a mouse infarct. After 8 weeks, cells were viable within the matrix and improved cardiac function and vascularization; however, there was no evidence that transplanted cells differentiated directly into vascular or myocardial cells (Ke et al., 2005). In a similar study, poly(lactide-*co*-caprolactone) was used to deliver MSC into the infarct, where the addition of the polymer improved cell survival and cardiac function (Jin et al., 2009). These studies suggest that similar to natural polymers, both adult and ES cells can be delivered with synthetic polymers with positive *in vivo* outcomes. However, more than just a delivery vehicle, synthetic materials offer the ability to introduce biological cues.

Molecular self-assembly of peptide amphiphiles in hydrogels has also been explored for CSC transplantation. Upon self-assembly peptide amphiphiles form nanofibers that are ~5–10 nm in length and mimic the filamentous structure of naturally occurring ECM fibers. Amphiphilic peptides thermodynamically self assemble upon exposure to physiological osmolarity and pH and produce scaffolds that can be applied for stem cell delivery into the infarcted heart (Davis et al., 2006). Davis et al. injected RAD16-II peptides (peptide sequence: AcN-RARADADARARADADA-CNH2) into the myocardium of mice, and the self-assembled nanofibers recruited progenitor cells that expressed endothelial markers and vascular smooth muscle cells, despite the fact that the peptides themselves have no known biological signaling sequence (Davis et al., 2005, 2006). In another study by the same group, insulin-like growth factor 1 (IGF-1), a cardiomyocyte growth and survival factor, was tethered to the same peptide (RAD16-II) nanofibers and the IGF-1 nanofibers improved cardiac function in an infarct model. Moreover, neonatal cardiomyocytes delivered within the IGF-1 biotinylated nanofibers showed improved cardiac function and decreased apoptosis than cell injection alone (Davis et al., 2006). Additionally, IGF-1 nanofibers in combination with cardiac progenitor stem cells transplanted into a rat infarct model showed improved differentiation and survival of cardiac progenitor cells (CPC) into cardiomyocytes which led to significant improvements in cardiac muscle regeneration (Padin-Iruegas et al., 2009). The RAD16 peptide nanofibers have also been used to deliver skeletal myocytes and vascular promoting growth factors in myocardial infarcts (Davis et al., 2006, Dubois et al., 2008, Hsieh et al., 2006). These studies suggest that the high aspect ratio of self-assembling nanofibers can be exploited to enhance stem cell survival and function in the cardiac tissue particularly when cardioprotective and angiogeneic growth factors are codelivered.

Semi-interpenetrating polymer networks (sIPNs) composed of P(NIPAAm) have also been explored as biomimetic ECMs for CSC transplantation and myocardial stabilization (Wall et al., 2010). Green fluorescent protein (GFP)-positive MSC were entrained in environmentally responsive poly(*N*-isopropylacrylamide-*co*-acrylic acid) hydrogels, incorporated with matrix metalloproteinase labile crosslinkers (e.g., MMP-2, 9, and 13) and peptides containing integrin binding domains (i.e., Arg-Gly-Asp), and were directly injected into infarcted murine left ventricles without inducing arrhythmias. Wall et al. reported significant differences in comparisons of pooled data from treatment groups that received MSC. At 2 weeks, the addition of MSC resulted in a significantly higher fractional shortening and a numerically higher EF than other treatments, while at 6 weeks function was worse in hearts that received stem cells for both fractional shortening and EF. However, injection of the sIPNs without cells resulted in superior LV function at 6 weeks compared to groups with cells and saline controls. Donor GFP⁺ cells were detected 6 weeks after matrix-enhanced transplantation, but not without matrix support, and infarct thickness was increased in animal subgroups that had histologically confirmed matrix contained within the infarct border zone. Thus, sIPN hydrogels succeeded in both mechanically supporting the injured myocardium and modestly enhancing donor cell survival (Wall et al., 2010). These results are consistent with a theoretical mathematical model that reported that injection of biomaterials in the left ventricle might ameliorate both ventricular remodeling and infarct extension (Wall et al., 2006).

Combinations of natural and synthetic materials have also been explored. A gelatin-based scaffold was evaluated for surgical ventricular repair, a procedure in which a full thickness section of the infarcted muscle was removed and replaced with a scaffold (Miyagi et al., 2010). The addition of PCL to the gelatin scaffold improved its mechanical stability and biodegradability as a replacement scaffold. Moreover, rat MSC with/without angiogeneic cytokines delivered in a temperature sensitive triblock copolymer of poly δ-valerolactone-*b*-poly ethylene glycol-*b*-poly δ-valerolactone within this scaffold drastically improved heart function as compared to scaffold alone.

Collectively, these studies suggest that a wide range of biomaterials can be safely used for stem cell transplantation therapy and may improve cell retention and vascularization compared to naked cell delivery. However, there has not been strong evidence to date that the supportive biomaterials engage with the stem cells to promote differentiation into the appropriate functional tissues and electromechanical integration with the host heart. Also, longer-term studies are needed to fully understand the degradation rates and the relative impact on tissue inflammation. It is likely that by controlling mechanical properties and introducing cell specific biological cues, ideal biomaterials can be synthesized to drastically improve stem cell therapy.

6.8 Biomaterials Used to Engineer "Heart Patch" *In Vitro*

The myocardium comprises of a highly organized ECM in combination with cardiac, vascular, and nerve cells. With delivery of stem cells with or without a supporting matrix, there is little control over the type and structure of regenerated tissue. On the other hand, engineering a full thickness myocardial equivalent *in vitro* with appropriate structure and cellular components is expected to replace damaged tissue. Much of the work within "heart patch" engineering has been explored with neonatal rat cardiomyocytes, and recent work has emerged applying those techniques toward stem cells. The following summarizes the structure of native myocardial ECM and efforts to engineer it using natural and synthetic-based biomaterials.

6.8.1 Engineering Thick Myocardial Tissue Replacements

One of the most common approaches to engineer a thick 3D "cardiac" patch is to seed cells within a porous scaffold for a number of days *in vitro*, allow the cells to remodel the matrix, and then implant the tissue into the host. However, the complex nature of the myocardium requires efforts to engineer materials in a precise and controlled manner. Several groups have attempted to engineer such a thick viable myocardial tissue. Initially, Eschenhagen et al. pioneered the use of collagen for cardiac tissue engineering by forming a 3D gel of collagen with embryonic chick cardiomyocytes and demonstrated that cardiomyocytes survive and beat in culture with considerable force (Eschenhagen et al., 1997). Since then, over the last 10 years, they have developed a robust cardiac muscle model *in vitro* termed "engineered heart tissue" (EHT). Neonatal rat cardiomyocytes within a mixture of collagen and Matrigel combined with growth factors are cast in various geometries and stretched under controlled strain for 1 week. In this manner, synchronously beating thick composites (1–4 mm diameter) can be fabricated that display mature electrophysiological properties and respond to mechanical and pharmacological interventions appropriately. This model has been used to delineate the cardiac tissue development *in vitro*. For example, it was discovered that EHT was more robust with the addition of cardiac fibroblasts as opposed to pure cardiomyocytes alone. During culture, cardiac cells remodel the underlying matrix to a mature native cardiac ECM. Also, EHT applied to an infarct region improved cardiac function for 8 weeks, and transplanted cardiomyocytes matured rapidly *in vivo*. Another group has applied a similar approach with ES cell-derived cardiomyocytes (Guo et al., 2006). *In vitro* studies showed synchronously beating tissues with good electromechanical coupling and mechanical properties, albeit the contractile activity is similar to immature cardiomyocytes. Preliminary *in vivo* data suggest that transplanted stem cell engineered tissue survive subcutaneously in nude mice (Guo et al., 2006). Yet another group has cultured human MSCs within type I collagen gels for weeks and *in vivo* performance indicates improved function within collagen gels (Simpson et al., 2007).

Another approach to engineer 3D tissues is via cell-sheet engineering, which applies layers of whole sheets of myocardial tissues to create a multicellular layered structure. Okano et al. pioneered the technique of grafting thermoreversible P(NIPAAm) on tissue culture dishes and culturing cells on this hydrophobic surface (Okano et al., 1993). Below 32°C, the polymer becomes hydrophilic, swells and is no longer cell adhesive, allowing the cultured cells including cardiac cells to be harvested (Shimizu et al., 2001). In this manner, harvested cell sheets comprised of neonatal cardiomyocytes, skeletal myoblasts, and MSCs have been transplanted in rat infarct models and shown to improve cardiac function (Shimizu et al., 2009). Also, multiple layers of neonatal CM sheets were combined to generate thick tissues *in vitro* (~80 µm thick) and these thick sheets survived long term in rat infarct models. Interestingly, owing to a vascularization challenge, multilayer sheets more than 4 layer thick developed a necrotic core, and the authors performed multiple surgeries to achieve a thicker tissue graft. Although cell-sheet engineering is an interesting technique, multiple surgeries are not clinically viable, and other strategies must be employed to achieve a full thickness cell sheet graft (Haraguchi et al., 2006). Similar to the Okano group, Sung et al. have applied the cell sheet engineering approach using a thermoreversible methylcellulose hydrogel to generate layered multiple rat MSC cell sheets. The sheet stack was then inserted into a porous decellularized bovine pericardium to create a thick patch with evenly distributed cells. Transplantation of the MSC patch in cardiac infarct region improved left ventricular function at 12 weeks, and some transplanted cells survived and differentiated into vascular cells, albeit very few cells differentiated into mature cardiomyocytes (Wei et al., 2008). In another study, the same group demonstrated that MSC in bovine pericardium scaffolds implanted into the infarct also improved cardiac function and promoted vascular cell differentiation (Wei et al., 2006). The above techniques to engineer thick heart tissues appear promising, and further optimization of scaffolds in combination with stem cells should yield reliable tissue replacements.

6.8.2 Improving Vascularization of Thick Tissues

A major challenge with engineering thick tissues is to maintain sufficient vascularization such that all cells within the tissue are viable. Cells in culture or *in vivo* rely on diffusion of oxygen, typically a distance of 100–200 μm, for metabolic needs. Engineering thick (>1 mm), 3D functional tissues is challenging as cells in the interior of thick tissues do not receive adequate nutrients (Griffith et al., 2005, Nomi et al., 2002). In light of this, several groups have explored the transplantation of primary endothelial cells in conjunction with functional tissue to engineer a vascular network (Enis et al., 2005, Koike et al., 2004). Levenberg's group created a vascularized construct by seeding a tri-culture of hES cell-derived cardiomyocytes with primary human endothelial cells (EC) (or hES derived EC) and embryonic fibroblasts in a copolymer of PLG and PLL. This copolymer was optimized to provide a mechanically stable porous scaffold with fast degradation to enable cell infiltration required for vascularization. The triple cell culture combination showed the greatest vascularization potential *in vitro* as assessed by EC lumen area and density and promoted cardiomyocyte differentiation. Moreover, transplantation of the multicellular grafts in immune compromised rats resulted in improved angiogenesis (Lesman et al., 2010). Recently, the Murry group demonstrated a scaffold-free equivalent of the vascularized cardiac tissue in which hES cell-derived EC and cardiomyocytes were combined with fibroblasts. The "vascularized" grafts showed markedly improved cell survival compared to cardiomyocyte only grafts in ischemic hearts (Stevens et al., 2009). These studies highlight the need for incorporating vascularization strategies for cardiac patch engineering. Future studies employing these ideas to delineate transplanted cardiomyocyte cell survival and function within infarct regions are still needed. Several groups have also developed innovative strategies to culture thick cardiac tissue constructs *in vitro* while maintaining optimal perfusion and electrical and mechanical stimulation (Radisic et al., 2004, 2006). Applicability of these innovative bioreactors for culturing stem cell-based cardiac tissues will be exciting.

6.9 Aligned Biomaterials for CSC

Perhaps the most innovative application of biomaterials is to provide nanotopographic cues for cell and tissue organization. As discussed earlier, cardiac ECM consists of highly organized filamentous network that guide tissue organization. *In vitro* work has shown that nanoscale topography in 2D affects anisotropic action potential propagation and tissue contractility of cultured cardiomyocyte monolayers (Kim et al., 2010). Also, electrospun fibers have shown that cell alignment can impact myogenic induction of mesenchymal stem cells (Dang and Leong, 2007). The main strategy to obtain a fibrous polymer structure of cardiac appropriate dimensions has been electrospinning.

In electrospinning, an electrified polymer solution is ejected from a needle and collected on a grounded target. The polymer stretches and forms nanometer to micron-sized diameter fibers on the collector. In this manner, a nonwoven thick mat of randomly aligned fibers can be collected. Fibers of varying shape (ribbon like vs. cylindrical), diameter, alignment, and mat porosity can be produced by varying electrospinning parameters such as polymer, solvent, type of needle/collector, flow rate, and distance between needle and collector. Aligned fibers are typically achieved by collecting the polymer solution between two parallel plates or on a rotating mandrel. One limitation of electrospinning is that the low porosity of electrospun mats limits cell infiltration and does not create 3D tissues, hence the resulting aligned tissue are thin monolayers.

For cardiac tissue engineering, the most commonly used biomaterials for electrospinning have been biocompatible materials such as PLL, PG, and polyurethanes. The majority of the studies have evaluated the proliferation of neonatal cardiomyocytes, MSC or resident CSC *in vitro* and demonstrated that cells align well along the electrospun fibers (generally around 1 μm fiber diameter) and display mature cell-to-cell connection (Hosseinkhani et al., 2010, Li et al., 2006, Zong et al., 2005). A common theme within these studies is that while synthetic materials provide control over electrospinning conditions, natural materials such as gelatin must be incorporated to promote cell attachment and migration within the scaffolds (Ifkovits et al., 2009, Li et al., 2006, Kenar et al., 2010, Zong et al., 2005). Additionally, culturing stem cells or neonatal cardiomyocytes on aligned matrices improves their morphological elongation relative to nonaligned hydrogels. Human MSCs on aligned hydroxybutyl chitosan matrices show both improved elongation and expressed more myogeneic markers than when cultured on nonaligned hydrogels (Dang and Leong, 2007). Meanwhile, mouse ES cell derived cardiomyocytes on PCL-based polyurethane electrospun scaffolds (nonaligned) had improved morphological appearance when compared to scaffolds prepared through thermally induced phase separation, however, no functional differences in junction proteins and contractile properties were noted between the two fabrication methods (Fromstein et al., 2008). It is plausible that if the scaffolds above were aligned, functional improvements in the cardiac tissue may have been observed.

Only a few studies have investigated the degradation of electrospun scaffolds *in vivo*. Recently, acrylated poly(glycerol sebacate) (PGS) was electrospun and crosslinked by UV curing and free radical polymerization. A ratio of the polymers with gelatin was optimized to electrospin a semipenetrating network that promoted human mesenchymal cell attachment and spreading. Interestingly, *in vivo* studies of the scaffolds sutured onto the epicardium indicated that a high degree of alkylation correlated with higher inflammation into the tissue (Ifkovits et al., 2009). This data provides evidence that to determine optimal patch properties, materials must be investigated *in vivo*.

Electrospun fibers have also been used to generate thicker 3D tissues. For example, a unique blend of polyesters: poly(3-hydroxybutyrate-*co*-3-hydroxyvalerate), poly(L-D,L-lactic acid), and PGS were electrospun to produce aligned fiber mats. Human umbilical cord derived MSC aligned onto these mats for 14 days, and these sheets were further assembled around porous tubes to form a 3D construct. While the cells aligned and attached well onto the scaffolds, there was no improved in the cardiogeneic potential of these stem cells (Kenar et al., 2010). In another example, PCL was electrospun to produce thin (~10 μm) nonwoven nanofiber meshes that when coated with collagen supported neonatal cardiomyocyte attachment. The cardiomyocyte sheets on these PCL meshes were stacked to create a 3D tissue *in vitro*; cell–cell contact and electrical integration were maintained within the layers and synchronized beating was observed within the thick tissue graft (Ishii et al., 2005). Combining electrospun aligned matrices to produce thick 3D tissue offers a unique approach to cardiac tissue engineering.

The above studies indicate that electrospinning is a viable technique to create fibrous scaffolds that mimic the cardiac ECM structure and support rodent neonatal and ES cell-derived cardiomyocytes. Although a number of polymers appear promising for this application, the ideal biomaterial has still not been demonstrated. It appears that cardiomycoytes require ligands presented by biological ECM to attach and proliferate. However, natural ECM proteins do not offer adequate mechanical stability and flexibility (i.e., with selecting fiber diameter) for electrospinning. Therefore, a majority of the strategies employ synthetic polymers coated with animal-derived ECM proteins. From a clinical perspective, animal derived proteins are not desirable as they can elicit an immunogeneic response. Ideally, synthetic polymers that provide appropriate biological cues and ligands to support human cardiomyocytes would be used for engineering fibrous scaffolds. Another major concern with electrospinning mats is that they lack porosity, and thus cells cannot truly invade into the mats and create a full thick 3D tissue. Future strategies would improve porosity to enable cell infiltration within the tissue.

6.10 Summary

HF due to ischemic heart disease is one of the leading causes of worldwide mortality. While current clinical therapies can improve hemodynamics in HF, currently there is no viable option for replacing damaged cardiac muscle cells. Stem cell transplantation therapy offers tremendous potential to regenerate the myocardium and improve overall quality of life. However, there are several critical challenges with stem cell transplantation such as poor cell retention at the site of transplantation, survival, and eventual functional integration into the diseased tissue. Various natural and synthetic biomaterials have been explored to enhance cell retention and survival in the ischemic myocardium, and ultimately cardiac function. Moreover, stem cells and biomaterials are employed to engineer artificial heart patches that could potentially replace the diseased tissue. A cohort of biomaterials used as both delivery vehicles and within cardiac patches have shown promising results in cardiac tissue engineering. Future work within this field will yield ideal platforms that structurally and functionally promote the viability and differentiation of stem cells for treating the damaged myocardium.

References

Agarkova, I. and Perriard, J. C. 2005. The M-band: An elastic web that crosslinks thick filaments in the center of the sarcomere. *Trends Cell Biol.*, 15, 477–485.

Alper, J. 2009. Geron gets green light for human trial of ES cell-derived product. *Nat. Biotechnol.*, 27, 213–214.

Alves, N. M. and Mano, J. F. 2008. Chitosan derivatives obtained by chemical modifications for biomedical and environmental applications. *Int. J. Biol. Macromol.*, 43, 401–414.

Amass, W., Amass, A., and Tighe, B. 1998. A review of biodegradable polymers: Uses, current developments in the synthesis and characterization of biodegradable polyesters, blends of biodegradable polymers and recent advances in biodegradation studies. *Polym. Int.*, 47, 89–144.

Anderson, R. H., HO, S. Y., Redmann, K., Sanchez-Quintana, D., and Lunkenheimer, P. P. 2005. The anatomical arrangement of the myocardial cells making up the ventricular mass. *Eur. J. Cardiothorac. Surg.*, 28, 517–525.

Assmus, B., Honold, J., Schachinger, V., Britten, M. B., Fischer-Rasokat, U., Lehmann, R., Teupe, C. et al. 2006. Transcoronary transplantation of progenitor cells after myocardial infarction. *N. Engl. J. Med.*, 355, 1222–1232.

Bearzi, C., Rota, M., Hosoda, T., Tillmanns, J., Nascimbene, A., De Angelis, A., Yasuzawa-Amano, S. et al. 2007. Human cardiac stem cells. *Proc. Natl. Acad. Sci. U.S.A.*, 104, 14068–14073.

Braunwald, E., Ross, J., Jr., and Sonnenblick, E. H. 1967. Mechanisms of contraction of the normal and failing heart. *N. Engl. J. Med.*, 277, 1012–1022.

Camelliti, P., Borg, T. K., and Kohl, P. 2005. Structural and functional characterisation of cardiac fibroblasts. *Cardiovasc. Res.*, 65, 40–51.

Caspi, O., Huber, I., Kehat, I., Habib, M., Arbel, G., Gepstein, A., Yankelson, L., Aronson, D., Beyar, R., and Gepstein, L. 2007. Transplantation of human embryonic stem cell-derived cardiomyocytes improves myocardial performance in infarcted rat hearts. *J. Am. Coll. Cardiol.*, 50, 1884–1893.

Chachques, J. C., Trainini, J. C., Lago, N., Masoli, O. H., Barisani, J. L., Cortes-Morichetti, M., Schussler, O., and Carpentier, A. 2007. Myocardial assistance by grafting a new bioartificial upgraded myocardium (Magnum clinical trial): One year follow-up. *Cell Transplant.*, 16, 927–934.

Chen, P. S., Chen, L. S., Cao, J. M., Sharifi, B., Karagueuzian, H. S., and Fishbein, M. C. 2001. Sympathetic nerve sprouting, electrical remodeling and the mechanisms of sudden cardiac death. *Cardiovasc. Res.*, 50, 409–416.

Christman, K. L., Fok, H. H., Sievers, R. E., Fang, Q. H., and Lee, R. J. 2004a. Fibrin glue alone and skeletal myoblasts in a fibrin scaffold preserve cardiac function after myocardial infarction. *Tissue Eng.*, 10, 403–409.

Christman, K. L., Vardanian, A. J., Fang, Q. Z., Sievers, R. E., Fok, H. H., and Lee, R. J. 2004b. Injectable fibrin scaffold improves cell transplant survival, reduces infarct expansion, and induces neovasculature formation in ischemic myocardium. *J. Am. Coll. Cardiol.*, 44, 654–660.

Dai, W., Hale, S. L., Kay, G. L., Jyrala, A. J., and Kloner, R. A. 2009. Delivering stem cells to the heart in a collagen matrix reduces relocation of cells to other organs as assessed by nanoparticle technology. *Regen. Med.*, 4, 387–395.

Dang, J. M. and Leong, K. W. 2007. Myogenic induction of aligned mesenchymal stem cell sheets by culture on thermally responsive electrospun nanofibers. *Adv. Mater.*, 19, 2775–2779.

Davis, M. E., Hsieh, P. C. H., Takahashi, T., Song, Q., Zhang, S. G., Kamm, R. D., Grodzinsky, A. J., Anversa, P., and Lee, R. T. 2006. Local myocardial insulin-like growth factor 1 (IGF-1) delivery with biotinylated peptide nanofibers improves cell therapy for myocardial infarction. *Proc. Natl. Acad. Sci. U.S.A.*, 103, 8155–8160.

Davis, M. E., Motion, J. P. M., Narmoneva, D. A., Takahashi, T., Hakuno, D., Kamm, R. D., Zhang, S. G., and Lee, R. T. 2005. Injectable self-assembling peptide nanofibers create intramyocardial microenvironments for endothelial cells. *Circulation*, 111, 442–450.

Dill, T., Schachinger, V., Rolf, A., Mollmann, S., Thiele, H., Tillmanns, H., Assmus, B., Dimmeler, S., Zeiher, A. M., and Hamm, C. 2009. Intracoronary administration of bone marrow-derived progenitor cells improves left ventricular function in patients at risk for adverse remodeling after acute ST-segment elevation myocardial infarction: Results of the reinfusion of enriched progenitor cells and infarct remodeling in acute myocardial infarction study (Repair-Ami) cardiac magnetic resonance imaging substudy. *Am. Heart J.*, 157, 541–547.

Dong, Y. X., Liao, S., Ngiam, M., Chan, C. K., and Ramakrishna, S. 2009. Degradation behaviors of electrospun resorbable polyester nanofibers. *Tissue Eng. Part B-Rev.*, 15, 333–351.

Drury, J. L. and Mooney, D. J. 2003. Hydrogels for tissue engineering: Scaffold design variables and applications. *Biomaterials*, 24, 4337–4351.

Dubois, G., Segers, V. F. M., Bellamy, V., Sabbah, L., Peyrard, S., Bruneval, P., Hagege, A. A., Lee, R. T., and Menasche, P. 2008. Self-assembling peptide nanofibers and skeletal myoblast transplantation in infarcted myocardium. *J. Biomed. Mater. Res. Part B-Appl. Biomater.*, 87B, 222–228.

Enis, D. R., Shepherd, B. R., Wang, Y. N., Qasim, A., Shanahan, C. M., Weissberg, P. L., Kashgarian, M., Pober, J. S., and Schechner, J. S. 2005. Induction, differentiation, and remodeling of blood vessels after transplantation of Bcl-2-transduced endothelial cells. *Proc. Natl. Acad. Sci. U.S.A*, 102, 425–430.

Eschenhagen, T., Fink, C., Remmers, U., Scholz, H., Wattchow, J., Weil, J., Zimmermann, W., Dohmen, H. H., Schafer, H., Bishopric, N., Wakatsuki, T., and Elson, E. L. 1997. Three-dimensional reconstitution of embryonic cardiomyocytes in a collagen matrix: A new heart muscle model system. *Faseb J.*, 11, 683–694.

Fromstein, J. D., Zandstra, P. W., Alperin, C., Rockwood, D., Rabolt, J. F., and Woodhouse, K. A. 2008. Seeding bioreactor-produced embryonic stem cell-derived cardiomyocytes on different porous, degradable, polyurethane scaffolds reveals the effect of scaffold architecture on cell morphology. *Tissue Eng. Part A*, 14, 369–378.

Gelse, K., Poschl, E., and Aigner, T. 2003. Collagens—structure, function, and biosynthesis. *Adv. Drug Deliv. Rev.*, 55, 1531–1546.

Gordon, M. K. and Hahn, R. A. 2010. Collagens. *Cell Tissue Res.*, 339, 247–257.

Griffith, C. K., Miller, C., Sainson, R. C. A., Calvert, J. W., Jeon, N. L., Hughes, C. C. W., and George, S. C. 2005. Diffusion limits of an *in vitro* thick prevascularized tissue. *Tissue Eng.*, 11, 257–266.

Griffith, L. G. 2002. Emerging design principles in biomaterials and scaffolds for tissue engineering. *Reparative Med.: Growing Tissues Organs*, 961, 83–95.

Guelcher, S. A. 2008. Biodegradable polyurethanes: Synthesis and applications in regenerative medicine. *Tissue Eng. Part B Rev.*, 14, 3–17.

Guo, X. M., Zhao, Y. S., Chang, H. X., Wang, C. Y., Ll, E., Zhang, X. A., Duan, C. M. et al. 2006. Creation of engineered cardiac tissue *in vitro* from mouse embryonic stem cells. *Circulation*, 113, 2229–2237.

Haraguchi, Y., Shimizu, T., Yamato, M., Kikuchi, A., and Okano, T. 2006. Electrical coupling of cardiomyocyte sheets occurs rapidly via functional gap junction formation. *Biomaterials*, 27, 4765–4774.

Heydarkhan-Hagvall, S., Schenke-Layland, K., Dhanasopon, A. P., Rofail, F., Smith, H., Wu, B. M., Shemin, R., Beygui, R. E., and Maclellan, W. R. 2008. Three-dimensional electrospun ECM-based hybrid scaffolds for cardiovascular tissue engineering. *Biomaterials*, 29, 2907–2914.

Hosseinkhani, H., Hosseinkhani, M., Hattori, S., Matsuoka, R., and Kawaguchi, N. 2010. Micro and nanoscale *in vitro* 3D culture system for cardiac stem cells. *J. Biomed. Mater. Res. Part A*, 94A, 1–8.

Hsieh, P. C. H., Davis, M. E., Gannon, J., Macgillivray, C., and Lee, R. T. 2006. Controlled delivery of PDGF-BB for myocardial protection using injectable self-assembling peptide nanofibers. *J. Clin. Invest.*, 116, 237–248.

Hughes, C. S., Postovit, L. M., and Lajoie, G. A. 2010. Matrigel: A complex protein mixture required for optimal growth of cell culture. *Proteomics*, 10, 1886–1890.

Huikuri, H. V., Kervinen, K., Niemela, M., Ylitalo, K., Saily, M., Koistinen, P., Savolainen, E. R. et al. 2008. Effects of intracoronary injection of mononuclear bone marrow cells on left ventricular function, arrhythmia risk profile, and restenosis after thrombolytic therapy of acute myocardial infarction. *Eur. Heart J.*, 29, 2723–2732.

Ieda, M., Fu, J. D., Delgado-Olguin, P., Vedantham, V., Hayashi, Y., Bruneau, B. G., and Srivastava, D. 2010. Direct reprogramming of fibroblasts into functional cardiomyocytes by defined factors. *Cell*, 142, 375–386.

Ifkovits, J. L., Devlin, J. J., Eng, G., Martens, T. P., Vunjak-Novakovic, G., and Burdick, J. A. 2009. Biodegradable fibrous scaffolds with tunable properties formed from photo-cross-linkable poly(glycerol sebacate). *ACS Appl. Mater. Interfaces*, 1, 1878–1886.

Irion, S., Nostro, M. C., Kattman, S. J., and Keller, G. M. 2008. Directed differentiation of pluripotent stem cells: From developmental biology to therapeutic applications. *Cold Spring Harb. Symp. Quant. Biol.*, 73, 101–110.

Ishii, O., Shin, M., Sueda, T., and Vacanti, J. P. 2005. *In vitro* tissue engineering of a cardiac graft using a degradable scaffold with an extracellular matrix-like topography. *J. Thorac. Cardiovasc. Surg.*, 130, 1358–1363.

Janssens, S., Dubois, C., Bogaert, J., Theunissen, K., Deroose, C., Desmet, W., Kalantzi, M. et al. 2006. Autologous bone marrow-derived stem-cell transfer in patients with ST-segment elevation myocardial infarction: Double-blind, randomised controlled trial. *Lancet*, 367, 113–121.

Jin, J., Jeong, S. I., Shin, Y. M., Lim, K. S., Shin, H., Lee, Y. M., Koh, H. C., and Kim, K. S. 2009. Transplantation of mesenchymal stem cells within a poly(lactide-*co*-epsilon-caprolactone) scaffold improves cardiac function in a rat myocardial infarction model. *Eur. J. Heart Fail.*, 11, 147–153.

Ke, Q., Yang, Y., Rana, J. S., Chen, Y., Morgan, J. P., and Xiao, Y. F. 2005. Embryonic stem cells cultured in biodegradable scaffold repair infarcted myocardium in mice. *Sheng Li Xue. Bao.*, 57, 673–681.

Kenar, H., Kose, G. T., and Hasirci, V. 2010. Design of a 3D aligned myocardial tissue construct from biodegradable polyesters. *J. Mater. Sci.-Mater. Med.*, 21, 989–997.

Kim, D.-H., Lipke, E. A., Kim, P., Cheong, R., Thompson, S., Delannoy, M., Suh, K.-Y., Tung, L., and Levchenko, A. 2010. Nanoscale cues regulate the structure and function of macroscopic cardiac tissue constructs. *Proc. Natl. Acad. Sci. U.S.A*, 107, 565–570.

Kleinman, H. K. and Martin, G. R. 2005. Matrigel: Basement membrane matrix with biological activity. *Semin. Cancer Biol.*, 15, 378–386.

Kocher, A. A., Schuster, M. D., Szabolcs, M. J., Takuma, S., Burkhoff, D., Wang, J., Homma, S., Edwards, N. M., and Itescu, S. 2001. Neovascularization of ischemic myocardium by human bone-marrow-derived angioblasts prevents cardiomyocyte apoptosis, reduces remodeling and improves cardiac function. *Nat. Med.*, 7, 430–436.

Kofidis, T., De Bruin, J. L., Hoyt, G., Ho, Y., Tanaka, M., Yamane, T., Lebl, D. R., Swijnenburg, R. J., Chang, C. P., Quertermous, T., and Robbins, R. C. 2005a. Myocardial restoration with embryonic stem cell bioartificial tissue transplantation. *J. Heart Lung Transplant.*, 24, 737–744.

Kofidis, T., Lebl, D. R., Martinez, E. C., Hoyt, G., Tanaka, M., and Robbins, R. C. 2005b. Novel injectable bioartificial tissue facilitates targeted, less invasive, large-scale tissue restoration on the beating heart after myocardial injury. *Circulation*, 112, I173–I177.

Koike, N., Fukumura, D., Gralla, O., Au, P., Schechner, J. S., and Jain, R. K. 2004. Creation of long-lasting blood vessels. *Nature*, 428, 138–139.

Kutschka, I., Chen, I. Y., Kofidis, T., Arai, T., Von Degenfeld, G., Sheikh, A. Y., Hendry, S. L. et al. 2006. Collagen matrices enhance survival of transplanted cardiomyoblasts and contribute to functional improvement of ischemic rat hearts. *Circulation*, 114, I167–I173.

Laflamme, M. A., Chen, K. Y., Naumova, A. V., Muskheli, V., Fugate, J. A., Dupras, S. K., Reinecke, H. et al. 2007. Cardiomyocytes derived from human embryonic stem cells in pro-survival factors enhance function of infarcted rat hearts. *Nat. Biotechnol.*, 25, 1015–1024.

Laflamme, M. A. and Murry, C. E. 2005. Regenerating the heart. *Nat. Biotechnol.*, 23, 845–856.

Le Blanc, K. and Ringden, O. 2005. Immunobiology of human mesenchymal stem cells and future use in hematopoietic stem cell transplantation. *Biol. Blood Marrow Transplant*, 11, 321–334.

Leor, J., Aboulafia-Etzion, S., Dar, A., Shapiro, L., Barbash, I. M., Battler, A., Granot, Y., and Cohen, S. 2000. Bioengineered cardiac grafts: A new approach to repair the infarcted myocardium? *Circulation*, 102, Iii56–Iii61.

Leor, J., Gerecht, S., Cohen, S., Miller, L., Holbova, R., Ziskind, A., Shachar, M., Feinberg, M. S., Guetta, E., and Itskovitz-Eldor, J. 2007. Human embryonic stem cell transplantation to repair the infarcted myocardium. *Heart*, 93, 1278–1284.

Lesman, A., Habib, M., Caspi, O., Gepstein, A., Arbel, G., Levenberg, S., and Gepstein, L. 2010. Transplantation of a tissue-engineered human vascularized cardiac muscle. *Tissue Eng. Part A*, 16, 115–125.

Li, M., Mondrinos, M. J., Chen, X., Gandhi, M. R., Ko, F. K., and Lelkes, P. I. 2006. Co-electrospun poly(lactide-*co*-glycolide), gelatin, and elastin blends for tissue engineering scaffolds. *J. Biomed. Mater. Res. Part A*, 79A, 963–973.

Lloyd-Jones, D., Adams, R. J., Brown, T. M., Carnethon, M., Dai, S., De Simone, G., Ferguson, T. B. et al. 2010. Executive summary: Heart disease and stroke statistics-2010 update a report from the American Heart Association. *Circulation*, 121, 948–954.

Lloyd-Jones, D. M., Larson, M. G., Beiser, A., and Levy, D. 1999. Lifetime risk of developing coronary heart disease. *Lancet*, 353, 89–92.

Lu, W. N., Lu, S. H., Wang, H. B., Li, D. X., Duan, C. M., Liu, Z. Q., Hao, T. etal. 2009. Functional improvement of infarcted heart by co-injection of embryonic stem cells with temperature-responsive chitosan hydrogel. *Tissue Eng. Part A*, 15, 1437–1447.

Lunde, K. and Aakhus, S. 2008. Intracoronary injection of mononuclear bone marrow cells after acute myocardial infarction: Lessons from the Astami trial. *Eur. Heart J. Suppl.*, 10, K35–K38.

Macchiarelli, G., Ohtani, O., Nottola, S. A., Stallone, T., Camboni, A., Prado, I. M., and Motta, P. M. 2002. A micro-anatomical model of the distribution of myocardial endomysial collagen. *Histol. Histopathol.*, 17, 699–706.

McMurray, J. J. V., Petrie, M. C., Murdoch, D. R., and Davie, A. P. 1998. Clinical epidemiology of heart failure: Public and private health burden. *Eur. Heart J.*, 19, 9–16.

Menasche, P., Alfieri, O., Janssens, S., Mckenna, W., Reichenspurner, H., Trinquart, L., Vilquin, J. T. et al. 2008. The myoblast autologous grafting in ischemic cardiomyopathy (Magic) trial: First randomized placebo-controlled study of myoblast transplantation. *Circulation*, 117, 1189–200.

Meyer, G. P., Wollert, K. C., Lotz, J., Pirr, J., Rager, U., Lippolt, P., Hahn, A., Fichtner, S., Schaefer, A., Arseniev, L., Ganser, A., and Drexler, H. 2009. Intracoronary bone marrow cell transfer after myocardial infarction: 5-year follow-up from the randomized-controlled Boost trial. *Eur. Heart J.*, 30, 2978–2984.

Miettinen, J. A., Ylitalo, K., Hedberg, P., Jokelainen, J., Kervinen, K., Niemela, M., Saily, M. et al. 2010. Determinants of functional recovery after myocardial infarction of patients treated with bone marrow-derived stem cells after thrombolytic therapy. *Heart*, 96, 362–367.

Miyagi, Y., Zeng, F., Huang, X. P., Foltz, W. D., Wu, J., Mihic, A., Yau, T. M., Weisel, R. D., and Li, R. K. 2010. Surgical ventricular restoration with a cell- and cytokine-seeded biodegradable scaffold. *Biomaterials*, 31, 7684–7694.

Murry, C. E. and Keller, G. 2008. Differentiation of embryonic stem cells to clinically relevant populations: Lessons from embryonic development. *Cell*, 132, 661–680.

Nomi, M., Atala, A., Coppi, P. D., and Soker, S. 2002. Principals of neovascularization for tissue engineering. *Mol. Aspects Med.*, 23, 463–483.

Noorman, M., Van der Heyden, M. A., Van veen, T. A., Cox, M. G., Hauer, R. N., DE Bakker, J. M., and Van Rijen, H. V. 2009. Cardiac cell-cell junctions in health and disease: Electrical versus mechanical coupling. *J. Mol. Cell Cardiol.*, 47, 23–31.

Okano, T., Yamada, N., Sakai, H., and Sakurai, Y. 1993. A novel recovery system for cultured cells using plasma-treated polystyrene dishes grafted with poly(*N*-isopropylacrylamide). *J. Biomed. Mater. Res.*, 27, 1243–1251.

Olsen, D., Yang, C., Bodo, M., Chang, R., Leigh, S., Baez, J., Carmichael, D., Perala, M., Hamalainen, E. R., Jarvinen, M., and Polarek, J. 2003. Recombinant collagen and gelatin for drug delivery. *Adv. Drug Deliv. Rev.*, 55, 1547–1567.

Orlic, D., Hill, J. M., and Arai, A. E. 2002. Stem cells for myocardial regeneration. *Circ. Res.*, 91, 1092–1102.

Orlic, D., Kajstura, J., Chimenti, S., Bodine, D. M., Leri, A., and Anversa, P. 2003. Bone marrow stem cells regenerate infarcted myocardium. *Pediatr. Transplant*, 7(Suppl 3), 86–88.

Padin-Iruegas, M. E., Misao, Y., Davis, M. E., Segers, V. F., Esposito, G., Tokunou, T., Urbanek, K. et al. 2009. Cardiac progenitor cells and biotinylated insulin-like growth factor-1 nanofibers improve endogenous and exogenous myocardial regeneration after infarction. *Circulation*, 120, 876–887.

Pittenger, M. F. and Martin, B. J. 2004. Mesenchymal stem cells and their potential as cardiac therapeutics. *Circ. Res.*, 95, 9–20.

Psaltis, P. J., Zannettino, A. C., Worthley, S. G., and Gronthos, S. 2008. Concise review: Mesenchymal stromal cells: Potential for cardiovascular repair. *Stem Cells*, 26, 2201–2210.

Radisic, M., Park, H., Chen, F., Salazar-Lazzaro, J. E., Wang, Y., Dennis, R., Langer, R., Freed, L. E., and Vunjak-Novakovic, G. 2006. Biomimetic approach to cardiac tissue engineering: Oxygen carriers and channeled scaffolds. *Tissue Eng.*, 12, 2077–2091.

Radisic, M., Yang, L., Boublik, J., Cohen, R. J., Langer, R., Freed, L. E., and Vunjak-Novakovic, G. 2004. Medium perfusion enables engineering of compact and contractile cardiac tissue. *Am. J. Physiol. Heart Circ. Physiol.*, 286, H507–H516.

Radosevich, M., Goubran, H. A., and Burnouf, T. 1997. Fibrin sealant: Scientific rationale, production methods, properties, and current clinical use. *Vox Sanguinis*, 72, 133–143.

Reyes, C. D. and Garcia, A. J. 2004. Alpha2beta1 integrin-specific collagen-mimetic surfaces supporting osteoblastic differentiation. *J. Biomed. Mater. Res. A*, 69, 591–600.

Rowley, J. A., Madlambayan, G., and Mooney, D. J. 1999. Alginate hydrogels as synthetic extracellular matrix materials. *Biomaterials*, 20, 45–53.

Ryu, J. H., Kim, I. K., Cho, S. W., Cho, M. C., Hwang, K. K., Piao, H., Piao, S., Lim, S. H., Hong, Y. S., Choi, C. Y., Yoo, K. J., and Kim, B. S. 2005. Implantation of bone marrow mononuclear cells using injectable fibrin matrix enhances neovascularization in infarcted myocardium. *Biomaterials*, 26, 319–326.

Saha, K., Pollock, J. F., Schaffer, D. V., and Healy, K. E. 2007. Designing synthetic materials to control stem cell phenotype. *Curr. Opin. Chem. Biol.*, 11, 381–387.

Santerre, J. P., Woodhouse, K., Laroche, G., and Labow, R. S. 2005. Understanding the biodegradation of polyurethanes: From classical implants to tissue engineering materials. *Biomaterials*, 26, 7457–7470.

Schachinger, V., Assmus, B., Honold, J., Lehmann, R., Hofmann, W. K., Martin, H., Dimmeler, S., and Zeiher, A. M. 2006a. Normalization of coronary blood flow in the infarct-related artery after intracoronary progenitor cell therapy: Intracoronary Doppler substudy of the Topcare-Ami trial. *Clin. Res. Cardiol.*, 95, 13–22.

Schachinger, V., Erbs, S., Elsasser, A., Haberbosch, W., Hambrecht, R., Holschermann, H., Yu, J. et al. 2006b. Intracoronary bone marrow-derived progenitor cells in acute myocardial infarction. *N. Engl. J. Med.*, 355, 1210–1221.

Schachinger, V., Erbs, S., Elsasser, A., Haberbosch, W., Hambrecht, R., Holschermann, H., Yu, J. et al. 2006c. Improved clinical outcome after intracoronary administration of bone-marrow-derived progenitor cells in acute myocardial infarction: Final 1-year results of the Repair-Ami trial. *Eur. Heart J.*, 27, 2775–2783.

Shimizu, T., Sekine, H., Yamato, M., and Okano, T. 2009. Cell sheet-based myocardial tissue engineering: New hope for damaged heart rescue. *Curr. Pharm. Des.*, 15, 2807–2814.

Shimizu, T., Yamato, M., Kikuchi, A., and Okano, T. 2001. Two-dimensional manipulation of cardiac myocyte sheets utilizing temperature-responsive culture dishes augments the pulsatile amplitude. *Tissue Eng.*, 7, 141–51.

Simpson, D., Liu, H., Fan, T. H., Nerem, R., and Dudley, S. C., Jr. 2007. A tissue engineering approach to progenitor cell delivery results in significant cell engraftment and improved myocardial remodeling. *Stem Cells*, 25, 2350–2357.

Spaan, J., Kolyva, C., Van den, W. J., Ter Wee, R., Van Horssen, P., Piek, J., and Siebes, M. 2008. Coronary structure and perfusion in health and disease. *Philos. Transact. A Math. Phys. Eng. Sci.*, 366, 3137–3153.

Stevens, K. R., Kreutziger, K. L., Dupras, S. K., Korte, F. S., Regnier, M., Muskheli, V., Nourse, M. B., Bendixen, K., Reinecke, H., and Murry, C. E. 2009. Physiological function and transplantation of scaffold-free and vascularized human cardiac muscle tissue. *Proc. Natl. Acad. Sci. U.S.A.*, 106, 16568–16573.

Tendera, M., Wojakowski, W., Ruzyllo, W., Chojnowska, L., Kepka, C., Tracz, W., Musialek, P. et al. 2009. Intracoronary infusion of bone marrow-derived selected Cd34+ Cxcr4+ cells and non-selected mononuclear cells in patients with acute Stemi and reduced left ventricular ejection fraction: Results of randomized, multicentre myocardial regeneration by intracoronary infusion of selected population of stem cells in acute myocardial infarction (Regent) trial. *Eur. Heart J.*, 30, 1313–1321.

Torella, D., Ellison, G. M., Mendez-Ferrer, S., Ibanez, B., and Nadal-Ginard, B. 2006. Resident human cardiac stem cells: Role in cardiac cellular homeostasis and potential for myocardial regeneration. *Nat. Clin. Pract. Cardiovasc. Med.*, 3(Suppl 1), S8–S13.

Van der Laan, A., Hirsch, A., Nijveldt, R., Van der Vleuten, P. A., Van der Giessen, W. J., Doevendans, P. A., Waltenberger, J. et al. 2008. Bone marrow cell therapy after acute myocardial infarction: The HEBE trial in perspective, first results. *Neth. Heart J.*, 16, 436–439.

Velagaleti, R. S., Pencina, M. J., Murabito, J. M., Wang, T. J., Parikh, N. I., D'agostino, R. B., Levy, D., Kannel, W. B., and Vasan, R. S. 2008. Long-term trends in the incidence of heart failure after myocardial infarction. *Circulation*, 118, 2057–2062.

Wall, S., Yeh, C., Yu, R., Mann, M., and Healy, K. E. 2010. Biomimetic matrices for myocardial stabilization and stem cell transplantation. *J. Biomed. Mater. Res.: Part A*, 95(4), 1055–1066.

Wall, S. T., Walker, J. C., Healy, K. E., Ratcliffe, M. B., and Guccione, J. M. 2006. Theoretical impact of the injection of material into the myocardium: A finite element model simulation. *Circulation*, 114, 2627–2635.

Wallace, D. G. and Rosenblatt, J. 2003. Collagen gel systems for sustained delivery and tissue engineering. *Adv. Drug Deliv. Rev.*, 55, 1631–1649.

Wang, Y., Liu, X. C., Zhao, J., Kong, X. R., Shi, R. F., Zhao, X. B., Song, C. X., Liu, T. J., and Lu, F. 2009. Degradable PLGA scaffolds with basic fibroblast growth factor experimental studies in myocardial revascularization. *Texas Heart Inst. J.*, 36, 89–97.

Wei, H. J., Chen, S. C., Chang, Y., Hwang, S. M., Lin, W. W., Lai, P. H., Chiang, H. H. K., Hsu, L. F., Yang, H. H., and Sung, H. W. 2006. Porous acellular bovine pericardia seeded with mesenchymal stem cells as a patch to repair a myocardial defect in a syngeneic rat model. *Biomaterials*, 27, 5409–5419.

Wei, H. J., Chen, C. H., Lee, W. Y., Chiu, I., Hwang, S. M., Lin, W. W., Huang, C. C., Yeh, Y. C., Chang, Y., and Sung, H. W. 2008. Bioengineered cardiac patch constructed from multilayered mesenchymal stem cells for myocardial repair. *Biomaterials*, 29, 3547–3556.

Wollert, K. C., and Drexler, H. 2005. Clinical applications of stem cells for the heart. *Circ. Res.*, 96, 151–163.

Wollert, K. C. and Drexler, H. 2010. Cell therapy for the treatment of coronary heart disease: A critical appraisal. *Nat. Rev. Cardiol.*, 7, 204–215.

Wollert, K. C., Meyer, G. P., Lotz, J., Ringes-Lichtenberg, S., Lippolt, P., Breidenbach, C., Fichtner, S. et al. 2004. Intracoronary autologous bone-marrow cell transfer after myocardial infarction: The Boost randomised controlled clinical trial. *Lancet*, 364, 141–148.

Wu, J., Zeng, F., Weisel, R. D., and Li, R. K. 2009. Stem cells for cardiac regeneration by cell therapy and myocardial tissue engineering. *Adv. Biochem. Eng. Biotechnol.*, 114, 107–128.

Xiang, Z., Liao, R., Kelly, M. S., and Spector, M. 2006. Collagen-Gag scaffolds grafted onto myocardial infarcts in a rat model: A delivery vehicle for mesenchymal stem cells. *Tissue Eng.*, 12, 2467–2478.

Ye, W. P., Du, F. S., Jin, J. Y., Yang, J. Y., and Xu, Y. 1997. *In vitro* degradation of poly(caprolactone), poly(lactide) and their block copolymers: Influence of composition, temperature and morphology. *Reactive Funct. Polym.*, 32, 161–168.

Yoshida, Y. and Yamanaka, S. 2010. Recent stem cell advances: Induced pluripotent stem cells for disease modeling and stem cell-based regeneration. *Circulation*, 122, 80–87.

Zhang, G., Hu, Q., Braunlin, E. A., Suggs, L. J., and Zhang, J. 2008. Enhancing efficacy of stem cell transplantation to the heart with a PEGylated fibrin biomatrix. *Tissue Eng. Part A*, 14, 1025–1036.

Zhang, J., Wilson, G. F., Soerens, A. G., Koonce, C. H., Yu, J., Palecek, S. P., Thomson, J. A., and Kamp, T. J. 2009. Functional cardiomyocytes derived from human induced pluripotent stem cells. *Circ. Res.*, 104, e30–e41.

Zhang, Y., Sievers, R. E., Prasad, M., Mirsky, R., Shih, H., Wong, M. L., Angeli, F. S. et al. 2010. Timing of bone marrow cell therapy is more important than repeated injections after myocardial infarction. *Cardiovasc. Pathol*, 20(4), 204–212.

Zimmermann, W. H., Melnychenko, I., and Eschenhagen, T. 2004. Engineered heart tissue for regeneration of diseased hearts. *Biomaterials*, 25, 1639–1647.

Zong, X. H., Bien, H., Chung, C. Y., Yin, L. H., Fang, D. F., Hsiao, B. S., Chu, B., and Entcheva, E. 2005. Electrospun fine-textured scaffolds for heart tissue constructs. *Biomaterials*, 26, 5330–5338.

7

Stem Cells and Hematopoiesis

Krista M. Fridley
University of Texas, Austin

Krishnendu Roy
University of Texas, Austin

7.1 Introduction

Controlled differentiation of stem and progenitor cells into lineage-specific, functional, and transplantable cells could provide new directions in cell therapy. Although the biological mechanisms for stem cell differentiation into various tissue types are widely studied by cell and molecular biologists, quantitative manipulation of these cells under engineered microenvironments as well as strategies to produce therapeutic cells in large scale are increasingly being studied by engineers, especially biomedical engineers. The use of biomaterials, bioreactors, and process-control tools along with quantitative studies of mechanical properties of stem cells is producing new insights in basic biology while transforming stem cell research into clinical possibilities.

Stem cell-derived blood cells could provide potentially unlimited and on-demand source of therapeutic cells for a variety of clinical applications, including bone marrow transplantation, adoptive T cell and dendritic cell therapies, as well as for blood transfusions, for example, platelet or red blood cell therapy. Some of these applications, for example, bone marrow transplantation, has been used for decades and have revolutionized modern medicine. Yet, current paradigms of isolating donor or patient cells for acute or future transplantation, with or without *in vitro* modification, are unsustainable in the face of high demand and immediate needs. The true impact of these therapies could only be realized if blood lineage cells (hematopoietic stem and progenitor cells, dendritic cells, T cells, red blood cells, platelets, etc.) are available on-demand and as ready-to-use therapeutics.

This chapter provides a fundamental understanding of hematopoietic development, both in the embryo and in adults, and describes recent advances made by stem cell biologists and engineers in further understanding hematopoiesis to generate potentially therapeutic blood cells.

7.2 Hematopoietic Development and Sources of Hematopoietic Stem Cells

Hematopoiesis is the development of blood lineage cells from stem and progenitor cells, including both red blood cells (erythrocytes) and white blood cells (leukocytes). There are two types of hematopoietic tissue, myeloid and lymphoid tissue. Myeloid tissue is found in the bone marrow and produces red and white blood cells. Lymphoid tissue functions to mature lymphocytes and is found in the lymph nodes, thymus, spleen, and mucosa of respiratory and digestive tracts.

7.2.1 Hematopoietic Cells: *In Vivo* Development

In embryonic development, hematopoiesis begins in the blood islands of the yolk sac. This beginning stage of hematopoiesis is termed "primitive" and functions to produce red blood cells for oxygenation of the rapidly growing embryo. Primitive hematopoiesis is transient and is eventually replaced by definitive or adult-type hematopoiesis. The site of hematopoiesis shifts to an intraembryonic region known as the aorta–gonad–mesonephros (AGM), followed by the fetal liver. Although experimental evidence confirms all of these locations as sites for hematopoietic cells, the precise origin where hematopoietic stem cells (HSCs) are first generated is believed to be the AGM region (Medvinsky and Dzierzak 1996). Additionally, the placenta has been identified as a location for human hematopoietic development (Barcena et al. 2009) and as a source of murine HSCs with adult reconstitution abilities (Ottersbach and Dzierzak 2005). The hemangioblast is a multipotent cell believed to be a common origin for both blood and vascular cells (Choi et al. 1998). The experimental evidence for the origin and locations of HSCs as well as a comparison of hematopoiesis in various vertebrate animals has been reviewed (Cumano and Godin 2007). In addition, the discovery that umbilical cord blood (UCB) cells contain HSCs has led to the harvest of these cells.

At the time of birth and throughout adulthood, the bone marrow is the primary site for hematopoiesis and is the location of most of the HSC population. A small percentage of HSCs are mobilized and found in circulating or peripheral blood (PB). Additionally, other hematopoietic sites function to further differentiate and mature hematopoietic stem and progenitor cells (HSPCs). The thymus differentiates HSPCs into T cells, and the spleen functions to differentiate into B cells (in mice and humans).

HSCs are adult multipotent stem cells that are defined by the ability to self-renew and develop all cellular components of the blood, including progenitor cells for both the lymphoid and myeloid lineages. On the other hand, HSPCs are classified primarily by their reconstitution potential and self-renewing ability (Morrison and Weissman 1994; Morrison et al. 1997). Long-term self-renewing HSCs (LT–HSCs) have self-renewal characteristics throughout the lifetime of the organism and are able to repopulate a host's hematopoietic system. LT–HSCs develop into short-term self-renewing HSCs (ST–HSCs) and subsequently into hematopoietic progenitor cells (HPCs) or multipotent progenitor cells, which have the ability to differentiate into all cellular components of the blood; however, these ST–HSC and progenitor cells only have a limited ability to self-renew. Therefore, identifying methods for the production and maintenance of LT–HSC is essential for the use of HSCs in cellular therapies.

Terminally differentiated hematopoietic cells develop from two types of progenitor cells, lymphoid and myeloid. Figure 7.1 illustrates the differentiation from HSCs into lineage-specific cells. Lymphoid progenitors develop into lymphocytes, which are T cells, B cells, and natural killer (NK) cells. Myeloid progenitors differentiate into dendritic cells, monocytes, macrophages, neutrophils, eosinophils, mast cells, basophils, platelets, and erythrocytes. Although the mouse model of hematopoiesis has been extensively studied, significant gaps remain in our knowledge of lineage commitments during differentiation.

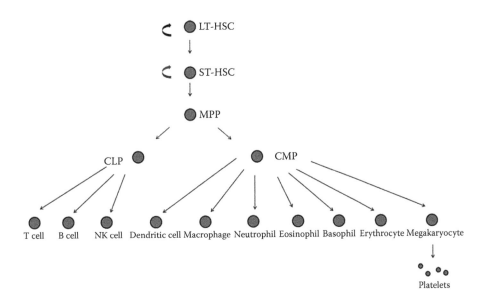

FIGURE 7.1 Long-term hematopoietic stem cells (LT-HSC) renew throughout life and give rise to short-term hematopoietic stem cells (ST-HSC) which have a limited capability for self-renewal. Multipotent progenitor (MPP) cells differentiate into common lymphoid progenitors (CLP) and common myeloid progenitors (CMP). Those progenitors then differentiate into the cell types of the blood, including T lymphocytes, B lymphocytes, NK cells, dendritic cells, macrophages, neutrophils, eosinophils, basophils, erythrocytes, megakaryocytes, and platelets.

7.2.2 Hematopoietic Cells from Pluripotent Cells

Embryonic stem (ES) cells can indefinitely self-renew and have the potential to differentiate to every cell in the body (Sato et al. 2003), including hematopoietic cells. ES cells were first isolated from the inner cell mass of developing mouse blastocysts (Sukoyan et al. 1993). Because of their renewal in the undifferentiated state and pluripotent properties, ES cells are a prospective cell source for clinical therapies (Kaji and Leiden 2001).

ES cells offer advantages over adult stem cells which can be difficult to isolate due to low frequency of adult HSCs (Wang et al. 1997). Additionally, adult stem cells can have decreased differentiation potentials and growth which reduces their utility for tissue engineering applications (Guillot et al. 2007).

More recently, somatic cells have been induced to form pluripotent cells which resemble ES cells. Since the derivation of ES cells is ethically controversial, induced pluripotent (iPS) cells could prove to be a novel cell source to replace ES cells. Cells are reprogrammed using genes important for pluripotency (OCT-3/4, SOX2, c-Myc, Klf4, NANOG, and/or LIN28) using viral transfection systems or recombinant proteins (Takahashi and Yamanaka 2006; Yu et al. 2007; Huangfu et al. 2008; Zhou et al. 2009). However, further studies must investigate the safety of reprogramming of adult cells to obtain iPS cells. These iPS cells have been increasingly studied as a source for hematopoietic cells (Choi et al. 2009; Lengerke et al. 2009).

During differentiation encouraged by suspension culture, ES and iPS cells typically form aggregates known as embryoid bodies (EBs). Similar to embryonic development, these EBs increase in complexity and differentiate into the three germ layers of embryonic development, which are the endoderm, ectoderm, and mesoderm. The mesoderm gives rise to the blood tissue, and hematopoietic cells including HSCs and HPC develop within the EB.

7.3 The HSC Niche

Stem cell niches are the microenvironments in which stem cells reside that regulate stem cell renewal and differentiation. The balance between self-renewal and differentiation is critical for stem cells, as the stem cell population could be depleted if differentiating cells surpassed those undergoing self-renewal or create tumors with unconstrained proliferation.

Two distinct niches have been identified in the bone marrow to support HSCs, which are the osteoblastic niche and the vascular niche. The role of osteoblastic niche is to support HSC maintenance. This claim is supported in mouse models by a gain in the number of HSCs when the number of osteoblastic cells is increased (Calvi et al. 2003; Zhang et al. 2003), as well as a decrease in hematopoietic progenitors in the bone marrow of mice with an induced osteoblast deficiency (Visnjic et al. 2004). The vascular niche has been reported to aid in self-renewal, differentiation, as well as migration of HSCs. This statement is supported by translocation of megakaryocyte progenitors to the bone marrow vascular sinusoids-induced megakaryocyte maturation as well as disruption of bone marrow endothelial cell (BMEC) VE-cadherin-mediated intercellular adhesion interactions results in an inability of the vascular niche to support megakaryocyte differentiation (Avecilla et al. 2004). Additionally, sinusoidal endothelium cells may create a niche that sustains HSCs in extramedullary tissues (Kiel et al. 2005).

The bone marrow contains extracellular matrix (ECM) as well as stromal cells which contribute to hematopoiesis. The ECM is composed of various proteins, including fibronectin and laminin, structural macromolecules, such as proteoglycans and glycosaminoglycan. Stromal cells are various cells at different developmental stages, including fibroblasts, osteocytes, adipocytes, chondrocytes, osteoblasts, and endothelial cells. Additionally, mesenchymal stem cells (MSCs) reside in the bone marrow and give rise to the majority of these stromal cells. These stromal cells provide secreted and membrane-bound cytokines to support HSC maintenance and differentiation. Stromal cells in combination with the blood vessels of the bone marrow create what is termed the hematopoietic inductive microenvironment (HIM). Additionally, the stem cell niches within the bone marrow are low oxygen environments, and the proliferation of hematopoietic progenitors has been shown to be regulated by a hypoxia-mediated signaling pathway in culture (Adelman et al. 1999).

7.4 Identification of HSCs

In order to effectively utilize HSCs and their derivatives in a clinical setting, reliable methods must be employed to identify and examine the function of hematopoietic cell populations. Cell surface markers can be used to help identify HSCs, and functional assays aim to measure cellular proliferation and differentiation both *in vitro* and *in vivo*.

7.4.1 Cell Surface Markers

Hematopoietic cells can be identified by cell surface markers via flow cytometry. If testing of a purified population of either HSCs or differentiated cells is desired, hematopoietic cells can be isolated by fluorescence activated cell sorting (FACS) or magnetic separation by either positive or negative selection of known hematopoietic surface markers. Continued study of the markers which distinguish HSCs and their derivatives to clearly identify HSCs must be done in order to further examine and utilize these various cell populations.

There are no definitive markers expressed on HSCs; however, many markers on both mouse and human HSC have been identified. HSPCs lack expression of lineage (lin) markers found on mature hematopoietic cells, and hence are classified as lin⁻ cells. HSC are identified in mouse by the expression cell surface markers c-kit (CD117) and sca-1 (Spangrude et al. 1988; Ogawa et al. 1991; Ikuta and Weissman 1992). Other markers that have been used to identify mouse HSC are Thy-1.1(CD90), CD150, CD244, CD48, CD45, and CD41 (Kiel et al. 2005; McKinney-Freeman et al. 2009). Human HSCs are most often characterized with CD34 (Baum et al. 1992) and CD38 (Muench et al. 1994) but CD133, and c-kit (CD117) markers are also used (Ogawa et al. 1991; Wognum et al. 2003). Additionally, HSCs have been identified by their ability to efflux various fluorescent dyes, such as Rhodamine-123 (Rho) and Hoechst 33342 (Ho). Cells identified using this method are referred to as side population (SP) cells because they form a unique cluster of events when identified using flow cytometry (Goodell et al. 1996). Further identification of HSCs as well as terminally differentiated hematopoietic cells is desired so that these populations can be isolated and used for cellular therapies.

7.4.2 Functional Assays

Both *in vitro* and *in vivo* assays have been developed to characterize and measure the function of HSCs. Cobble stone area-forming cell (CAFC) assays measure both progenitors and stem cells. Cells are cultured on a stromal layer, and the number of hematopoietic colonies, which appear underneath the stromal layer, are counted. Colony-forming units (CFU) or colony-forming cells (CFC) assays measure the ability of progenitor cells to rapidly produce colonies, including erythroid, granulocyte, megakaryocyte, or combinations of these precursors. The long-term culture-initiating cell (LTC–IC) assay determines the prevalence of stem cells, and as the name suggests, measures the ability to form colonies after longer culture than the CFU assay.

Although *in vitro* assays provide an indication of *in vivo* activity, *in vivo* assays are a better demonstration of HSC function. *In vivo* assays involve reconstitution of the hematopoietic system in immunodeficient or irradiated animal models. Spleen colony-forming unit (CFU-S) measures the ability of HSCs to repopulate the spleens of lethally irradiated mice. However, CFU-S assays are short-term assays (approximately 12 days). Therefore, *in vivo* assays which examine the reconstitution of the entire hematopoietic system in immunodeficient or irradiated animal models for the lifetime of the animal are the only assays which truly identify HSCs.

7.5 Plasticity of HSCs

Recent research has indicated that adult stem cells can differentiate to cells of another lineage, which is referred to as plasticity or transdifferentiation. Plasticity refers to differentiation to cells within the same germ layer, while transdifferentiation designates crossing barriers to another germ layer. For example, bone marrow transplants have demonstrated both plasticity by differentiating into other cells of the mesoderm and transdifferentiation into cells originating in the ectoderm and endoderm.

Mouse bone marrow has demonstrated hepatic regeneration by rescuing the function of a mouse liver (Lagasse et al. 2000), regenerated myocardium in infarcted mice (Orlic et al. 2001), developed into cells expressing neuronal proteins in the central nervous system (Brazelton et al. 2000), and remyelination of the spinal cord (Akiyama et al. 2002). Additionally, adult bone marrow cells have shown differentiation into epithelial cells of the liver, lung, gastrointestinal (GI) tract, and skin (Krause et al. 2001), contribution to the myofibers of skeletal muscle (Doyonnas et al. 2004; Palermo et al. 2005).

However, this concept of plasticity has been questioned. One alternative is that fusion of stem cells with other cells gives the appearance of differentiation (Vassilopoulos et al. 2003; Wang et al. 2003). Regardless of the mechanism, HSCs may have the ability to not only repopulate the hematopoietic system but also to support the regeneration of other failing tissues.

7.6 Clinical Therapies with HSCs

Hematopoietic stem cell transplantation (HSCT) was originally referred to as bone marrow transplantation, as bone marrow was the cell source for these transplants, but now may also involve transplanting adult stem cells from the PB or UCB. Hematopoietic cell transplants can be self-transplants (autologous) or donor transplants (allologous). Hematopoietic cells have the potential to treat several diseases (Dang et al. 2002), including hematopoietic malignancies (Hsu et al. 1996; Nestle et al. 1998; Reichardt et al. 1999; Galea-Lauri et al. 2002; Buchler et al. 2003) (e.g., leukemia, lymphoma, and myeloma), certain cancers (Murphy et al. 1996), and immunodeficiency (Rideout et al. 2002).

Disadvantages exist with current methods, including isolation difficulties of rare cell populations, problems with long-term expansion of HSPCs *in vitro*, and availability of matched human leukocyte antigen (HLA) donor marrows (Daley 2003). In addition, autologous grafts require the expansion of the patient's own hematopoietic cells which is a time-consuming process. The high-throughput generation of HSPCs or terminally differentiated hematopoietic cells could provide a renewable, readily available cell source for cell-based therapies.

7.7 Generation of Hematopoietic Cells in Culture

One difficulty associated with the use of adult HSCs for therapeutic applications is maintaining and expanding these cells *in vitro*; therefore, researchers have attempted to optimize culture environments for the expansion and differentiation of these cells. One common practice to influence hematopoietic cell expansion and differentiation is the addition of various cytokine and growth factor combinations *in vitro* to help replicate the signals provided by the HSC niche. The expansion and maturation of hematopoietic lineages from murine ES cells has been reviewed (Mohle and Kanz 2007). Briefly, cytokines that have been studied included thrombopoietin (TPO), interleukin (IL)-6, and IL-11 for megakaryocytes and platelets, erythropoietin (EPO), c-Kit ligand (KL), insulin, insulin-like growth factor (IGF)-1, and IL-3 for erythroid cells, IL-3 and KL for mast cells, IL-1, IL-3, macrophage colony stimulating factor (M-CSF), and granulocyte/macrophage colony stimulating factor (GM–CSF) for macrophages, FMS-like tyrosine kinase-3 ligand (Flt-3L) and IL-7 for T and B lymphocytes, IL-5, IL-3, GM-CSF, eotaxin for eosinophils, oncostatin M (OSM), basic fibroblast growth factor (b-FGF), IL-6, IL-11, leukemia inhibitory factor (LIF), KL, granulocyte colony stimulating factor (G-CSF), GM–CSF, and IL-6 for neutrophils, GM–CSF, IL-3, IL-4, tumor necrosis factor alpha (TNF-α), lipopolysaccharide (LPS), and anti-CD40 for dendritic cells, Flt-3L, IL-15, IL-6, IL-7, and KL for NK cells, and M-CSF, receptor activator for nuclear factor ligand (RANKL) for osteoclasts. Additionally, cytokine growth factors for human UCB HSC expansion have employed various combinations of stem cell factor (SCF), GM-CSF, IL-3, TPO, Flt-3L, IL-6, G-CSF, and EPO and have been reviewed (Mohle and Kanz 2007; Andrade-Zaldivar et al. 2008).

As previously mentioned, ES and iPS cells provide a potentially unlimited cell source; therefore, embryoid body (EB)-based differentiation has been studied for the generation of hematopoietic cells. EB development has been studied using a variety of static techniques, including liquid suspension, hanging drop, methylcellulose, and liquid attached cultures, to encourage ES cell aggregation and hematopoietic differentiation. Liquid suspension, methylcellulose, and hanging drop cultures produce no difference in hematopoietic differentiation of EBs produced in these systems. However, EBs created in the liquid-attached culture show decreased hematopoietic differentiation (Dang et al. 2002). Interestingly, different stem cell lines may have varying differentiation potentials. Testing of five different human ES cell lines have demonstrated that although all ES cells form the three germ layers of embryonic development, they were marked differences in their hematopoietic differentiation potential (Chang et al. 2008).

Definitive blood lineages can be developed from pluripotent cells. Studies which have involved EB or stromal cell coculture-based differentiation with the addition of different cytokines to produce terminal differentiated hematopoietic cells are reviewed by Olsen et al. (2006).

Several biomedical engineering techniques have been employed to generate a clinically applicable number of therapeutic cells. For the generation of hematopoietic cells, these techniques often involve mimicking the bone marrow environment for the enhanced expansion or differentiation of stem cells. Another important concern is the ability to scale up culture methods to provide clinically relevant numbers of therapeutic cell populations. In addition to the use of cytokines and growth factors, culture techniques include manipulation of oxygen tension, stromal cell coculture, biomaterial culture, and bioreactor culture. These methods will be discussed in further detail in the following sections.

7.7.1 Hypoxia

As mentioned previously, the HSC niche is a low oxygen or hypoxic environment, also referred to as hypoxia. Low oxygen tension has shown increases in the expansion and differentiation potentials of stem cells. In stirred culture systems, recent studies have shown increased HPCs from encapsulated ES cells when differentiated at low (3–4%) oxygen (Dang et al. 2004) as well as examined hypoxic response in EB-based hematopoietic differentiation (Cameron et al. 2008). Additionally, the expansion of human cord blood progenitors showed significantly higher cell increases in hypoxic (5% oxygen) compared to normoxic conditions (Koller et al. 1992). Hypoxic effects may further increase hematopoiesis, including the expansion of HSPCs and differentiation into hematopoietic cells.

7.7.2 Stromal Cell Line Coculture and Conditioned Medium

Stromal cell lines have been established to support hematopoietic cell maintenance and differentiation in culture, comparable to stromal cells support *in vivo*. Several stromal lines have been utilized from the mouse bone marrow (Pessina et al. 1992; Qiu et al. 2003). Additionally, murine stromal cell lines have been established from the AGM region (Ohneda et al. 1998; Xu et al. 1998; Weisel et al. 2006) and the fetal liver (Moore et al. 1997) for the maintenance and differentiation of HSPCs.

OP9 is a commonly used murine stromal cell line from the bone marrow which is deficient in M-CSF and known to support hematopoiesis (Nakano et al. 1994). Although the mechanism by which these cells support hematopoietic differentiation from ES cells is largely unclear, the importance of M-CSF deficiency was demonstrated when recombinant M-CSF reduced differentiation of ES cells into meso-dermal cells and subsequent development to hematopoietic cells (Nakano et al. 1994). OP9 cells have been studied for generating HPCs as well as definitive hematopoietic lineages from ES cells (Suzuki and Nakano 2001; Kitajima et al. 2003; Schmitt et al. 2004; Umeda et al. 2004; Vieira and Cumano 2004; La Motte-Mohs et al. 2005).

Most previous work with OP9 cells has involved coculture with ES cells. However, disadvantages exist with a coculture system; for example, the coculture of cells may require sorting to obtain and analyze the specific cell population of interest. Taqvi et al. demonstrated that cell–cell contact may not be required for OP9 support of hematopoiesis from ES cells by culturing OP9 and ES cells on separate scaffolds that did not contact but were placed in the same culture so that soluble factors could penetrate throughout (Taqvi and Roy 2006). OP9 conditioned medium supplemented with cytokines and growth factors has recently shown increased hematopoietic differentiation of ES cells (Zhang et al. 2006).

Potential problems for the use of the xenogeneic stromal cells in a clinical setting may include transfer of infectious diseases or rejection of transplanted HSCs. Therefore, a human bone marrow (hBM) stro-mal cell line has been developed, which demonstrated potential support for multilineage differentiation (Bertolini et al. 1997). Furthermore, microencapsulated feeder cells have been investigated and demon-strated effective expansion of human UCB (hUCB) cells (Fujimoto et al. 2007).

7.7.3 Biomaterials for Hematopoietic Cell Culture

To mimic the biological microenvironment of cells, a variety of three-dimensional (3D) biomaterials have been used as substitutes for the ECM. Unlike conventional two-dimensional (2D) culture systems where cells generally grow and proliferate as a horizontal monolayer, 3D scaffolds provide a physical support matrix thus increasing cell–cell and cell–substrate interactions (Martin et al. 1998; Tan et al. 2001). Therefore, bioengineered 3D culture systems have become a promising experimental approach for the differentiation of both adult and ES cells (Martin et al. 1997; Solchaga et al. 1999; Dawson et al. 2008). Table 7.1 summarizes biomaterials used for the expansion and differentiation of hematopoietic cells that are discussed in detail below.

TABLE 7.1 Biomaterials for Hematopoietic Cell Culture

Scaffold Material	Application	Reference(s)
Aluminum oxide (Al$_2$O$_3$)	hPB cell expansion and differentiation	Schubert et al. (2004)
Apatite	hPB cell expansion and differentiation	Schubert et al. (2004)
Cellulose	mBM cell expansion	Tomimori et al. (2000)
Chitosan	CD34$^+$ hUCB cell expansion	Cho et al. (2008)
Collagen	CD34$^+$ hUCB cell expansion	Kim et al. (2003); Oswald et al. (2006)
Poly(lactic acid) (PLA)	ESC differentiation to HSPCs	Taqvi et al. (2006)
Polyester	mBM cell expansion	Tomimori et al. (2000); Sasaki et al. (2002)
Polyethersulfone	CD34$^+$ hUCB cell expansion	Chua et al. (2007)
Polyvinyl formal (PVF)	mBM cell expansion	Tun et al. (2002)
Polyethylene terephthalate (PET)	hCD34$^+$ cell expansion	Feng et al. (2006)
Tantalum (Cytomatrix™)	hCD34$^+$ cell differentiation to T cells	Poznansky et al. (2000)
	ESC differentiation to HSPCs	Liu et al. (2005)
	HSC expansion	Banu et al. (2001); Ehring et al. (2003)
Tantalum-coated porous biomaterial (TCPB)	HSC expansion	Bagley et al. (1999)

Several natural materials have been investigated as scaffolds for hematopoietic cell culture. Collagen microbeads have been reported to improve human CD34$^+$ UCB cell expansion compared to traditional 2D culture (Kim et al. 2003). Using a fibrillar collagen matrix, human CD34$^+$ UCB cells showed higher levels of expressed growth factors and cytokines compared to cells grown in suspension (Oswald et al. 2006). Additionally, human CD34$^+$ UCB cells seeded on chitosan scaffold with immobilized heparin demonstrated higher percentages of progenitors and increased CFUs as compared to cells from static cultures (Cho et al. 2008).

Additionally, synthetic materials have been employed as a hematopoietic culture environment. One common scaffold material used for hematopoietic culture is tantalum, which is an inert metal. Poznansky et al. (2000) have reported the successful generation of human T cells from CD34$^+$ progenitors *in vitro* by coculturing with murine thymic stromal cells on 3D tantalum-based Cytomatrix™. The critical cell–cell associations provided by the 3D architectures resulted in more efficient T cell production than that in a monolayer culture system. Tantalum 3D culture systems facilitate and enhance maintenance and multipotency of HPCs in long-term cultures where low concentrations or no exogenous cytokines need to be added, while under the same condition, 2D systems are less capable of supporting progenitor viability and multipotency (Bagley et al. 1999; Banu et al. 2001; Ehring et al. 2003). The tantalum-based Cytomatrix scaffolds have also been used for the successful hematopoietic differentiation from ES cells (Liu and Roy 2005; Liu et al. 2006a).

Polyethylene terephthalate (PET) scaffolds demonstrated successful expansion of CD34$^+$ cells, with conjugated fibronectin (FN) resulting in a higher expansion compared to adsorbed or soluble FN in PET scaffolds (Feng et al. 2006). Additionally, aluminum oxide and apatite ceramics provide support for the proliferation and differentiation of human PB (hPB) cells (Schubert et al. 2004). Polyvinyl formal (PVF) scaffolds demonstrated enhanced mouse bone marrow (BM) cell proliferation (Tun et al. 2002).

Yoshida and colleagues investigated medium pore (100 μm) sized cellulose beads, large pore (500 μm) sized cellulose cubes, and nonwoven polyester disks to create a hematopoietic microenvironment for mouse bone marrow cultures. Although the cellulose carriers showed a decrease in progenitor cells, the polyester disks maintained the progenitor cells over the 4 week cultivation (Tomimori et al. 2000). Furthermore, the nonwoven fabric disks (Fribra-cel) exhibited superior expansion of HPCs from mouse bone marrow without the addition of cytokines compared to 2D culture (Sasaki et al. 2002).

In addition to various scaffold materials, the physical properties of scaffold structure (i.e., pore size, polymer concentration) have also been studied. Specifically, decreasing pore size and increasing polymer concentration (which increased the compression modulus) significantly increases the efficiency of HPC generation from ES cells in poly(lactic acid) (PLA) scaffolds (Taqvi et al. 2006). Poly(ethersulfone) nanofiber scaffolds generated significantly higher numbers of total CFU, CFU–GEMM units, and LTC–IC counts on aminated nanofiber scaffolds in contrast to unmodified nanofiber scaffolds. Further highlighting the importance of scaffold topography, the length of the different spacer groups affected the expansion outcome (Chua et al. 2007).

Furthermore, differentiation and maturation of hematopoietic cells can be achieved by creating an artificial signaling system. Biomaterials can be used to immobilize ligands, which may be important for efficient activity compared to soluble factors. For example, immobilized Delta1-Fc protein was more effective at expanding human CD133$^+$ cord blood cells than as a soluble growth factor (Suzuki et al. 2006), and functionalized microbeads have demonstrated efficient notch signaling for the differentiation of murine BM–HSCs into T cells (Taqvi et al. 2006). Additionally, immobilized ligands can allow for more controlled presentation, including creating signaling gradients. Using cellulose binding domain fusion protein, bioactive SCF stimulated receptor polarization in the cell membrane of mouse progenitor cells and adherence to the cellulose matrix (Jervis et al. 2005). Furthermore, major histocompatibility (MHC) tetramers have been shown to produce T cells from mouse ES cells with significant cytotoxic T lymphocyte activity against antigen-loaded target cells, which demonstrates the potential to produce antigen-specific T cells from stem cells without the use of stromal cells (Lin et al. 2010).

Again, HSC expansion and the hematopoietic differentiation potential of ES cells have been mostly studied in static cultures. The culture of stem cells in a more biomimetic environment may greatly increase the expansion and differentiation potential of hematopoietic cells. Additionally, these 3D microenvironments may reduce the dependence on cytokines and stromal cell cultures compared to traditional 2D culture methods.

7.7.4 Bioreactors for Hematopoietic Cell Culture

Traditional 2D static methods can culture only a limited number of cells and is generally considered time-consuming and labor-intensive. Unlike traditional static 2D culture methods, bioreactor systems have the ability to achieve scale-up, which makes bioreactors critical for potential clinical applications. Additionally, the dynamic flow of bioreactors creates a more homogenous environment and increases nutrient availability when compared to traditional static culture (Nielsen 1999). Stem cell expansion and differentiation has been typically performed in static cultures, but recently the expansion of adult HPCs and efficiency of ES cell differentiation into various lineages has been studied in several different types of bioreactors, including stirred flasks, rotary wall, perfusion cultures, and packed bed bioreactors. Bioreactors and their application in the culture of hematopoietic cells are summarized in Table 7.2 and discussed in detail in the following sections.

7.7.4.1 Stirred Bioreactors

Stirred-tank bioreactors have been used for culturing a variety of suspension cells, including HSCs, as well as adherent cells using microcarriers. The spinner flask system has shown expansion of human bone marrow progenitors (Zandstra et al. 1994; Kim 1998), UCB (Collins et al. 1998), CD34$^+$ PB (Collins et al. 1998), mouse bone marrow cells (Kwon et al. 2003), and human T cells (Carswell and Papoutsakis 2000). Spinner flask culture showed no detrimental effects on primary T cell expansion; however, a significant downregulation of interleukin-2 receptor (IL-2R) occurred compared to static culture (Carswell and Papoutsakis 2000). Additionally, a T cell line had severally reduced growth rates showing extensive sensitivity to agitation at much lower speeds than the primary T cells, demonstrating the potential culture differences between model cell lines and the hematopoietic cells of interest (Carswell and Papoutsakis 2000). Spinner flask cultures have also been used for successful EB-based differentiation,

TABLE 7.2 Bioreactors for Hematopoietic Cell Culture

Bioreactor	Application	Reference(s)
Spinner flask	Expansion of hBM, PB, UCB, mBM cells	Zandstra et al. (1994); Collins et al. (1998); Kim (1998); Kwon et al. (2003)
	Expansion of T cells	Carswell and Papoutsakis (2000)
	Differentiation of ES cells	Dang et al. (2004); Fok and Zandstra (2005); Cameron et al. (2006); Fridley et al. (2010)
Rotating wall (Synthecon)	Expansion of hBM, mBM, and UCB cells	Plett et al. (2001); Konstantinov et al. (2004); Liu et al. (2006a,b)
	Differentiation of ES cells	Gerecht-Nir et al. (2004); Fridley et al. (2010)
Perfusion chamber	Proliferation of mBM, hPB, hUCB cells	Koller et al. (1993); Peng and Palsson (1996); Sandstrom et al. (1996); Jaroscak et al. (2003)
Packed bed bioreactors	Expansion of mBM, hBM, hUCB, hPB cells	Mantalaris et al. (1998); Meissner et al. (1999)
Airlift packed bed bioreactor	Expansion of mBM cells	Highfill et al. (1996)

demonstrating potential hematopoietic-specific differentiation of ES cells (Dang et al. 2004; Fok and Zandstra 2005; Cameron et al. 2006), and higher cell seeding densities of ES cells have been shown to improve hematopoietic differentiation from ES cells in spinner flasks (Fridley et al. 2010).

7.7.4.2 Rotary Wall Vessels

Rotary wall vessels, also referred to as microgravity bioreactors, have also been used for suspension and microcarrier cell culture. Previous studies have employed rotary wall cultures for the expansion of human CD34+ bone marrow, UCB, and mouse bone marrow cells (Plett et al. 2001; Konstantinov et al. 2004; Liu et al. 2006a,b). Originally developed by NASA, the Synthecon, Inc. (Houston, TX) Rotary Cell Culture System with a slow turning lateral vessel (STLV) has been reported to increase the efficiency of EB formation and differentiation of stem cells into the three germ layers of embryonic development; however, the differentiation specifically into hematopoietic lineages was not investigated (Gerecht-Nir et al. 2004). Exploration of ES cell differentiation specifically into hematopoietic lineages using the Synthecon system demonstrated that hematopoietic differentiation could be improved using an optimal cell seeding density and rotation speed (Fridley et al. 2010). However, differentiation of ES cells into HSCs and their derivatives needs to be explored in this system.

7.7.4.3 Perfusion Cultures

In perfusion cultures, fresh medium flows through the system continuously, which can mimic the *in vivo* environment and increase cell productivity by providing fresh nutrients and removing waste products. Perfusion chambers have been used for the expansion of HSCs and have shown increased expansion and colony formation compared to static culture (Koller et al. 1993). The geometry of these bioreactors has shown to affect the proliferation of bone marrow cells (Peng and Palsson 1996). The greatest challenge with this system is the retention of cells as medium flows through the chamber. Therefore, flatbed perfusion chambers can be modified with grooves perpendicular to the flow to retain nonadherent hematopoietic progenitors (Sandstrom et al. 1996).

The potential use of perfusion bioreactors in a clinical setting has been recently demonstrated in a clinical study. UCB cells expanded in perfusion cultures demonstrated durable engraftment within the follow-up time of approximately 47 months in a phase I clinical study (Jaroscak et al. 2003).

7.7.4.4 Packed or Fixed Bed Bioreactor

In a packed or fixed bed bioreactor system, cells are seeded on or encapsulated within packed particles that do not move within the culture medium. Packed bed bioreactors support high cell densities in a compact volume; however, the volume of these systems is limited due to the requirement for uniform flow over the entire cross-section (Meuwly et al. 2007). A bioreactor packed with porous microspheres generated a higher percentage of erythroid cells than traditional flask cultures (Mantalaris et al. 1998). Another fixed bed bioreactor with immobilized stromal cells in porous glass carriers demonstrated expansion of both early and late progenitor cells from human mononuclear cells derived from UCB or PB (Meissner et al. 1999). In an airlift packed bed bioreactor system, stromal cells established in a fiber-glass matrix followed by seeding of fresh bone marrow cells demonstrated sustained cell production (Highfill et al. 1996).

Bioreactor cultures have been increasingly studied for stem cell expansion and differentiation; however, much work remains to study lineage-specific differentiation and generate blood cell lineages. The varying hydrodynamics of these bioreactor systems may have different effects on hematopoietic differentiation. For example, spinner flasks are generally characterized by turbulent flows and high shear stress whereas rotating vessels produce laminar flow and low shear forces (Vunjak-Novakovic et al. 1999, 2006). Most hematopoietic stem and progenitor cell expansion studies have cultured mononuclear cell populations containing accessory cells which may contribute to the expansion of progenitor cells; therefore, isolated progenitor populations may have different requirements for expansion. Furthermore, ES cell differentiation in bioreactor systems has focused on EB formation and not lineage-specific differentiation into hematopoietic cells. Optimizing bioreactor systems for the production of a homogeneous, therapeutic cell population is critical for potential clinical applications, particularly because these systems have the ability to achieve scale-up.

7.8 Summary

One goal of tissue engineering or regenerative medicine is the replacement of damaged tissue and organs by transplanting therapeutic cells. Hematopoietic cells have the potential to treat hematopoietic malignancies, certain cancers, and immunodeficiencies. In order for hematopoietic cell-based therapies to be used in clinical applications, expansion and differentiation of stem cells must be further studied to provide clinically relevant numbers of therapeutic cell populations. The use of bioengineering techniques including biomaterial and bioreactor cell culture may provide the capability to generate these therapeutic cells with standard, clinical-grade production techniques.

References

Adelman, D. M., E. Maltepe, and M. C. Simon. 1999. Multilineage embryonic hematopoiesis requires hypoxic ARNT activity. *Genes Dev* 13(19): 2478–83.

Akiyama, Y., C. Radtke, O. Honmou, and J. D. Kocsis. 2002. Remyelination of the spinal cord following intravenous delivery of bone marrow cells. *Glia* 39(3): 229–36.

Andrade-Zaldivar, H., L. Santos, and A. D. Rodriguez. 2008. Expansion of human hematopoietic stem cells for transplantation: Trends and perspectives. *Cytotechnology* 56(3): 151–60.

Avecilla, S. T., K. Hattori, B. Heissig et al. 2004. Chemokine-mediated interaction of hematopoietic progenitors with the bone marrow vascular niche is required for thrombopoiesis. *Nat Med* 10(1): 64–71.

Bagley, J., M. Rosenzweig, D. F. Marks, and M. J. Pykett. 1999. Extended culture of multipotent hematopoietic progenitors without cytokine augmentation in a novel three-dimensional device. *Exp Hematol* 27(3): 496–504.

Banu, N., M. Rosenzweig, H. Kim, J. Bagley, and M. Pykett. 2001. Cytokine-augmented culture of haematopoietic progenitor cells in a novel three-dimensional cell growth matrix. *Cytokine* 13(6): 349–58.

Barcena, A., M. Kapidzic, M. O. Muench et al. 2009. The human placenta is a hematopoietic organ during the embryonic and fetal periods of development. *Dev Biol* 327(1): 24–33.

Baum, C. M., I. L. Weissman, A. S. Tsukamoto, A. M. Buckle, and B. Peault. 1992. Isolation of a candidate human hematopoietic stem-cell population. *Proc Natl Acad Sci USA* 89(7): 2804–8.

Bertolini, F., M. Battaglia, D. Soligo et al. 1997. "Stem cell candidates" purified by liquid culture in the presence of Steel factor, IL-3, and 5FU are strictly stroma-dependent and have myeloid, lymphoid, and megakaryocytic potential. *Exp Hematol* 25(4): 350–6.

Brazelton, T. R., F. M. V. Rossi, G. I. Keshet, and H. M. Blau. 2000. From marrow to brain: Expression of neuronal phenotypes in adult mice. *Science* 290(5497): 1775–9.

Buchler, T., J. Michalek, L. Kovarova, R. Musilova, and R. Hajek .2003. Dendritic cell-based immunotherapy for the treatment of hematological malignancies. *Hematology* 8(2): 97–104.

Calvi, L. M., G. B. Adams, K. W. Weibrecht et al. 2003. Osteoblastic cells regulate the haematopoietic stem cell niche. *Nature* 425(6960): 841–6.

Cameron, C. M., F. Harding, W. S. Hu, and S. Kaufman. 2008. Activation of hypoxic response in human embryonic stem cell-derived embryoid bodies. *Exp Biol Med* 233(8): 1044–57.

Cameron, C. M., W. S. Hu, and D. S. Kaufman. 2006. Improved development of human embryonic stem cell-derived embryoid bodies by stirred vessel cultivation. *Biotechnol Bioeng* 94(5): 938–48.

Carswell, K. S. and E. T. Papoutsakis. 2000. Culture of human T cells in stirred bioreactors for cellular immunotherapy applications: Shear, proliferation, and the IL-2 receptor. *Biotechnol Bioeng* 68(3): 328–38.

Chang, K. H., A. M. Nelson, P. A. Fields et al. 2008. Diverse hematopoietic potentials of five human embryonic stem cell lines. *Exp Cell Res* 314(16): 2930–40.

Cho, C. H., J. F. Eliason, and H. W. T. Matthew. 2008. Application of porous glycosaminoglycan-based scaffolds for expansion of human cord blood stem cells in perfusion culture. *J Biomed Mater Res Part A* 86A(1): 98–107.

Choi, K., M. Kennedy, A. Kazarov, J. C. Papadimitriou, and G. Keller. 1998. A common precursor for hematopoietic and endothelial cells. *Development* 125(4): 725–32.

Choi, K. D., J. Yu, K. Smuga-Otto et al. 2009. Hematopoietic and endothelial differentiation of human induced pluripotent stem cells. *Stem Cells* 27(3): 559–67.

Chua, K. N., C. Chai, P. C. Lee et al. 2007. Functional nanofiber scaffolds with different spacers modulate adhesion and expansion of cryopreserved umbilical cord blood hematopoietic stem/progenitor cells. *Exp Hematol* 35(5): 771–81.

Collins, P. C., W. M. Miller, and E. T. Papoutsakis. 1998. Stirred culture of peripheral and cord blood hematopoietic cells offers advantages over traditional static systems for clinically relevant applications. *Biotechnol Bioeng* 59(5): 534–43.

Cumano, A. and I. Godin. 2007. Ontogeny of the hematopoietic system. *Annu Rev Immunol* 25:745–85.

Daley, G. Q. 2003. From embryos to embryoid bodies: generating blood from embryonic stem cells. *Ann NY Acad Sci* 996:122–31.

Dang, S. M., S. Gerecht-Nir, J. Chen, J. Itskovitz-Eldor, and P. W. Zandstra. 2004. Controlled, scalable embryonic stem cell differentiation culture. *Stem Cells* 22(3): 275–82.

Dang, S. M., M. Kyba, R. Perlingeiro, G. Q. Daley, and P. W. Zandstra. 2002. Efficiency of embryoid body formation and hematopoietic development from embryonic stem cells in different culture systems. *Biotechnol Bioeng* 78(4): 442–53.

Dawson, E., G. Mapili, K. Erickson, S. Taqvi, and K. Roy. 2008. Biomaterials for stem cell differentiation. *Adv Drug Deliv Rev* 60(2): 215–28.

Doyonnas, R., M. A. LaBarge, A. Sacco, C. Charlton, and H. M. Blau. 2004. Hematopoietic contribution to skeletal muscle regeneration by myelomonocytic precursors. *Proc Natl Acad Sci USA* 101(37): 13507–12.

Ehring, B., K. Biber, T. M. Upton et al. 2003. Expansion of HPCs from cord blood in a novel 3D matrix. *Cytotherapy* 5(6): 490–9.

Feng, Q., C. Chai, X. S. Jiang, K. W. Leong, and H. Q. Mao. 2006. Expansion of engrafting human hematopoietic stem/progenitor cells in three-dimensional scaffolds with surface-immobilized fibronectin. *J Biomed Mater Res Part A* 78A(4): 781–91.

Fok, E. Y. L. and P. W. Zandstra. 2005. Shear-controlled single-step mouse embryonic stem cell expansion and embryoid body-based differentiation. *Stem Cells* 23(9): 1333–42.

Fridley, K. M., I. Fernandez, M. T. A. Li, R. B. Kettlewell, and K. Roy. 2010. Unique differentiation profile of mouse embryonic stem cells in rotary and stirred tank bioreactors. *Tissue Engineering Part A* 16(11): 3285–98.

Fujimoto, N., S. Fujita, T. Tsuji et al. 2007. Microencapsulated feeder cells as a source of soluble factors for expansion of CD34(+) hematopoietic stem cells. *Biomaterials* 28(32): 4795–805.

Galea-Lauri, J., D. Darling, G. Mufti, P. Harrison, and F. Farzaneh. 2002. Eliciting cytotoxic T lymphocytes against acute myeloid leukemia-derived antigens: Evaluation of dendritic cell-leukemia cell hybrids and other antigen-loading strategies for dendritic cell-based vaccination. *Cancer Immunol Immunother* 51(6): 299–310.

Gerecht-Nir, S., S. Cohen, and J. Itskovitz-Eldor. 2004. Bioreactor cultivation enhances the efficiency of human embryoid body (hEB) formation and differentiation. *Biotechnol Bioeng* 86(5): 493–502.

Goodell, M. A., K. Brose, G. Paradis, A. S. Conner, and R. C. Mulligan. 1996. Isolation and functional properties of murine hematopoietic stem cells that are replicating *in vivo*. *J Exp Med* 183(4): 1797–806.

Guillot, P. V., W. Cui, N. M. Fisk, and D. J. Polak. 2007. Stem cell differentiation and expansion for clinical applications of tissue engineering. *J Cell Mol Med* 11(5): 935–44.

Highfill, J. G., S. D. Haley, and D. S. Kompala. 1996. Large-scale production of murine bone marrow cells in an airlift packed bed bioreactor. *Biotechnol Bioeng* 50(5): 514–20.

Hsu, F. J., C. Benike, F. Fagnoni et al. 1996. Vaccination of patients with B-cell lymphoma using autologous antigen-pulsed dendritic cells. *Nat Med* 2(1): 52–8.

Huangfu, D. W., K. Osafune, R. Maehr et al. 2008. Induction of pluripotent stem cells from primary human fibroblasts with only Oct4 and Sox2. *Nat Biotechnol* 26(11): 1269–75.

Ikuta, K. and I. L. Weissman. 1992. Evidence that hematopoietic stem-cells express mouse C-Kit but do not depend on steel factor for their generation. *Proc Natl Acad Sci USA* 89(4): 1502–6.

Jaroscak, J., K. Goltry, A. Smith et al. 2003. Augmentation of umbilical cord blood (UCB) transplantation with *ex vivo*-expanded UCB cells: Results of a phase 1 trial using the AastromReplicell System. *Blood* 101(12): 5061–7.

Jervis, E. J., M. M. Guarna, J. G. Doheny, C. A. Haynes, and D. G. Kilburn. 2005. Dynamic localization and persistent stimulation of factor-dependent cells by a stem cell factor/cellulose binding domain fusion protein. *Biotechnol Bioeng* 91(3): 314–24.

Kaji, E. H. and J. M. Leiden. 2001. Gene and stem cell therapies. *JAMA* 285(5): 545–50.

Kiel, M. J., O. H. Yilmaz, T. Iwashita et al. 2005. SLAM family receptors distinguish hematopoietic stem and progenitor cells and reveal endothelial niches for stem cells. *Cell* 121(7): 1109–21.

Kim, B. S. 1998. Production of human hematopoietic progenitors in a clinical-scale stirred suspension bioreactor. *Biotechnol Lett* 20(6): 595–601.

Kim, H. S., J. B. Lim, Y. H. Min et al. 2003. ex vivo expansion of human umbilical cord blood CD34(+) cells in a collagen bead-containing 3-dimensional culture system. *Int J Hematol* 78(2): 126–32.

Kitajima, K., M. Tanaka, J. Zheng, E. Sakai-Ogawa, and T. Nakano. 2003. *In vitro* differentiation of mouse embryonic stem cells to hematopoietic cells on an OP9 stromal cell monolayer. *Methods Enzymol* 365:72–83.

Koller, M. R., J. G. Bender, W. M. Miller, and E. T. Papoutsakis. 1993. Expansion of primitive human hematopoietic progenitors in a perfusion bioreactor system with Il-3, Il-6, and stem-cell factor. *Bio-Technology* 11(3): 358–63.

Koller, M. R., J. G. Bender, E. T. Papoutsakis, and W. M. Miller. 1992. Effects of synergistic cytokine combinations, low oxygen, and irradiated stroma on the expansion of human cord blood progenitors. *Blood* 80(2): 403–11.

Konstantinov, S. M., M. M. Mindova, P. T. Gospodinov, and P. I. Genova. 2004. Three-dimensional bioreactor cultures: A useful dynamic model for the study of cellular interactions. *Ann NY Acad Sci* 1030:103–15.

Krause, D. S., N. D. Theise, M. I. Collector et al. 2001. Multi-organ, multi-lineage engraftment by a single bone marrow-derived stem cell. *Cell* 105(3): 369–77.

Kwon, J., B. S. Kim, M. J. Kim, and H. W. Park. 2003. Suspension culture of hematopoietic stem cells in stirred bioreactors. *Biotechnol Lett* 25(2): 179–82.

La Motte-Mohs, R. N., E. Herer, and J. C. Zuniga-Pflucker. 2005. Induction of T-cell development from human cord blood hematopoietic stem cells by Delta-like 1 *in vitro*. *Blood* 105(4): 1431–9.

Lagasse, E., H. Connors, M. Al-Dhalimy et al. 2000. Purified hematopoietic stem cells can differentiate into hepatocytes *in vivo*. *Nat Med* 6(11): 1229–34.

Lengerke, C., M. Grauer, N. I. Niebuhr et al. 2009. Hematopoietic development from human induced pluripotent stem cells. In *Hematopoietic Stem Cells VII*, eds. L. Kanz, K. C. Weisel, J. E. Dick and W. E. Fibbe, 1176:219–27. Oxford: Blackwell Publishing.

Lin, J., H. Nie, P. W. Tucker, and K. Roy. 2010. Controlled major histocompatibility complex-T cell receptor signaling allows efficient generation of functional, antigen-specific CD8+ T cells from embryonic stem cells and thymic progenitors. *Tissue Engineering Part A* 16(9): 2709–20.

Liu, H., J. Lin, and K. Roy. 2006a. Effect of 3D scaffold and dynamic culture condition on the global gene expression profile of mouse embryonic stem cells. *Biomaterials* 27(36): 5978–89.

Liu, H. and K. Roy. 2005. Biomimetic three-dimensional cultures significantly increase hematopoietic differentiation efficacy of embryonic stem cells. *Tissue Eng* 11(1–2): 319–30.

Liu, Y., T. Q. Liu, X. B. Fan, X. H. Ma, and Z. F. Cui. 2006b. *Ex vivo* expansion of hematopoietic stem cells derived from umbilical cord blood in rotating wall vessel. *J Biotechnol* 124(3): 592–601.

Mantalaris, A., P. Keng, P. Bourne, A. Y. C. Chang, and J. H. D. Wu. 1998. Engineering a human bone marrow model: A case study on *ex vivo* erythropoiesis. *Biotechnol Prog* 14(1): 126–33.

Martin, I., A. Muraglia, G. Campanile, R. Cancedda, and R. Quarto. 1997. Fibroblast growth factor-2 supports ex vivo expansion and maintenance of osteogenic precursors from human bone marrow. *Endocrinology* 138(10): 4456–62.

Martin, I., R. F. Padera, G. Vunjak-Novakovic, and L. E. Freed. 1998. *In vitro* differentiation of chick embryo bone marrow stromal cells into cartilaginous and bone-like tissues. *J Orthop Res* 16(2): 181–9.

McKinney-Freeman, S. L., O. Naveiras, F. Yates et al. 2009. Surface antigen phenotypes of hematopoietic stem cells from embryos and murine embryonic stem cells. *Blood* 114(2): 268–78.

Medvinsky, A. and E. Dzierzak. 1996. Definitive hematopoiesis is autonomously initiated by the AGM region. *Cell* 86(6): 897–906.

Meissner, P., B. Schroder, C. Herfurth, and M. Biselli. 1999. Development of a fixed bed bioreactor for the expansion of human hematopoietic progenitor cells. *Cytotechnology* 30(1–3): 227–34.

Meuwly, F., P. A. Ruffieux, A. Kadouri, and U. von Stockar. 2007. Packed-bed bioreactors for mammalian cell culture: Bioprocess and biomedical applications. *Biotechnol Adv* 25(1): 45–56.

Mohle, R. and L. Kanz. 2007. Hematopoietic growth factors for hematopoietic stem cell mobilization and expansion. *Semin Hematol* 44(3): 193–202.

Moore, K. A., H. Ema, and I. R. Lemischka. 1997. *In vitro* maintenance of highly purified, transplantable hematopoietic stem cells. *Blood* 89(12): 4337–47.

Morrison, S. J., A. M. Wandycz, H. D. Hemmati, D. E. Wright, and I. L. Weissman. 1997. Identification of a lineage of multipotent hematopoietic progenitors. *Development* 124(10): 1929–39.

Morrison, S. J. and I. L. Weissman. 1994. The long-term repopulating subset of hematopoietic stem-cells is deterministic and isolatable by phenotype. *Immunity* 1(8): 661–73.

Muench, M. O., J. Cupp, J. Polakoff, and M. G. Roncarolo. 1994. Expression of Cd33, Cd38, and Hla-Dr on Cd34+ human fetal liver progenitors with a high proliferative potential. *Blood* 83(11): 3170–81.

Murphy, G., B. Tjoa, H. Ragde, G. Kenny, and A. Boynton. 1996. Phase I clinical trial: T-cell therapy for prostate cancer using autologous dendritic cells pulsed with HLA-A0201-specific peptides from prostate-specific membrane antigen. *Prostate* 29(6): 371–80.

Nakano, T., H. Kodama, and T. Honjo. 1994. Generation of lymphohematopoietic cells from embryonic stem-cells in culture. *Science* 265(5175): 1098–101.

Nestle, F. O., S. Alijagic, M. Gilliet et al. 1998. Vaccination of melanoma patients with peptide- or tumor lysate-pulsed dendritic cells. *Nat Med* 4(3): 328–32.

Nielsen, L. K. 1999. Bioreactors for hematopoietic cell culture. *Annu Rev Biomed Eng* 1:129–52.

Ogawa, M., Y. Matsuzaki, S. Nishikawa et al. 1991. Expression and function of C-Kit in hematopoietic progenitor cells. *J Exp Med* 174(1): 63–71.

Ohneda, O., C. Fennie, Z. Zheng et al. 1998. Hematopoietic stem cell maintenance and differentiation are supported by embryonic aorta-gonad-mesonephros region-derived endothelium. *Blood* 92(3): 908–19.

Olsen, A. L., D. L. Stachura, and M. J. Weiss. 2006. Designer blood: Creating hematopoietic lineages from embryonic stem cells. *Blood* 107(4): 1265–75.

Orlic, D., J. Kajstura, S. Chimenti et al. 2001. Bone marrow cells regenerate infarcted myocardium. *Nature* 410(6829): 701–5.

Oswald, J., C. Steudel, K. Salchert et al. 2006. Gene-expression profiling of CD34(+) hematopoietic cells expanded in a collagen I matrix. *Stem Cells* 24(3): 494–500.

Ottersbach, K. and E. Dzierzak. 2005. The murine placenta contains hematopoietic stem cells within the vascular labyrinth region. *Dev Cell* 8(3): 377–87.

Palermo, A. T., M. A. LaBarge, R. Doyonnas, J. Pomerantz, and H. M. Blau. 2005. Bone marrow contribution to skeletal muscle: A physiological response to stress. *Dev Biol* 279(2): 336–44.

Peng, C. A. and B. O. Palsson. 1996. Cell growth and differentiation on feeder layers is predicted to be influenced by bioreactor geometry. *Biotechnol Bioeng* 50(5): 479–92.

Pessina, A., E. Mineo, M. G. Neri et al. 1992. Establishment and Characterization of a New Murine Cell-Line (Sr-4987) Derived from Marrow Stromal Cells. *Cytotechnology* 8(2): 93–102.

Plett, R. A., S. M. Frankovitz, R. Abonour, and C. M. Orschell-Traycoff. 2001. Proliferation of human hematopoietic bone marrow cells in simulated microgravity. *In Vitro Cell Dev Biol Anim* 37(2): 73–8.

Poznansky, M. C., R. H. Evans, R. B. Foxall et al. 2000. Efficient generation of human T cells from a tissue-engineered thymic organoid. *Nat Biotechnol* 18(7): 729–34.

Qiu, H. Y., Y. Fujimori, S. R. Kai et al. 2003. Establishment of mouse embryonic fibroblast cell lines that promote *ex vivo* expansion of human cord blood CD34(+) hematopoietic progenitors. *J Hematother Stem Cell Res* 12(1): 39–46.

Reichardt, V. L., C. Y. Okada, A. Liso et al. 1999. Idiotype vaccination using dendritic cells after autologous peripheral blood stem cell transplantation for multiple myeloma—A feasibility study. *Blood* 93(7): 2411–9.

Rideout, W. M. 3rd, K. Hochedlinger, M. Kyba, G. Q. Daley, and R. Jaenisch. 2002. Correction of a genetic defect by nuclear transplantation and combined cell and gene therapy. *Cell* 109(1): 17–27.

Sandstrom, C. E., J. G. Bender, W. M. Miller, and E. T. Papoutsakis. 1996. Development of novel perfusion chamber to retain nonadherent cells and its use for comparison of human "mobilized" peripheral blood mononuclear cell cultures with and without irradiated bone marrow stroma. *Biotechnol Bioeng* 50(5): 493–504.

Sasaki, T., M. Takagi, T. Soma, and T. Yoshida. 2002. 3D culture of murine hematopoietic cells with spatial development of stromal cells in nonwoven fabrics. *Cytotherapy* 4(3): 285–91.

Sato, N., I. M. Sanjuan, M. Heke et al. 2003. Molecular signature of human embryonic stem cells and its comparison with the mouse. *Dev Biol* 260(2): 404–13.

Schmitt, T. M., M. Ciofani, H. T. Petrie, and J. C. Zuniga-Pflucker. 2004. Maintenance of T cell specification and differentiation requires recurrent notch receptor-ligand interactions. *J Exp Med* 200(4): 469–79.

Schubert, H., I. Garrn, A. Berthold et al. 2004. Culture of haematopoietic cells in a 3-D bioreactor made of Al_2O_3 or apatite foam. *J Mater Sci Mater Med* 15(4): 331–4.

Solchaga, L. A., J. E. Dennis, V. M. Goldberg, and A. I. Caplan. 1999. Hyaluronic acid-based polymers as cell carriers for tissue-engineered repair of bone and cartilage. *J Orthop Res* 17(2): 205–13.

Spangrude, G. J., S. Heimfeld, and I. L. Weissman. 1988. Purification and characterization of mouse hematopoietic stem-cells. *Science* 241(4861): 58–62.

Sukoyan, M. A., S. Y. Vatolin, A. N. Golubitsa et al. 1993. Embryonic stem cells derived from morulae, inner cell mass, and blastocysts of mink: comparisons of their pluripotencies. *Mol Reprod Dev* 36(2): 148–58.

Suzuki, A. and T. Nakano. 2001. Development of hematopoietic cells from embryonic stem cells. *Int J Hematol* 73(1): 1–5.

Suzuki, T., Y. Yokoyama, K. Kumano et al. 2006. Highly efficient *ex vivo* expansion of human hematopoietic stem cells using Delta1-Fc chimeric protein. *Stem Cells* 24(11): 2456–65.

Takahashi, K. and S. Yamanaka. 2006. Induction of pluripotent stem cells from mouse embryonic and adult fibroblast cultures by defined factors. *Cell* 126(4): 663–76.

Tan, W., R. Krishnaraj, and T. A. Desai. 2001. Evaluation of nanostructured composite collagen—Chitosan matrices for tissue engineering. *Tissue Eng* 7(2): 203–10.

Taqvi, S., L. Dixit, and K. Roy. 2006. Biomaterial-based notch signaling for the differentiation of hematopoietic stem cells into T cells. *J Biomed Mater Res Part A* 79A(3): 689–97.

Taqvi, S. and K. Roy. 2006. Influence of scaffold physical properties and stromal cell coculture on hematopoietic differentiation of mouse embryonic stem cells. *Biomaterials* 27(36): 6024–31.

Tomimori, Y., M. Takagi, and T. Yoshida. 2000. The construction of an *in vitro* three-dimensional hematopoietic microenvironment for mouse bone marrow cells employing porous carriers. *Cytotechnology* 34(1–2): 121–30.

Tun, T., H. Miyoshi, T. Aung et al. 2002. Effect of growth factors on ex vivo bone marrow cell expansion using three-dimensional matrix support. *Artif Organs* 26(4): 333–9.

Umeda, K., T. Heike, M. Yoshimoto et al. 2004. Development of primitive and definitive hematopoiesis from nonhuman primate embryonic stem cells in vitro. *Development* 131(8): 1869–79.

Vassilopoulos, G., P. R. Wang, and D. W. Russell. 2003. Transplanted bone marrow regenerates liver by cell fusion. *Nature* 422(6934): 901–4.

Vieira, P. and A. Cumano. 2004. Differentiation of B lymphocytes from hematopoietic stem cells. *Methods Mol Biol* 271: 67–76.

Visnjic, D., Z. Kalajzic, D. W. Rowe et al. 2004. Hematopoiesis is severely altered in mice with an induced osteoblast deficiency. *Blood* 103(9): 3258–64.

Vunjak-Novakovic, G., I. Martin, B. Obradovic et al. 1999. Bioreactor cultivation conditions modulate the composition and mechanical properties of tissue-engineered cartilage. *J Orthop Res* 17(1): 130–8.

Vunjak-Novakovic, G., M. Radisic, and B. Obradovic. 2006. Cardiac tissue engineering: Effects of bioreactor flow environment on tissue constructs. *J Chem Technol Biotechnol* 81(4): 485–90.

Wang, J. C. Y., M. Doedens, and J. E. Dick. 1997. Primitive human hematopoietic cells are enriched in cord blood compared with adult bone marrow or mobilized peripheral blood as measured by the quantitative *in vivo* SCID-repopulating cell assay. *Blood* 89(11): 3919–24.

Wang, X., H. Willenbring, Y. Akkari et al. 2003. Cell fusion is the principal source of bone-marrow-derived hepatocytes. *Nature* 422(6934): 897–901.

Weisel, K. C., Y. Gao, J. H. Shieh, and M. A. S. Moore. 2006. Stromal cell lines from the aorta-gonado-mesonephros region are potent supporters of murine and human hematopoiesis. *Exp Hematol* 34(11): 1505–16.

Wognum, A. W., A. C. Eaves, and T. E. Thomas. 2003. Identification and isolation of hematopoietic stem cells. *Arch Med Res* 34(6): 461–75.

Xu, M. J., K. Tsuji, T. Ueda et al. 1998. Stimulation of mouse and human primitive hematopoiesis by murine embryonic aorta-gonad-mesonephros-derived stromal cell lines. *Blood* 92(6): 2032–40.

Yu, J. Y., M. A. Vodyanik, K. Smuga-Otto et al. 2007. Induced pluripotent stem cell lines derived from human somatic cells. *Science* 318(5858): 1917–20.

Zandstra, P. W., C. J. Eaves, and J. M. Piret. 1994. Expansion of hematopoietic progenitor cell populations in stirred suspension bioreactors of normal human bone marrow cells. *Biotechnology* 12(9): 909–14.

Zhang, H., K. Saeki, A. Kimura et al. 2006. Efficient and repetitive production of hematopoietic and endothelial cells from feeder-free monolayer culture system of primate embryonic stem cells. *Biol Reprod* 74(2): 295–306.

Zhang, J. W., C. Niu, L. Ye et al. 2003. Identification of the haematopoietic stem cell niche and control of the niche size. *Nature* 425(6960): 836–41.

Zhou, H. Y., S. L. Wu, J. Y. Joo et al. 2009. Generation of induced pluripotent stem cells using recombinant proteins. *Cell Stem Cell* 4(5): 381–4.

8

Synthetic Biomaterials and Stem Cells for Connective Tissue Engineering

Ameya Phadke
University of California, San Diego

Shyni Varghese
University of California, San Diego

8.1 Emergence of Stem Cells in Regenerative Medicine

Stem cells have been an extremely valuable asset to researchers in understanding the complex molecular and cellular events underlying the early tissue development, tissue repair, disease progression, epigenetics, and pathophysiology. However, the most exciting applications of stem cells are in regenerative medicine, wherein the ability of the stem cells to differentiate and contribute to tissue repair for the treatment of diseased and damaged tissue is harnessed. Several approaches have utilized directed differentiation of stem cells into tissue-specific lineages; one of the most widely explored approaches involves utilizing biomaterials with defined biochemical and physical properties to direct stem cell proliferation and/or lineage-specific differentiation, in conjunction with other soluble factors.

Stem cells have been isolated from nearly every tissue in the human body and also from embryonic sources. Embryonic stem cells (ESCs) are pluripotent (i.e., they can differentiate into all three germ layers) and were first isolated from preimplantation blastocysts (Thomson et al., 1998). Mesenchymal stem cells (MSCs) are multipotent progenitors which give rise to tissues originating in the mesoderm. As connective tissue such as bone and cartilage develop from this germ layer, these cells have commonly been used to great effect for regeneration of these tissues; in fact there are several stem cell-based therapies currently under clinical trials (www.clinicaltrials.gov). While MSCs are typically isolated from bone marrow and adipose, they have been successfully isolated from almost all adult tissues (Hwang et al., 2009). A recent advance is the development of induced pluripotent stem cells (iPSCs) (Okita et al., 2007, Takahashi and Yamanaka, 2006, Takahashi et al., 2007, Yu et al., 2007). These cells are obtained by effectively dedifferentiating terminally differentiated cells into a phenotype resembling that of ESCs and have immense potential in obtaining autologous pluripotent cells for therapeutic applications such as personalized regenerative medicine.

8.2 Role of the Extracellular Microenvironment

The extracellular environment plays an important role in regulating cell behavior. The main components of the local extracellular environment are extracellular matrix (ECM), soluble factors (chemicals, growth factors, and chemokines), and neighboring cells. The ECM is a three-dimensional hydrophilic network consisting of fibrous proteins such as collagens, fibronectin and elastin, and specialized tissue-specific components, such as increased content of glycosaminoglycans in cartilage or apatite crystals in bone. Current research in regenerative medicine has given a substantial amount of importance to mimicking these tissue-specific characteristics of the ECM using a new generation of novel biomaterials. Figure 8.1 is a schematic representation of the dynamic interactions between cells and the extracellular microenvironment.

In addition to the structural and chemical specificity, the ECM is multifunctional and dynamic in nature. It is this dynamic nature, which allows for the presentation of different cues to tissue-specific cells during development and morphogenesis in a spatio-temporal manner. An example of this was reported by Hoshiba et al., who observed that ECM obtained from early stages of osteogenesis of hMSCs had a greater ability to induce osteogenic differentiation of freshly seeded hMSCs than those obtained from later stages of osteogenesis (Hoshiba et al., 2009). It is thus evident that a highly sought after goal in regenerative medicine is the recapitulation of spatio-temporal characteristics of native ECM.

The multifunctionality of ECM along with its dynamic nature makes the development of artificial ECM extremely challenging; however, several newly developed synthetic biomaterials have utilized precoded instructive signals inspired by the ECM, to modulate various stem cell functions such as *ex vivo* expansion and tissue-specific differentiation. In order to develop synthetic biomaterials with defined chemical and physical properties for directing tissue-specific cellular response, it is crucial to understand the tissue-specific structure and composition of ECM and how they play a role in maintaining tissue-specific cellular function. In this review, we focus on the development of synthetic ECM-mimicking biomaterials, their use in guided stem cell differentiation and the subsequent impact on connective tissue engineering.

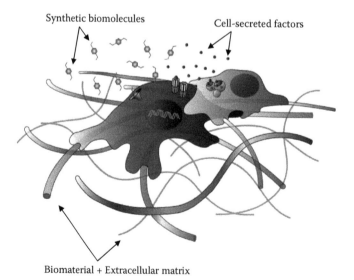

FIGURE 8.1   (See color insert.) Reciprocal interactions between cells and the extracellular microenvironment, comprising the extracellular matrix, cell-secreted factors, and other cells. (From Phadke, A., Chang, C.-W., and Varghese, S. 2010a. Functional biomaterials for controlling stem cell differentiation. In Roy, K. (Ed.) *Biomaterials as Stem Cell Niche*, Berlin/Heidelberg, Springer-Verlag, pp. 19–44.)

8.3 Biomaterial-Mediated Repair of Connective Tissue

8.3.1 Bone

8.3.1.1 Structure and Composition

Bone is a predominantly load-bearing tissue and has specialized structure to optimally perform this function. The ECM of bone is a composite, consisting of an organic phase called the osteoid, integrated with an inorganic calcium phosphate mineral phase (Weiner and Traub, 1992). The osteoid consists of chiefly collagen type I (COL1). Additionally, bone matrix is known to consist of bone sialoprotein (BSP) and osteopontin (OPN). Both these proteins contain pockets rich in anionic Asp and Glu residues, believed to promote apatite nucleation. The inorganic calcium phosphate phase appears to be semicrystalline and shows a structure similar to that of hydroxyapatite ($Ca_5(PO_4)_3(OH)$); however, Raman spectroscopic analyses of bone mineral suggest the substitution of hydroxyl groups with carbonate groups within the lattice (Gamsjäger et al., 2009).

Bone consists of three kinds of cells: osteoblasts, osteoclasts, and osteocytes. Osteoblasts are derived from mesoderm-specific progenitor cells and synthesize ECM associated with bone. Osteocytes are osteoblasts that become embedded within the bone matrix. Osteoclasts (from hematopoietic lineage) degrade bone matrix using a combination of proteases and are instrumental in resorption of bone and regulation of Ca^{2+} balance in serum. The antagonistic activity of osteoblasts and osteoclasts is critical in the remodeling of bone tissue.

8.3.1.2 Osteogenic Differentiation Mediated by Synthetic Biomaterials

8.3.1.2.1 Effects of Substrate Chemistry on Osteogenesis

A variety of biomaterials composed of both synthetic and natural polymers have been extensively employed to promote osteogenic differentiation (Hing, 2004). One of these approaches utilizes biomaterials with specific chemical groups to direct tissue-specific differentiation. Functionalization of substrates with the appropriate chemical groups ($-NH_2$, $-OH$, $-PO_4$, for example) allows for mimicking the interactions between these cells and the osteoid. A recent study by Benoit et al. demonstrates the ability of hydrogel matrices functionalized with $-PO_4$ groups to promote osteogenesis of hMSCs under two-dimensional and three-dimensional culture conditions in growth medium (without the addition of osteogenesis-inducing soluble supplements) (Benoit et al., 2008). In two-dimensional culture conditions, hMSCs cultured on phosphate-functionalized surface assumed spread morphology similar to that seen in osteoblasts. Moreover, these cells showed a significant upregulation of CBFA1, a known marker for osteogenesis. In fact, cell morphology has been shown to affect the lineages into which stem cells differentiate (Huang and Ingber, 2000, McBeath et al., 2004). In order to determine whether material chemistry promoted osteogenesis solely by influencing cell shape, cells were encapsulated in poly(ethylene glycol) (PEG) diacrylate hydrogels functionalized with $-PO_4$ groups. Interestingly, these cells also showed evidence of osteogenic differentiation (through the upregulation of CBFA1); as encapsulated cells are restricted to a rounded morphology, the authors concluded that material chemistry can affect stem cell lineage independent of any effect on morphology. The authors attributed this effect of material chemistry to the possible sequestering of osteogenesis-specific signals by the $-PO_4$ groups or effect of chemistry on serum protein conformation. In a similar study, Wang et al. demonstrated enhanced osteogenic differentiation of encapsulated goat MSCs within hydrogels containing phosphoester groups (Wang et al., 2005) Synthetic biomaterials with other functional groups have also been shown to support osteogenic differentiation of stem cells through preferential adsorption of serum proteins such as fibronectin. These groups include $-NH_2$ (Curran et al., 2005, 2006, Keselowsky et al., 2005, Phillips et al., 2009), $-SH$ groups (Curran et al., 2005, 2006), and $-OH$ groups (Keselowsky et al., 2005).

Studies have demonstrated that modifications in surface chemistry can change the conformation of adsorbed fibronectin, exposing binding domains corresponding to different cellular integrins, which in turn can selectively promote osteogenic differentiation (Keselowsky et al., 2005, Michael et al., 2003). The conformations of fibronectin adsorbed on surfaces functionalized with amine and hydroxyl groups, respectively, were found to promote activation, binding, and subsequent upregulation of integrin $\alpha5\beta1$ in immature MC3T3-E1 osteoblasts; surface carboxyl groups on the other hand, were found to induce fibronectin conformations favoring the binding and upregulation of integrin $\alpha v \beta 3$. Previous studies have shown that increased activity of integrin $\alpha5\beta1$ is associated with increase in osteoblastic activity and osteogenic differentiation (Moursi et al., 1997) while integrin $\alpha v \beta 3$ is associated with the suppression of osteoblastic phenotype (Cheng et al., 2006).

Indeed, increased osteoblastic activity was observed in cells seeded on hydroxyl- and amine-enriched surfaces when compared to carboxyl functionalized surfaces, as evidenced by increased ALP activity, matrix mineralization, and expression of osteocalcin (OCN) and BSP (Keselowsky et al., 2005). Use of $\beta1$-blocking antibodies was found to suppress matrix mineralization even for surfaces functionalized with $-OH$ and $-NH_2$, while $\beta3$-blocking antibodies promoted matrix mineralization on $-COOH$ surfaces and surprisingly, even $-OH$- and $-NH_2$-modified surfaces. In other words, this selective binding of integrins (as controlled by fibronectin conformation) was sufficient to both promote and suppress osteogenic differentiation. These findings clearly demonstrate the ability of material chemistry-mediated protein adsorptions at the matrix interface in determining the extracellular and intracellular pathways to direct lineage-specific differentiation of stem cells. Figure 8.2 shows evidence supporting the ability of the ECM to induce osteogenic differentiation.

The aforementioned studies thus clearly suggest that selective integrin activation can promote increased osteoblast activity and differentiation of immature osteoblasts. It is important to determine whether this phenomenon can trigger osteogenic differentiation of uncommitted multipotent progenitor cells. To address this question, a recent study by Hamidouche et al. reported the vital role of integrin $\alpha5$ in the osteogenic differentiation of hMSCs, induced using the glucocorticosteroid dexamathasone (Hamidouche et al., 2009). It was found that integrin $\alpha5$ is substantially upregulated in the early stages of osteogenic differentiation. Additionally, it was found that silencing the expression of integrins $\alpha5$ and $\beta1$ individually caused a substantial downregulation in the expression of osteogenic biomarkers, namely RUNX-2, ALP, and COL1A1, suggesting that silencing of these integrins led to suppression of pathways pertaining to osteogenic differentiation. Moreover, it was found that endogenous priming of integrin $\alpha5\beta1$ was sufficient to induce osteogenic differentiation; additionally, overexpression of this integrin in MSCs was found to promote osteogenesis *in vivo* upon implantation into ectopic sites in immunocompromised mice. This study thus provides overwhelming support to the hypothesis that selective activation of integrins, mediated by material chemistry, is an efficient method for directing the differentiation of mesenchymal progenitors into osteoblasts.

The adsorption of proteins onto synthetic substrates can also be influenced by alterations in surface hydrophobicity. We have recently demonstrated that small changes in hydrophobicity can have substantial effects on the adhesion and differentiation of hMSCs (Ayala et al., 2011). Polyacrylamide hydrogel matrices modified with *N*-acryloyl derivatives of aliphatic amino acids with varying length, thereby yielding pendant side chains of varying length and terminal carboxyl group were found to differentially regulate adhesion and differentiation of hMSCs. The length of the pendant side chain (and as a result, material hydrophobicity) was varied through a number of $-CH_2$ groups by synthesizing acrylamide hydrogels containing acryloyl-2-glycine or C1 (one methylene group), acryloyl-4-aminobutyric acid or C3 (three methylene groups), acryloyl-6-aminocaproic acid or C5 (five methylene groups), and acryloyl-8-aminocaprylic acid or C7 (seven methylene groups). We found that the amount of fibronectin adsorbed on these surfaces increased with increases in side chain length; however, when the number of methylene groups in the pendant side chain was increased beyond five, the amount of fibronectin adsorption showed a sharp decrease. The ability of these matrices to support adhesion and proliferation of hMSCs was found to follow the same trend, with C5 hydrogels proving to be optimally suited as an hMSC substrate. Figure 8.3 shows evidence that surface hydrophobicity can profoundly affect differentiation into osteogenic lineage.

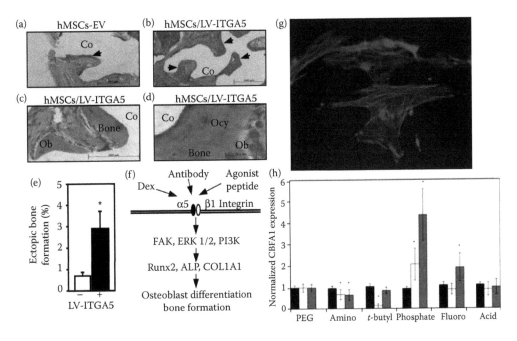

FIGURE 8.2 (**See color insert.**) (a–f) Coral/hydroxyapatite (labeled Co) scaffolds seeded with hMSCs transduced with either empty vector (EV) or LV-ITGA5 (overexpression of integrin α5), implanted into ectopic sites in nude mice. (From Hamidouche, Z. 2009. *Proc Natl Acad Sci*, 106, 18587–18591.) (a) Control cells (EV), (b–d) cells transduced with LV-ITGA5, showing forced overexpression of integrin α5. Ectopic bone is labeled pink with plump osteoblasts (Ob) and osteocytes (Ocy). (e) Quantification of ectopic bone formation of LV-ITGA5 hMSCs. Error bars represent standard deviation. (f) Proposed mechanism by which activity of α5β1 promotes osteogenic differentiation. (g, h) Osteogenic differentiation of hMSCs on phosphate-functionalized surfaces. (From Benoit, D. S. W. et al. 2008. *Nat Mater*, 7, 816–823.) (g) hMSCs cultured on phosphate functionalized hydrogels (red: actin filaments, blue: nuclei) (h) Expression of osteogenic marker CBFA1 at day 0 (black), day 4 (white), and day 5 (gray) on gels functionalized with various groups. Error bars represent standard deviation. * Indicates statistical significance as compared with PEG.

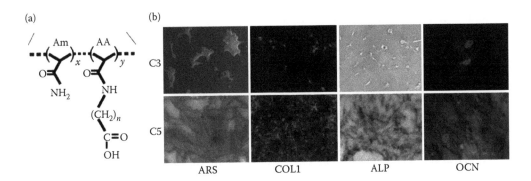

FIGURE 8.3 (**See color insert.**) Effect of tunable matrix hydrophobicity on adhesion, proliferation, and osteogenic differentiation of hMSCs. (a) Schematic representation of polymeric surfaces used. To obtain hydrogels with varying matrix hydrophobicity but identical functionality, *n* was varied from 1 to 7. (b) Histological staining hMSCs cultured on C3 and C5 gels for alkaline phosphatase (ALP) and matrix mineralization as evidence by alizarin red-S (ARS), along with immunohistochemical staining for COL1 and OCN. (From Ayala, R. et al. 2011. *Biomaterials*, 32, 3700–3711.)

C5 hydrogels were also found to optimally support osteogenic and myogenic differentiation of hMSCs in the presence of appropriate soluble factors, as evidenced by the expression of the osteogenic markers OCN, ALP, BSP, and CBFA1 in osteogenic medium, and myogenic markers MyoD, Myf5, and MHC in myogenic medium. The changes in regulation of these genes were supported by immunohistochemical staining. Our findings thus suggest that interfacial hydrophobicity can indeed affect the stem cell response to biomaterials, supported by a report by Jansen et al. which reported an effect of scaffold hydrophobicity on bone ingrowth upon implantation (Jansen et al., 2005).

In addition to small functional groups, incorporation of specific cell-binding ligands within synthetic matrices has been shown to promote osteogenic differentiation of progenitor cells (Chastain et al., 2006, Hu et al., 2003, Shin et al., 2005). In fact, Shin et al. demonstrated the ability of oligo-PEG fumerate scaffolds functionalized with arginine–glycine–aspartate peptides to induce osteogenic differentiation of rat bone marrow stromal cells even in the absence of osteogenesis-inducing components such as dexamethasone and β-glycerol phosphate (Shin et al., 2005). The authors suggested that the presence of this well-documented cell adhesion peptide in the matrix stimulated the activity of integrins specific to osteogenic commitment in a manner similar to that described above. Several other studies have also made use of RGD peptide coupled with polymer scaffolds to induce osteogenic differentiation of progenitor cells, particularly in three-dimensional culture conditions (Paletta et al., 2009, Schofer et al., 2009, Yang et al., 2001, 2005).

8.3.1.2.2 *Effect of Bone-Like Minerals on Osteogenic Differentiation*

In addition to the effect of functional groups (chemical groups, peptides, etc.) and the resultant influence on protein adsorption on osteogenic differentiation, studies have also evaluated the effect of scaffold materials containing inorganic crystalline/semicrystalline calcium phosphate minerals on osteogenic differentiation of stem cells.

As calcium phosphates mimic the mineralized microstructure of bone, it is possible that they would be capable of directing differentiation of progenitor cells into bone-specific lineages. Several studies have demonstrated the ability of calcium phosphate scaffolds to promote osteoinduction, that is, differentiation of progenitor cells (typically marrow stromal cells) into osteoblastic phenotype (Koc et al., 2008, Ohgushi et al., 1993, 1996). Interestingly, a study by Marino et al. reported the ability of β-tricalcium phosphate (TCP) scaffolds to induce osteogenic differentiation of adipose-derived stem cells in the absence of osteogenesis-inducing additives to culture medium (Marino et al., 2010). Briefly, MSCs were extracted from human adipose tissue and cultured on porous β-TCP scaffolds as well as tissue culture dishes in both growth and osteogenic medium, respectively. While cells cultured in growth medium on tissue culture dishes showed no evidence of osteogenic differentiation, cells cultured in porous β-TCP scaffolds showed evidence of osteogenic differentiation in both growth and osteogenic medium, through ALP activity as well as production of OCN and OPN. That these scaffolds can induce osteogenic differentiation in the absence of medium supplements illustrates the immense potential for use of mineralized scaffolds in the bone tissue engineering. Depending on the scaffold material in question, there are several of methods by which such an apatite-like crystalline phase can be incorporated, ranging from direct embedding to templated mineralization. For example, several ceramics have shown the ability to form a surface layer of apatite upon immersion in simulated body fluid, under conditions mimicking those observed *in vivo* (Chen et al., 2006, Kokubo, 1990, Kokubo et al., 1990). Such modification of materials with a bone-mimetic apatite layer has been shown to greatly improve their osseointegration (Ducheyne and Cuckler, 1992, Geesink et al., 1987, Hench and Paschall, 1973, Karabatsos et al., 2001). Cowan et al. have shown that apatite-coated poly(lactic-*co*-glycolic acid) (PLGA) materials promoted osteogenic differentiation of adipose-derived MSCs *in vivo* where more than 84% of implanted cells contributed to the repair of critical size mouse calvarial defects (Cowan et al., 2004). Osathanon et al. also demonstrated the ability of composite fibrin/mineral scaffolds to induce osteogenic differentiation of murine calvarial cells (Osathanon et al., 2008). This study compared the osteoinductive capacity of two kinds of composite scaffolds: scaffolds wherein crystalline nanosize hydroxyapatite particles were incorporated directly into the scaffold material and scaffolds mineralized by immersion in simulated body fluid (SBF). Scaffolds mineralized by immersion in SBF were found to promote osteogenic differentiation to a greater degree than scaffolds incorporating nanosize hydroxyapatite as well as control fibrin scaffolds, as determined by measuring expression levels of osteogenic biomarkers such as OCN, CBFA1, COL1, BSP, and osterix. This was attributed to dissolution of the mineral phase in SBF-mineralized samples, leading to increased extracellular calcium and phosphate levels. The Ca^{2+} release from SBF-mineralized scaffolds was indeed found to be greater than that from scaffolds incorporating nanosize hydroxyapatite. It is believed that dissolution of mineralized scaffold matrices leads to increased Ca^{2+} and PO_4^{3-} concentration in the cellular microenvironment. The effect of increased extracellular Ca^{2+} concentrations on the osteoblastic activity of cells was studied in depth by Dvorak et al. (2004). This study revealed that increases in extracellular Ca^{2+} levels led to upregulation of osteogenic biomarkers such as OCN, OPN, CBFA1, and COL1. This suggests that the presence of a mineralized scaffold alone may be sufficient to direct differentiation of progenitor cells into bone-specific lineages, thereby showing agreement with the findings reported by Osathonon et al. There has also been substantial experimental evidence to suggest that polymer/mineral composite scaffolds are capable of inducing osteogenic differentiation of progenitor cells (Schantz et al., 2005, Yu et al., 2009).

Mineral–polymer composite materials can also be fabricated through templated mineralization. This process promotes mineral nucleation by incorporation of specific functional groups into materials (Ball et al., 2005, Ngankam et al., 2000, Nuttelman et al., 2006, Shkilnyy et al., 2008). Mineralization of these synthetic materials often yields composites that incorporate characteristics of not only the apatite phase but also the osteoid. Moreover, such composites often exhibit a great deal of affinity between the polymer and mineral phases, thereby providing a more effective mimic of the microstructure of osseous tissue. Kretlow and Mikos have discussed the various methods by which polymers can be mineralized in great detail (Kretlow and Mikos, 2007). A popular method of nucleation of inorganic crystalline phases on polymer substrates is through modification with charged functional groups (Ball et al., 2005, Ngankam et al., 2000, Nuttelman et al., 2006, Shkilnyy et al., 2008). Additionally, our studies have shown that interfacial hydrophobicity can affect the extent of templated apatite formation on PEG-based hydrogel matrices and the topology of the mineralized phase (Phadke et al., 2010b). This is especially important as surface topology has recently been shown to affect osteogenic differentiation of progenitor cells (Dalby et al., 2007, Oh et al., 2009). Studies have also shown that poly(2-hydroxyethyl methacrylate) matrices can nucleate hydroxyapatite through thermal decomposition of urea (Song et al., 2005), exposure to simulated body fluid supplemented with serum proteins (Zainuddin et al., 2006) and mixing with COL1 (Cífková et al., 1987). The synthesis of mineral/polymer composites through templated mineralization thus represents a fairly new but extremely exciting development in the fabrication of bone-mimetic materials for directed stem cell differentiation.

8.3.1.2.3 *Effect of Substrate Mechanical Properties on Osteogenesis*

There is substantial evidence indicating that mechanical properties of material substrates can influence the lineage specificity of hMSCs (Engler et al., 2006, 2009). In other words, tissue-specific mechanical cues are capable of directing differentiation of stem cells into tissue-specific lineages. Briefly, polyacrylamide hydrogels with varying stiffness were synthesized by varying cross-linking density. Rigid matrices (elastic modulus = 34 kPa) were found to promote osteogenic differentiation of hMSCs, while moderately stiff (elastic modulus = 11 kPa) and soft matrices (elastic modulus = 0.1–1 kPa) were found to promote myogenic differentiation and neuronal differentiation, respectively.

Effects of mechanotransduction can be observed from several hierarchical levels, ranging from single cells to complex tissues. A complex pathway mediated by the triggering of transmembrane adhesion and signaling proteins allows for the translation of extracellular mechanical cues into intracellular changes (Ingber, 2006). Integrins have been shown to play a dominant role in mechanotransduction; dynamic interactions between mechanical cues from the ECM and various integrins affect cell adhesion, spreading, and subsequently differentiation (Schwartz and DeSimone, 2008). A study by Khatiwala et al. shed further light on the intricate pathways through which mechanotransduction influences osteogenesis, primarily by activation of extracellular signaling kinase (ERK) (Khatiwala et al., 2008). The authors implicated downstream ERK-mitogen activated protein kinase (MAPK) activation of RhoA-Rho associated protein kinase (ROCK) signaling pathway in the subsequent osteogenesis and found a role of integrin expression on the osteogenic differentiation of progenitors on stiffer matrices (Kundu et al., 2008). These studies amply demonstrate the effects of mechanical properties of matrices on intracellular transduction pathways and their resultant effect on differentiation of stem cells. Utilization of scaffold matrices mimicking mechanical properties of native tissue thus represents an additional mode through which stem lineage can be directed into tissue-specific lineages.

8.3.2 Cartilage

8.3.2.1 Structure and Composition

Cartilage is a load-bearing tissue that primarily functions to reduce friction at joints in the body. It consists of chondrocytes embedded within lacunae in the ECM, mainly consisting of collagen type II and proteoglycans. Collagen type II provides load-bearing capacity. Glycosaminoglycans are long unbranched polysaccharides and are highly negatively charged; this vastly increases their hydrophilicity, leading to a highly water swollen matrix. This highly swollen character lends compressive strength to the ECM. Additionally, cartilage is avascular and as a result, relies on diffusion as the main mass transfer mechanism. The avascular nature of cartilage also limits its self-repairing ability.

8.3.2.2 Biomaterial-Directed Chondrogenic Differentiation of Stem Cells

Recent efforts have borne fruit in the regeneration of cartilage using a combination of stem cells, biomaterials, and bioreactors. Due to structural similarities of hydrogels (three-dimensional elastic networks of macromolecules containing aqueous solution) with native cartilage tissue, they have been widely used as an artificial matrix for supporting cartilage tissue formation from stem cells. In addition to providing the necessary three-dimensional structural support required for the chondrogenic differentiation of progenitor cells, biomaterials have also been shown to promote chondrogenic differentiation by way of cell–material interactions (Chung and Burdick, 2008a). Both natural (mostly cartilage-specific ECM-based) and synthetic materials have been utilized for cartilage tissue engineering. For instance, a number of studies have used the ability of PEG hydrogels to render round morphology to the encapsulated cells to promote chondrogenic differentiation of both adult and ESCs (Hwang et al., 2008, Williams et al., 2003). In one such study, goat MSCs were encapsulated in PEG diacrylate-based hydrogels and cultured with and without transforming growth factor β1 (TGF-β1). Cells cultured in TGF β-positive conditions showed evidence of chondrogenic differentiation, as evidenced by production of glycosaminoglycan, type II collagen, aggrecan, and link protein. Interestingly, even the control group cultured without TGF β1 showed evidence of chondrogenic differentiation, suggesting that the similarity of the constructs to the native structure of cartilage stimulated spontaneous chondrogenic differentiation of the MSCs. However, PEG hydrogels are bioinert and nonadhesive to cells; a number of approaches have been used to develop artificial matrices with improved cell–matrix interactions by incorporating cell-adhesive moieties in synthetic materials (Elisseeff et al., 2006, Nicodemus and Bryant, 2008, Salinas and Anseth, 2008).

Some studies have incorporated charged functional groups into synthetic matrices and demonstrated their ability to induce chondrogenic differentiation. This could potentially be due to mimicking of the highly charged character of cartilage ECM components such as GAGs, by the charged synthetic matrices. For example, Benoit et al. demonstrated the ability of PEG hydrogels functionalized with –COOH to promote chondrogenic differentiation of hMSCs (Benoit et al., 2008). Cells cultured on these gels assumed a round morphology, similar to that observed for chondrocytes, Moreover, analysis of gene expression revealed that these cells showed a significant upregulation of the proteoglycan aggrecan, a chondrogenic biomarker. Similar upregulation of aggrecan was also observed upon encapsulation of these cells in –COOH-functionalized PEG hydrogels. A study by Guo et al. further evaluated the use of functionalized surfaces to promote chondrogenic differentiation of hMSCs (Guo et al., 2008). Cells were cultured on polystyrene surfaces modified with azodiphenyl derivatives of polyacrylic acid (PAAc), polyallylamine (PAAm), and PEG which presented negatively charged, positively charged, and neutral surface, respectively. Upon culturing in chondrogenic medium, these cells were found to aggregate and form pellets. When compared to control and PAAc-modified surfaces, the pellets that formed on PAAm- and PEG-modified surfaces were found to contain significantly high levels of sulfated glycosaminoglycans, type II collagen, and proteoglycans. It was suggested that in PAAm- and PEG-modified surfaces, cell–cell interactions were dominant over cell–matrix interactions, facilitating the formation of pellets conducive to chondrogenic differentiation. A similar study demonstrated the importance of cell–cell interactions in chondrogenic differentiation through seeding on an aragonite matrix derived from coral (Gross-Aviv and Vago, 2009). This matrix was coated with gold to suppress cell–matrix interactions while promoting cell–cell interactions, after which it was seeded with hMSCs; this was found to favor chondrogenesis and the maintenance of chondrogenic phenotype.

Another extensively utilized approach is the functionalization of synthetic materials with cartilage-specific ECM components as well as ECM-specific ligands (Hwang et al., 2006a,b, Park et al., 2009, Salinas et al., 2007, Varghese et al., 2008). Varghese et al. have shown that PEG hydrogels functionalized with chondroitin sulfate (CS) moieties promoted chondrogenic differentiation of encapsulated goat MSCs (Varghese et al., 2008). Interestingly, the CS moieties of the hydrogels promoted aggregation of encapsulated MSCs and mimicked various stages of cartilage *in vivo* development. Similarly, Lee et al. have demonstrated the effect of collagen-mimetic-peptide (CMP) moieties on chondrogenic differentiation of MSCs (Lee et al., 2008). The CMP moieties mimic the native collagen and have a unique collagen-like triple helical confirmation. Chung and Burdick have demonstrated the effect of hyaluronic acid-based scaffolds on the chondrogenesis of MSCs (Chung and Burdick, 2008b).

Studies have also demonstrated the effect of cell–cell interactions in the form of soluble factors on promoting chondrogenic differentiation (Levenberg et al., 2003). In a recent study by Hwang et al., mesoderm progenitor cells derived from ESCs were differentiated in this manner into chondrogenic phenotype (Hwang et al., 2008). This study consisted of two main stages: first, efficiently obtaining MSCs from ESCs and second, using these MSCs to produce functional cartilage tissue upon *in vivo* implantation. ESC-derived MSCs were cultured in medium conditioned with morphogenetic factors secreted by chondrocytes; this medium has been demonstrated to induce chondrogenic differentiation of multipotent progenitors (Hwang et al., 2007). These cells were encapsulated in PEG-based hydrogels and implanted into athymic mice. After 12 weeks postimplantation, histological sections of these constructs showed characteristics of neocartilage, such as the presence of collagen type II. Additionally, these cells were also implanted into articular cartilage defects of athymic mice; these cells seemed to induce complete cartilage repair as indicated by the absence of a border between the original defect area and neocartilage produced by the implanted cells.

A common limitation of hydrogel-based systems is their weak mechanical properties, thereby posing a challenge in their use for load-bearing applications. To address this, we have recently developed single-precursor-based biodegradable hydrogel which is more mechanically robust when compared to PEG-based hydrogels (Zhang et al., 2009). We used a triblock copolymer precursor, namely oligo (trimethylene carbonate)-block-PEG-block-oligo (trimethylene carbonate) diacrylate. Due to a critical balance of hydrophilicity of the copolymer with small hydrophobic moieties by way of oligo (trimethylene carbonate), hydrogels synthesized with this triblock copolymer showed enhanced toughness and elasticity while maintaining a swelling ratio similar to that observed in unmodified PEG hydrogels prepared from PEG diacrylate of similar weight. Moreover, this hydrogel was found to undergo both hydrolytic and enzymatic degradation with carbon dioxide being the chief degradation product. Indeed, these tough hydrogels were found to support growth of hMSCs in 2D monolayers; hMSCs encapsulated within these hydrogels also showed excellent viability, suggesting their vast potential for use as scaffold for repair of load-bearing tissues such as cartilage.

8.3.2.3 Integration of Tissue-Engineered Constructs with Native Cartilage

While several studies have demonstrated the significance of biomaterials in engineering functional cartilage tissues, it is extremely important to facilitate the integration of these engineered tissues into native cartilage upon implantation for their efficient repair. Wang et al. demonstrated the use of chondroitin sulfate, a major component of cartilage, as a bioadhesive to facilitate the integration of tissue-engineered polymer-cartilage constructs with defects in native articular cartilage (Wang et al., 2007). Figure 8.4 shows histological sections that provide evidence of this integration.

Briefly, chondroitin sulfate was functionalized with methacrylate and aldehyde groups. These functionalized molecules served as a method to link acrylate-based polymer matrices containing and the ECM of native cartilage, in combination with biocompatible initiators upon photopolymerization, wherein the aldehyde groups reacted with amines present in the native tissue while the acrylate groups reacted with cell-embedded polymer precursors. This bioadhesive was found to support growth of cells both within the construct and in the native cartilage adjacent to the defect surface. Such approaches allow for an *in vivo* application of traditional methods for synthesis of stem-cell-based polymeric cartilage constructs for effective therapeutic administration.

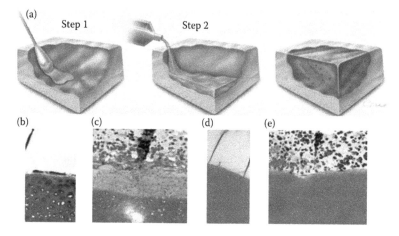

FIGURE 8.4   (See color insert.) (a–e) Functionalized chondroitin sulfate (CS) used as a bioadhesive for integration between polymeric cartilage constructs and native cartilage tissue. (a) Schematic diagram, representing (step 1) application of chondroitin sulfate adhesive and (step 2) application of polymeric precursor with suspended cells, ultimately resulting in the construct, integrated with native tissue. (b) Acellular hydrogel, attached to cartilage explant. (c) Cartilage explants attached to cell–polymer construct with formation of neocartilage. (Red: Safranin-O staining, indicating proteoglycans). (d) Acellular hydrogel attached to cartilage defect in athymic mice. (e) Cell-seeded hydrogel attached to native cartilage *in vivo*. (From Wang, D.-A. et al. 2007. *Nat Mater*, 6, 385–392.)

8.3.2.4 Maintenance of Chondrocytic Phenotype

A major challenge in current stem cell-based approaches to generate cartilaginous tissue *in vitro* is the loss of chondrocytic phenotype. Several studies have reported the terminal differentiation of chondrocytes into a phenotype resembling that of fibroblasts particularly during monolayer culture (Holtzer et al., 1960, Von Der Mark et al., 1977). Chondrocytes also have the tendency to undergo terminal differentiation. This leads to difficulty in long-term culture of chondrocytes *in vitro*. In fact, a study determined that chondrocytes obtained from *in vivo* implantation of MSCs had much lower COL10A1/COL2A1 and MMP13/COL2A1 ratios than those obtained through *in vitro* differentiation of MSCs, suggesting a much lower extent of hypertrophy in the former (Steck et al., 2009). In this study, the authors conclude that direct implantation of MSCs into cartilage defects may be a more efficient approach to cartilage repair, than their preimplantation differentiation.

As a result, there has been a surge in efforts to develop biomaterial substrates capable of maintaining collagen type X-negative (i.e., nonhypertrophic) phenotype of chondrocytes. Recently, Stokes et al. demonstrated a recovery of chondrocytic phenotype from dedifferentiated chondrocytes upon culturing on poly (2-hydroxyethyl methacrylate) (Stokes et al., 2001). In a study by Mwale et al., nitrogen doping of biaxially oriented polypropylene (BOPP) with nitrogen-rich plasma-polymerized ethylene was found to suppress collagen type X, an established marker for chondrocytic hypertrophy, in hMSCs extracted from patients suffering from osteoarthritis (Mwale et al., 2006). Plasma-treated BOPP was also found to suppress osteogenic markers such as ALP, BSP, and OCN; these markers were significantly expressed by hMSCs cultured on non-plasma-treated control surfaces. Moreover, this treatment did not affect collagen or aggrecan expression, suggesting that this treatment is an effective method of maintaining chondrocytic phenotype.

8.3.3 Tendons and Ligaments

Tendons and ligaments are tough fibrous connective tissues composed primarily of collagen fibers. Tendons serve to connect bones with muscles while ligaments primarily connect bones with other bones. Tendons and ligaments are frequently damaged during physical activity, leading to loss of function. Additionally, loss of ligament function could potentially cause joint instability, leading to osteoarthritis. As a result, there has been substantial research on the regeneration of these tissues using a combination of stem cells and specialized materials. Fibrous scaffolds have been found to efficiently mimic the structure and function of these tissues and as a result have been widely used to facilitate their repair (Wang et al., 2006). Ouyang et al. demonstrated the efficacy of knitted fibrous poly-lactide-*co*-glycolide (PLGA) scaffolds seeded with marrow stromal cells in the repair of Achilles tendons in adult white New Zealand rabbits (Ouyang et al., 2003). The scaffolds were seeded with autologous bone marrow stromal cells and implanted into Achilles tendon defects in rabbits; 8 and 12 weeks postimplantation, the newly generated tendons were found to be histologically similar to the native tendon structure and integrated well with the damaged tendons. It is important to note that there was mild inflammation due to the degradation products of the PLGA scaffolds at later time points.

Lin et al. demonstrated the use of synthetic fibrous scaffolds seeded with cells from anterior cruciate ligaments (ACL) and medial collateral ligaments (MCL) to generate ligament components *in vitro* (Lin et al., 1999). Bourke et al. also demonstrated the development of fibrous scaffolds prepared from desaminotyrosyl-tyrosamine derived polycarbonate (poly(DTE carbonate)) with potential use for engineering ligament tissue (Bourke et al., 2004). More recently, Hayami et al. demonstrated the potential of biodegradable polymeric scaffolds for use in ligament repair (Hayami et al., 2009). These scaffolds were composed of electrospun fibers of poly (ε-caprolactone-*co*-D,L-lactide) (PCLDLLA) embedded in photocrosslinked *N*-methacrylated glycol chitosan hydrogel. Fibroblasts cultured on these scaffolds were found to produce ligament markers COL1, collagen type III, and decorin along the fibers in these scaffolds, as evidenced by immunostaining. Synthetic fibrous scaffolds have thus shown immense promise in the repair of damaged tendons and ligaments.

8.3.4 Interfacial Tissue Engineering

As seen in this review, there has been considerable success in the use of biomaterial-mediated stem cell response for regeneration of separate tissues; engineering of interfaces between separate types of tissues is a much more challenging proposition. Examples of such interfaces are the osteochondral interface between bone and cartilage and the interface between bone and ligaments. This often requires the engineering of a multiphasic single construct incorporating properties of each tissue in separate regions; achieving integration tissues with markedly different properties is quite complex. In a recent review, Lu and Spalazzi have presented a detailed account of the various considerations that must be made for the successful design of scaffolds for interfacial tissue engineering (Lu and Spalazzi, 2009).

Gao et al. have reported the use of MSCs in the successful synthesis of an osteochondral construct (Gao et al., 2001). Briefly, MSCs were exposed to osteogenic medium and then seeded onto a porous calcium phosphate scaffold; another set of MSCs was exposed to TGF β1 and seeded in a porous hyaluronan-derived sponge. These two cell seeded scaffolds were then fused using fibrin sealant. Upon subcutaneous implantation into syngenic rats, this composite graft maintained its integrity as a single unit; additionally, evidence of bone ingrowth was observed in the ceramic component, while fibrocartilage was detected in the hyaluronan sponge. Significantly, observations with polarized light suggested continuity of newly synthesized collagen fibers between the two phases. Another study demonstrates the synthesis of an osteochondral graft molded into the shape of a human mandibular condoyle using sequential photoencapsulation of MSCs committed to chondrogenic and osteogenic lineages, respectively (Alhadlaq and Mao, 2005). Upon 12 weeks postimplantation into dorsum of immunodeficient mice, these constructs showed separate layers of cartilage-like and bone-like tissue, respectively. Interestingly, there was also significant cross-layer infiltration of both these tissues, indicating excellent integration between the cartilage and bone phases of the constructs. Other approaches utilized image-based design and solid free-form fabrication coupled with differential seeding of each region with cells of appropriate phenotype (Schek et al., 2004), generating scaffold anisotropy using chitosan particle aggregation (Malafaya et al., 2005) and computer-aided design followed by differential seeding (Hung et al., 2003). Several other studies have also reported the successful synthesis of osteochondral constructs. There has also been considerable research into the engineering of bone–ligament interfaces, as detailed by Moffat et al. (2009). Composite materials have been used to engineer this interface with considerable success. Recently, Paxton et al. demonstrated the potential use of a PEG diacrylate-hydroxyapatite composite functionalized with the RGD cell adhesion peptide for this purpose (Paxton et al., 2009).

8.4 Conclusions and Future Directions

The development of novel biomaterial-mediated stem cell therapies has clearly provided an enormous impetus to the advancement of regenerative medicine. Some of these advancements have led to clinical and preclinical trials. For instance, Atala et al. (2006) have tissue engineered functional bladders, using autologous cells seeded within a biodegradable scaffold made of polyglycolic acid and collagen. Similarly, vascularized bone grafts for critically sized mandible defects were tissue engineered using a custom-designed biomaterial and MSCs isolated from the patient (Warnke et al., 2004). While these achievements are encouraging, there are a number of challenges and engineering opportunities for clinically viable stem cell-based therapeutics. The development of biomaterial-based technologies will be an integral part of these advancements.

There is however much scope for the development of novel materials using novel methods such as high-throughput screening (Anderson et al., 2004, Peters et al., 2009) for optimal stem cell response. The development of self-healing biomimetic materials (Toohey et al., 2007) as well as materials exhibiting stimulus-mediated shape memory (Andreas et al., 2005, Behl and Lendlein, 2007, Feninat et al., 2002, Lendlein and Kelch, 2005, Ripamonti et al., 2007, Yakacki et al., 2007) are also promising areas for the growth of biomaterial technology; materials that exhibit some of these properties while producing a desired stem cell response could be immensely impactful on stem cell-mediated regenerative therapies. Finally, it is important to study the ability of biomaterial-presented cues to produce a favorable response in iPSCs, allowing for personalized regenerative therapies. It is thus evident that advances in biomaterial-mediated stem cell technologies have given medical professionals the tools to heal damaged tissues with unprecedented efficacy. These developments have ushered in a new exciting era of regenerative medicine with the ability to effect profound advances in healthcare in the near future.

Acknowledgments

The authors gratefully acknowledge the assistance from Dr. Nivedita Sangaj, Dr. Chao Zhang, and Dr. Ramses Ayala in the preparation of this manuscript.

References

Alhadlaq, A. and Mao, J. J. 2005. Tissue-engineered osteochondral constructs in the shape of an articular condyle. *J Bone Joint Surg Am*, 87, 936–944.

Anderson, D. G., Levenberg, S., and Langer, R. 2004. Nanoliter-scale synthesis of arrayed biomaterials and application to human embryonic stem cells. *Nat Biotech*, 22, 863–866.

Andreas, L., Annette, M. S., Michael, S., and Robert, L. 2005. Shape-memory polymer networks from oligo(ε- caprolactone)dimethacrylates. *J Poly Sci Part A: Poly Chem*, 43, 1369–1381.

Atala, A., Bauer, S. B., Soker, S., Yoo, J. J., and Retik, A. B. 2006. Tissue-engineered autologous bladders for patients needing cystoplasty. *Lancet*, 367, 1241–1246.

Ayala, R., Zhang, C., Yang, D., Hwang, Y., Aung, A., Shroff, S., Arce, F., Lal, R., Arya, G., and Varghese, S. 2011. Engineering the cell-material interface for controlling stem cell adhesion, migration, and differentiation. *Biomaterials*, 32, 3700–3711.

Ball, V., Michel, M., Boulmedais, F., Hemmerle, J., Haikel, Y., Schaaf, P., and Voegel, J. C. 2005. Nucleation kinetics of calcium phosphates on polyelectrolyte multilayers displaying internal secondary structure. *Crystal Growth Design*, 6, 327–334.

Behl, M. and Lendlein, A. 2007. Shape-memory polymers. *Mater Today*, 10, 20–28.

Benoit, D. S. W., Schwartz, M. P., Durney, A. R., and Anseth, K. S. 2008. Small functional groups for controlled differentiation of hydrogel-encapsulated human mesenchymal stem cells. *Nat Mater*, 7, 816–823.

Bourke, S. L., Kohn, J., and Dunn, M. G. 2004. Preliminary development of a novel resorbable synthetic polymer fiber scaffold for anterior cruciate ligament reconstruction. *Tissue Eng*, 10, 43–52.

Chastain, S. R., Kundu, A. K., Dhar, S., Calvert, J. W., and Putnam, A. J. 2006. Adhesion of mesenchymal stem cells to polymer scaffolds occurs via distinct ECM ligands and controls their osteogenic differentiation. *J Biomed Mater Res Part A*, 78A, 73–85.

Chen, Q. Z., Thompson, I. D., and Boccaccini, A. R. 2006. 45S5 Bioglass®-derived glass-ceramic scaffolds for bone tissue engineering. *Biomaterials*, 27, 2414–2425.

Cheng, S.-L., Lai, C.-F., Blystone, S. D., and Avioli, L. V. 2006. Bone mineralization and osteoblast differentiation are negatively modulated by integrin $\alpha_v\beta_3$. *J Bone Min Res*, 16, 277–288.

Chung, C. and Burdick, J. A. 2008a. Engineering cartilage tissue. *Adv Drug Deliv Rev*, 60, 243–262.

Chung, C. and Burdick, J. A. 2008b. Influence of three-dimensional hyaluronic acid microenvironments on mesenchymal stem cell chondrogenesis. *Tissue Eng Part A*, 15, 243–254.

Cífková, I., Stol, M., Holusa, R., and Adam, M. 1987. Calcification of poly(2-hydroxyethyl methacrylate)-collagen composites implanted in rats. *Biomaterials*, 8, 30–34.

Cowan, C. M., Shi, Y.-Y., Aalami, O. O., Chou, Y.-F., Mari, C., Thomas, R., Quarto, N., Contag, C. H., Wu, B., and Longaker, M. T. 2004. Adipose-derived adult stromal cells heal critical-size mouse calvarial defects. *Nat Biotech*, 22, 560–567.

Curran, J. M., Chen, R., and Hunt, J. A. 2005. Controlling the phenotype and function of mesenchymal stem cells *in vitro* by adhesion to silane-modified clean glass surfaces. *Biomaterials*, 26, 7057–7067.

Curran, J. M., Chen, R., and Hunt, J. A. 2006. The guidance of human mesenchymal stem cell differentiation *in vitro* by controlled modifications to the cell substrate. *Biomaterials*, 27, 4783–4793.

Dalby, M. J., Gadegaard, N., Tare, R., Andar, A., Riehle, M. O., Herzyk, P., Wilkinson, C. D. W., and Oreffo, R. O. C. 2007. The control of human mesenchymal cell differentiation using nanoscale symmetry and disorder. *Nat Mater*, 6, 997–1003.

Ducheyne, P. and Cuckler, J. M. 1992. Bioactive ceramic prosthetic coatings. *Clin Orth Rel Res*, 276, 102–114.

Dvorak, M. M., Siddiqua, A., Ward, D. T., Carter, D. H., Dallas, S. L., Nemeth, E. F., and Riccardi, D. 2004. Physiological changes in extracellular calcium concentration directly control osteoblast function in the absence of calciotropic hormones. *Proc Natl Acad Sci USA*, 101, 5140–5145.

Elisseeff, J., Ferran, A., Hwang, S., Varghese, S., and Zhang, Z. 2006. The role of biomaterials in stem cell differentiation: Applications in the musculoskeletal system. *Stem Cells Dev*, 15, 295–303.

Engler, A. J., Humbert, P. O., Wehrle-Haller, B., and Weaver, V. M. 2009. Multiscale modeling of form and function. *Science*, 324, 208–212.

Engler, A. J., Sen, S., Sweeney, H. L., and Discher, D. E. 2006. Matrix elasticity directs stem cell lineage specification. *Cell*, 126, 677–689.

Feninat, F. E., Laroche, G., Fiset, M., and Mantovani, D. 2002. Shape memory materials for biomedical applications. *Adv Eng Mater*, 4, 91–104.

Gamsjäger, S., Kazanci, M., Paschalis, E. P., and Fratzl, P. 2009. Raman application in bone imaging. In Amer, M. S. (Ed.) *Raman Spectroscopy for Soft Matter Applications*. Hoboken, New Jersey, John Wiley and Sons, Inc.

Gao, J., Dennis, J. E., Solchaga, L. A., Awadallah, A. S., Goldberg, V. M., and Caplan, A. I. 2001. Tissue-engineered fabrication of an osteochondral composite graft using rat bone marrow-derived mesenchymal stem cells. *Tissue Eng*, 7, 363–371.

Geesink, R. G. T., Groot, K. D., and Klein, C. P. A. T. 1987. Chemical implant fixation using hydroxyl-apatite coatings: The development of a human total hip prosthesis for chemical fixation to bone using hydroxyl-apatite coatings on titanium substrates. *Clin Orth Rel Res*, 225, 147–170.

Gross-Aviv, T. and Vago, R. 2009. The role of aragonite matrix surface chemistry on the chondrogenic differentiation of mesenchymal stem cells. *Biomaterials*, 30, 770–779.

Guo, L., Kawazoe, N., Fan, Y., Ito, Y., Tanaka, J., Tateishi, T., Zhang, X., and Chen, G. 2008. Chondrogenic differentiation of human mesenchymal stem cells on photoreactive polymer-modified surfaces. *Biomaterials*, 29, 23–32.

Hamidouche, Z., Fromigue, O., Ringe, J., Haupl, T., Vaudin, P., Pages, J.-C., Srouji, S., Livne, E., and Marie, P. J. 2009. Priming integrin a5 promotes human mesenchymal stromal cell osteoblast differentiation and osteogenesis. *Proc Natl Acad Sci USA*, 106, 18587–18591.

Hayami, J. W. S., Surrao, D. C., Waldman, S., and Amsden, B. G. 2009. Design and characterization of a biodegradable composite scaffold for ligament tissue engineering. *J Biomed Mater Res Part A*, 92A, 1407–1420.

Hench, L. L., and Paschall, H. A. 1973. Direct chemical bond of bioactive glass-ceramic materials to bone and muscle. *J Biomed Mater Res*, 7, 25–42.

Hing, K. A. 2004. Bone repair in the twenty-first century: Biology, chemistry or engineering? *Philos Trans R Soc Lond A Math Phys Eng Sci*, 362, 2821–2850.

Holtzer, H., Abbott, J., Lash, J., and Holtzer, S. 1960. The loss of phenotypic traits by differentiated cells *in vitro*, I. dedifferentiation of cartilage cells. *Proc Natl Acad Sci USA*, 46, 1533–1542.

Hoshiba, T., Kawazoe, N., Tateishi, T., and Chen, G. 2009. Development of stepwise osteogenesis-mimicking matrices for the regulation of mesenchymal stem cell functions. *J Biol Chem*, 284, 31164–31173.

Hu, Y., S. Winn, Krajbich, I., and Hollinger, J. O. 2003. Porous polymer scaffolds surface-modified with arginine-glycine-aspartic acid enhance bone cell attachment and differentiation *in vitro. J Biomed Mater Res*, 64A, 583–590.

Huang, S. and Ingber, D. E. 2000. Shape-dependent control of cell growth, differentiation, and apoptosis: Switching between attractors in cell regulatory networks. *Exp Cell Res*, 261, 91–103.

Hung, C. T., Lima, E. G., Mauck, R. L., Taki, E., Leroux, M. A., Lu, H. H., Stark, R. G., Guo, X. E., and Ateshian, G. A. 2003. Anatomically shaped osteochondral constructs for articular cartilage repair. *J Biomech*, 36, 1853–1864.

Hwang, N. S., Varghese, S., Lee, H. J., Zhang, Z., Ye, Z., Bae, J., Cheng, L., and Elisseeff, J. 2008. *In vivo* commitment and functional tissue regeneration using human embryonic stem cell-derived mesenchymal cells. *Proc Natl Acad Sci USA*, 105, 20641–20646.

Hwang, N. S., Varghese, S., Puleo, C., Zhang, Z., and Elisseeff, J. 2007. Morphogenetic signals from chondrocytes promote chondrogenic and osteogenic differentiation of mesenchymal stem cells. *J Cell Phy*, 212, 281–284.

Hwang, N. S., Varghese, S., Theprungsirikul, P., Canver, A., and Elisseeff, J. 2006a. Enhanced chondrogenic differentiation of murine embryonic stem cells in hydrogels with glucosamine. *Biomaterials*, 27, 6015–6023.

Hwang, N. S., Varghese, S., Zhang, Z., and Elisseeff, J. 2006b. Chondrogenic differentiation of human embryonic stem cell-derived cells in arginine-glycine-aspartate-modified hydrogels. *Tissue Eng*, 12, 2695–2706.

Hwang, N. S., Zhang, C., Hwang, Y.-S., and Varghese, S. 2009. Mesenchymal stem cell differentiation and roles in regenerative medicine. *Wiley Interdisciplinary Rev: Syst Biol Med*, 1, 97–106.

Ingber, D. E. 2006. Cellular mechanotransduction: Putting all the pieces together again. *FASEB J*, 20, 811–827.

Jansen, E. J. P., Sladek, R. E. J., Bahar, H., Yaffe, A., Gijbels, M. J., Kuijer, R., Bulstra, S. K., Guldemond, N. A., Binderman, I., and Koole, L. H. 2005. Hydrophobicity as a design criterion for polymer scaffolds in bone tissue engineering. *Biomaterials*, 26, 4423–4431.

Karabatsos, B., Myerthall, S. L., Fornasier, V. L., Binnington, A., and Maistrelli, G. L. 2001. Osseointegration of hydroxyapatite porous-coated femoral implants in a canine model. *Clin Orth Rel Res*, 392, 442–449.

Keselowsky, B. G., Collard, D. M., and Garcia, A. J. 2005. Integrin binding specificity regulates biomaterial surface chemistry effects on cell differentiation. *PNAS*, 102, 5953–5957.

Khatiwala, C. B., Kim, P. D., Peyton, S. R., and Putnam, A. J. 2008. ECM compliance regulates osteogenesis by influencing mapk signaling downstream of RhoA and rock. *J Bone Min Res*, 24, 886–898.

Koc, A., Emin, N., Elcin, A. E., and Elcin, Y. M. 2008. *In vitro* osteogenic differentiation of rat mesenchymal stem cells in a microgravity bioreactor. *J Bioactive Compatible Polym*, 23, 244–261.

Kokubo, T. 1990. Surface chemistry of bioactive glass-ceramics. *J Non-Crys Solids*, 120, 138–151.

Kokubo, T., Ito, S., Huang, Z. T., Hayashi, T., Sakka, S., Kitsugi, T., and Yamamuro, T. 1990. Ca, P-rich layer formed on high-strength bioactive glass-ceramic A-W. *J Biomed Mater Res*, 24, 331–343.

Kretlow, J. D. and Mikos, A. G. 2007. Review: Mineralization of synthetic polymer scaffolds for bone tissue engineering. *Tissue Eng*, 13, 927–938.

Kundu, A. K., Khatiwala, C. B., and Putnam, A. J. 2008. Extracellular matrix remodeling, integrin expression, and downstream signaling pathways influence the osteogenic differentiation of mesenchymal stem cells on poly(lactide-*co*-glycolide) substrates. *Tissue Eng Part A*, 15, 273–283.

Lee, H. J., Yu, C., Chansakul, T., Hwang, N. S., Varghese, S., Yu, S. M., and Elisseeff, J. H. 2008. Enhanced chondrogenesis of mesenchymal stem cells in collagen mimetic peptide-mediated microenvironment. *Tissue Eng Part A*, 14, 1843–1851.

Lendlein, A. and Kelch, S. 2005. Shape-memory polymers as stimuli-sensitive implant materials. *Clin Hemorheol Microcirculation*, 32, 105–116.

Levenberg, S., Huang, N. F., Lavik, E., Rogers, A. B., Itskovitz-Eldor, J., and Langer, R. 2003. Differentiation of human embryonic stem cells on three-dimensional polymer scaffolds. *Proc Natl Acad Sci USA*, 100, 12741–12746.

Lin, V. S., Lee, M. C., O'neal, S., Mckean, J., and Sung, K. L. P. 1999. Ligament tissue engineering using synthetic biodegradable fiber scaffolds. *Tissue Eng*, 5, 443–451.

Lu, H. H. and Spalazzi, J. P. 2009. Biomimetic stratified scaffold design for ligament-to-bone interface tissue engineering. *Combinatorial Chemistry & High Throughput Screening*, 12, 589–597.

Malafaya, P., Pedro, A., Peterbauer, A., Gabriel, C., Redl, H., and Reis, R. 2005. Chitosan particles agglomerated scaffolds for cartilage and osteochondral tissue engineering approaches with adipose tissue derived stem cells. *J Mater Sci Mater Med*, 16, 1077–1085.

Marino, G., Rosso, F., Cafiero, G., Tortora, C., Moraci, M., Barbarisi, M., and Barbarisi, A. 2010. β-Tricalcium phosphate 3D scaffold promote alone osteogenic differentiation of human adipose stem cells: *In vitro* study. *J Mater Sci Mater Med*, 21, 353–363.

McBeath, R., Pirone, D. M., Nelson, C. M., Bhadriraju, K., and Chen, C. S. 2004. Cell shape, cytoskeletal tension, and RhoA regulate stem cell lineage commitment. *Dev Cell*, 6, 483–495.

Michael, K. E., Vernekar, V. N., Keselowsky, B. G., Meredith, J. C., Latour, R. A., and Garcia, A. J. 2003. Adsorption-induced conformational changes in fibronectin due to interactions with well-defined surface chemistries. *Langmuir*, 19, 8033–8040.

Moffat, K. L., Wang, I. N. E., Rodeo, S. A., and Lu, H. H. 2009. Orthopedic interface tissue engineering for the biological fixation of soft tissue grafts. *Clin Sports Med*, 28, 157–176.

Moursi, A. M., Globus, R. K., and Damsky, C. H. 1997. Interactions between integrin receptors and fibronectin are required for calvarial osteoblast differentiation *in vitro*. *J Cell Sci*, 110, 2187–2196.

Mwale, F., Girard-Lauriault, P.-L., Wang, H. T., Lerouge, S., Antoniou, J., and Wertheimer, M. R. 2006. Suppression of genes related to hypertrophy and osteogenesis in committed human mesenchymal stem cells cultured on novel nitrogen-rich plasma polymer coatings. *Tissue Eng*, 12, 2639–2647.

Ngankam, P. A., Lavalle, P., Szyk, L., Decher, G., Schaaf, P., and Cuisinier, F. J. G. 2000. Influence of polyelectrolyte multilayer films on calcium phosphate nucleation. *J Am Chem Soc*, 122, 8998–9005.

Nicodemus, G. D., and Bryant, S. J. 2008. Cell encapsulation in biodegradable hydrogels for tissue engineering applications. *Tissue Eng Part B: Rev*, 14, 149–165.

Nuttelman, C. R., Benoit, D. S. W., Tripodi, M. C., and Anseth, K. S. 2006. The effect of ethylene glycol methacrylate phosphate in Peg hydrogels on mineralization and viability of encapsulated hMSCs. *Biomaterials*, 27, 1377–1386.

Oh, S., Brammer, K. S., Li, Y. S. J., Teng, D., Engler, A. J., Chien, S., and Jin, S. 2009. Stem cell fate dictated solely by altered nanotube dimension. *Proc Natl Acad Sci USA*, 106, 2130–2135.

Ohgushi, H., Dohi, Y., Tamai, S., and Tabata, S. 1993. Osteogenic differentiation of marrow stromal stem cells in porous hydroxyapatite ceramics. *J Biomed Mater Res*, 27, 1401–1407.

Ohgushi, H., Dohi, Y., Yoshikawa, T., Tamai, S., Tabata, S., Okunaga, K., and Shibuya, T. 1996. Osteogenic differentiation of cultured marrow stromal stem cells on the surface of bioactive glass ceramics. *J Biomed Mater Res*, 32, 341–348.

Okita, K., Ichisaka, T., and Yamanaka, S. 2007. Generation of germline-competent induced pluripotent stem cells. *Nature*, 448, 313–317.

Osathanon, T., Linnes, M. L., Rajachar, R. M., Ratner, B. D., Somerman, M. J., and Giachelli, C. M. 2008. Microporous nanofibrous fibrin-based scaffolds for bone tissue engineering. *Biomaterials*, 29, 4091–4099.

Ouyang, H. W., Goh, J. C. H., Thambyah, A., Teoh, S. H., and Lee, E. H. 2003. Knitted poly-lactide-*co*-glycolide scaffold loaded with bone marrow stromal cells in repair and regeneration of rabbit achilles tendon. *Tissue Eng*, 9, 431–439.

Paletta, J., Bockelmann, S., Walz, A., Theisen, C., Wendorff, J., Greiner, A., Fuchs-Winkelmann, S., and Schofer, M. 2009. RGD-functionalization of PLLA nanofibers by surface coupling using plasma treatment: Influence on stem cell differentiation. *J Mater Sci Mater Med*, 21, 1363–1369.

Park, H., Guo, X., Temenoff, J. S., Tabata, Y., Caplan, A. I., Kasper, F. K., and Mikos, A. G. 2009. Effect of swelling ratio of injectable hydrogel composites on chondrogenic differentiation of encapsulated rabbit marrow mesenchymal stem cells *in vitro*. *Biomacromolecules*, 10, 541–546.

Paxton, J. Z., Donnelly, K., Keatch, R. P., and Baar, K. 2009. Engineering the bone-ligament interface using polyethylene glycol diacrylate incorporated with hydroxyapatite. *Tissue Eng Part A*, 15, 1201–1209.

Peters, A., Brey, D. M., and Burdick, J. A. 2009. High-throughput and combinatorial technologies for tissue engineering applications. *Tissue Eng Part B: Rev*, 15, 225–239.

Phadke, A., Chang, C.-W., and Varghese, S. 2010a. Functional biomaterials for controlling stem cell differentiation. In Roy, K. (Ed.) *Biomaterials as Stem Cell Niche*, Berlin/Heidelberg, Springer-Verlag, pp. 19–44.

Phadke, A., Zhang, C., Hwang, Y., Vecchio, K., and Varghese, S. 2010b. Templated mineralization of synthetic hydrogels for bone-like composite materials: Role of matrix hydrophobicity. *Biomacromolecules*, 11, 2060–2068.

Phillips, J. E., Petrie, T. A., Creighton, F. P., and García, A. J. 2009. Human mesenchymal stem cell differentiation on self-assembled monolayers presenting different surface chemistries. *Acta Biomater*, 6, 12–20.

Ripamonti, U. M. D. P. D., Richter, P. W. P. D., and Thomas, M. E. P. D. 2007. Self-inducing shape memory geometric cues embedded within smart hydroxyapatite-based biomimetic matrices. *Plastic Recons Surg*, 120, 1796–1807.

Salinas, C. N., and Anseth, K. S. 2008. The enhancement of chondrogenic differentiation of human mesenchymal stem cells by enzymatically regulated Rgd functionalities. *Biomaterials*, 29, 2370–2377.

Salinas, C. N., Cole, B. B., Kasko, A. M., and Anseth, K. S. 2007. Chondrogenic differentiation potential of human mesenchymal stem cells photoencapsulated within poly(ethylene glycol)-arginine-glycine-aspartic acid-serine thiol-methacrylate mixed-mode networks. *Tissue Eng*, 13, 1025–1034.

Schantz, J.-T., Brandwood, A., Hutmacher, D., Khor, H., and Bittner, K. 2005. Osteogenic differentiation of mesenchymal progenitor cells in computer designed fibrin-polymer-ceramic scaffolds manufactured by fused deposition modeling. *J Mater Sci Mater Med*, 16, 807–819.

Schek, R. M., Taboas, J. M., Segvich, S. J., Hollister, S. J., and Krebsbach, P. H. 2004. Engineered osteochondral grafts using biphasic composite solid free-form fabricated scaffolds. *Tissue Eng*, 10, 1376–1385.

Schofer, M., Boudriot, U., Bockelmann, S., Walz, A., Wendorff, J., Greiner, A., Paletta, J., and Fuchs-Winkelmann, S. 2009. Effect of direct Rgd incorporation in Plla nanofibers on growth and osteogenic differentiation of human mesenchymal stem cells. *J Mater Sci Mater Med*, 20, 1535–1540.

Schwartz, M. A. and Desimone, D. W. 2008. Cell adhesion receptors in mechanotransduction. *Curr Opin Cell Biol*, 20, 551–556.

Shin, H., Temenoff, J. S., Bowden, G. C., Zygourakis, K., Farach-Carson, M. C., Yaszemski, M. J., and Mikos, A. G. 2005. Osteogenic differentiation of rat bone marrow stromal cells cultured on Arg-Gly-Asp modified hydrogels without dexamethasone and [beta]-glycerol phosphate. *Biomaterials*, 26, 3645–3654.

Shkilnyy, A., Friedrich, A., Tiersch, B., Schone, S., Fechner, M., Koetz, J., Schlapfer, C.-W., and Taubert, A. 2008. Poly(ethylene imine)-controlled calcium phosphate mineralization. *Langmuir*, 24, 2102–2109.

Song, J., Malathong, V., and Bertozzi, C. R. 2005. Mineralization of synthetic polymer scaffolds: A bottom-up approach for the development of artificial bone. *J Am Chem Soc*, 127, 3366–3372.

Steck, E., Fischer, J., Lorenz, H., Gotterbarm, T., Jung, M., and Richter, W. 2009. Mesenchymal stem cell differentiation in an experimental cartilage defect: Restriction of hypertrophy to bone-close neocartilage. *Stem Cells Dev*, 18, 969–978.

Stokes, D. G., Liu, G., Dharmavaram, R., Hawkins, D., Piera-Velazquez, S., and Jimenez, S. A. 2001. Regulation of type-II collagen gene expression during human chondrocyte de-differentiation and recovery of chondrocyte-specific *Biochem J*, 360, 461–470.

Takahashi, K. and Yamanaka, S. 2006. Induction of pluripotent stem cells from mouse embryonic and adult fibroblast cultures by defined factors. *Cell*, 126, 663–676.

Takahashi, K., Tanabe, K., Ohnuki, M., Narita, M., Ichisaka, T., Tomoda, K., and Yamanaka, S. 2007. Induction of pluripotent stem cells from adult human fibroblasts by defined factors. *Cell*, 131, 861–872.

Thomson, J. A., Itskovitz-Eldor, J., Shapiro, S. S., Waknitz, M. A., Swiergiel, J. J., Marshall, V. S., and Jones, J. M. 1998. Embryonic stem cell lines derived from human blastocysts. *Science*, 282, 1145–1147.

Toohey, K. S., Sottos, N. R., Lewis, J. A., Moore, J. S., and White, S. R. 2007. Self-healing materials with microvascular networks. *Nat Mater*, 6, 581–585.

Varghese, S., Hwang, N. S., Canver, A. C., Theprungsirikul, P., Lin, D. W., and Elisseeff, J. 2008. Chondroitin sulfate based niches for chondrogenic differentiation of mesenchymal stem cells. *Matr Biol*, 27, 12–21.

Von Der Mark, K., Gauss, V., Von Der Mark, H., and Muller, P. 1977. Relationship between cell shape and type of collagen synthesised as chondrocytes lose their cartilage phenotype in culture. *Nature*, 267, 531–532.

Wang, D.-A., Varghese, S., Sharma, B., Strehin, I., Fermanian, S., Gorham, J., Fairbrother, D. H., Cascio, B., and Elisseeff, J. H. 2007. Multifunctional chondroitin sulphate for cartilage tissue-biomaterial integration. *Nat Mater*, 6, 385–392.

Wang, D.-A., Williams, C. G., Yang, F., Cher, N., Lee, H., and Elisseeff, J. H. 2005. Bioresponsive phospho-ester hydrogels for bone tissue engineering. *Tissue Eng*, 11, 201–213.

Wang, Y., Kim, H.-J., Vunjak-Novakovic, G., and Kaplan, D. L. 2006. Stem cell-based tissue engineering with silk biomaterials. *Biomaterials*, 27, 6064–6082.

Warnke, P. H., Springer, I. N. G., Wiltfang, J., Acil, Y., Eufinger, H., Wehmöller, M., Russo, P. A. J., Bolte, H., Sherry, E., Behrens, E., and Terheyden, H. 2004. Growth and transplantation of a custom vascularised bone graft in a man. *Lancet*, 364, 766–770.

Weiner, S. and Traub, W. 1992. Bone structure: From angstroms to microns. *FASEB J.*, 6, 879–885.

Williams, C. G., Kim, T. K., Taboas, A., Malik, A., Manson, P., and Elisseeff, J. 2003. *In vitro* chondrogenesis of bone marrow-derived mesenchymal stem cells in a photopolymerizing hydrogel. *Tissue Eng*, 9, 679–688.

Yakacki, C. M., Shandas, R., Lanning, C., Rech, B., Eckstein, A., and Gall, K. 2007. Unconstrained recovery characterization of shape-memory polymer networks for cardiovascular applications. *Biomaterials*, 28, 2255–2263.

Yang, F., Williams, C. G., Wang, D.-A., Lee, H., Manson, P. N., and Elisseeff, J. 2005. The effect of incorporating Rgd adhesive peptide in polyethylene glycol diacrylate hydrogel on osteogenesis of bone marrow stromal cells. *Biomaterials*, 26, 5991–5998.

Yang, X. B., Roach, H. I., Clarke, N. M. P., Howdle, S. M., Quirk, R., Shakesheff, K. M., and Oreffo, R. O. C. 2001. Human osteoprogenitor growth and differentiation on synthetic biodegradable structures after surface modification. *Bone*, 29, 523–531.

Yu, H.-S., Hong, S.-J., and Kim, H.-W. 2009. Surface-mineralized polymeric nanofiber for the population and osteogenic stimulation of rat bone-marrow stromal cells. *Mater Chem Phys*, 113, 873–877.

Yu, J., Vodyanik, M. A., Smuga-Otto, K., Antosiewicz-Bourget, J., Frane, J. L., Tian, S., Nie, J. et al. 2007. Induced pluripotent stem cell lines derived from human somatic cells. *Science*, 318, 1917–1920.

Zainuddin, Hill, D. J. T., Chirila, T. V., Whittaker, A. K., and Kemp, A. 2006. Experimental calcification of HEMA-based hydrogels in the presence of albumin and a comparison to the *in vivo* calcification. *Biomacromolecules*, 7, 1758–1765.

Zhang, C., Aung, A., Liao, L., and Varghese, S. 2009. A novel single precursor-based biodegradable hydrogel with enhanced mechanical properties. *Soft Matter*, 5, 3831–3834.

9

Derivation and Expansion of Human Pluripotent Stem Cells

Sean P. Palecek
*University of Wisconsin,
Madison*
WiCell Research Institute

9.1 Applications of Human Pluripotent Stem Cells

Pluripotent stem cells, including embryonic stem cells (ESCs) and induced pluripotent stem cells (iPSCs), possess a unique combination of two properties, pluripotency and a theoretically unlimited self-renewal capacity. ESCs are derived from the inner cell mass (ICM) of a human blastocyst, while iPSCs are generated by expressing transcription factors that regulate pluripotency in somatic cells. Both ESCs and iPSCs have the potential to serve as a source of large quantities of any cell type, originating from a clonal cell source, for a diverse set of applications in basic biology and regenerative medicine.

The greatest impact of human pluripotent stem cells in the near term is their ability to provide a model system to study human development *in vitro*. For ethical and practical reasons, developmental biology relies on animal model systems and human cell lines or primary tissues. However, model systems do not always accurately recapitulate human development, and cell lines and primary tissue often behave differently *in vitro* than *in vivo*, and cannot access all stages of development. Human ESCs and iPSCs provide an *in vitro* model system to study development from the very earliest stages through terminal differentiation to specialized cells. Examples of disease modeling include generation of iPSCs from patients suffering from a familial form of amytrophic lateral sclerosis (ALS; Lou Gehrig's disease) and patients with inherited spinal muscular atrophy (Dimos et al. 2008; Ebert et al. 2009). These iPSCs were able to differentiate to motor neurons, the cell types afflicted by these two diseases, and provide a model for understanding the development of these diseases as well as a source of easily accessible human tissue for characterizing the effects of this disease and for screening or testing novel treatments. Disease-specific iPSCs have also been generated from patients with adenosine deaminase deficiency-related severe combined immunodeficiency (ADA-SCID), Shwachman–Bodian–Diamond syndrome (SBDS), Gaucher disease (GD) type III, Duchenne (DMD) and Becker muscular dystrophy (BMD), Parkinson disease (PD), Huntington disease (HD), juvenile-onset type 1 diabetes mellitus (JDM), Down syndrome (DS)/trisomy 21, the carrier state of Lesch–Nyhan syndrome, Fanconi anemia, various myeloproliferative disorders, and dyskeratosis congenita (Agarwal et al. 2010; Maehr et al. 2009; Park et al. 2008a; Raya et al. 2009; Ye et al. 2009).

Another near-term utility of human pluripotent stem cells is in drug screening or toxicology applications. Pluripotent stem cells themselves may serve as a model to assess the effects of compounds on development, or alternatively may provide an *in vitro* source of normal, human cells that are difficult to obtain from primary sources or cannot be maintained or expanded in culture. Cardiac myocytes, hepatocytes, neurons, and embryonic tissues offer particular commercialization prospects as model systems for toxicology testing (Cezar 2007; Freund and Mummery 2009; Snykers et al. 2009; Winkler et al. 2009).

A longer-term promise exists for using pluripotent-derived stem cells in regenerative medicine and tissue engineering applications. Pluripotent stem cells have the capacity to restore the structure and functions of cells, tissues, and organs. However, substantial technical hurdles face implementing these strategies. First, effective and efficient methods to differentiate the pluripotent stem cell *in vitro* or *in vivo* must be developed. The delivery of cells to the desired site and assembly of cells into the appropriate three-dimensional (3D) structure assemblies must be improved. The genetic stability of pluripotent stem cells and their progeny, and the teratomagenicity of undifferentiated pluripotent stem cells remain a concern.

The field of pluripotent stem cells is relatively young and rapidly changing. This chapter will summarize the current techniques to generate human pluripotent stem cells and describe common methods for expanding these cell populations. These technologies are crucial for translating advances in human pluripotent stem cell biology to applications *in vitro* and *in vivo*.

9.2 Deriving Human Embryonic Stem Cells

The first human pluripotent stem cells, hESCs, were derived by James Thomson in 1998 by isolation of the blastocyst ICM and subsequent selection and cultivation of the ICM cells in coculture with mouse embryonic fibroblast (MEF) feeder cells (Thomson et al. 1998). In this study, five hESC lines were established from 14 ICMs. Isolation and culture of the blastocyst ICM remains the gold standard for deriving new hESC lines, but hESCs have also been derived at earlier developmental stages, such as from blastomeres from the four-cell stage or the 16–32-cell morula (Geens et al. 2009; Klimanskaya et al. 2006; Strelchenko et al. 2004; Zhang et al. 2006). The efficiency of deriving hESC lines from the blastomeres appears to be lower than the efficiency from the ICM. It is not yet clear whether differences in the donor cell developmental stage affects growth or differentiation potential of the resulting hESC line; blastomeres and ICM cells possess distinct DNA methylation patterns and histone modifications (Fulka et al. 2008). A systematic study of the timing of ICM isolation from the blastocyst on the efficiency of hESC line derivation revealed an optimum yield of pluripotent cell lines at day 6 postfertilization (Chen et al. 2009). In general, the likelihood of success in obtaining an hESC line from a blastomere or blastocyst depends on the quality of the embryo and its storage. Embryos with a low morphological grade and arrest prior to the blastocyst stage do not efficiently yield hESC lines, but poor-quality embryos that become blastocysts still have the potential to generate hESC lines (Lerou et al. 2008).

As a result of ethical concerns regarding the destruction of human embryos during the establishment of hESC lines, substantial effort has been directed toward developing technologies that generate hESC lines without damaging the developmental potential of the embryo. Single blastomeres can give rise to hESC lines (Geens et al. 2009; Klimanskaya et al. 2006). Blastomeres isolated from a morula have been used to generate hESC lines, and the morula was able to further develop to the blastocyst stage *in vitro* (Chung et al. 2008). While none of these morulas was allowed to develop into an organism, the method of cell harvest is similar to that used to obtain cells for preimplantation genetic diagnosis (PGD) and is likely to be safe.

Most embryos used to generate blastomeres and blastocysts for hESC line derivation are produced by *in vitro* fertilization. Somatic cell nuclear transfer (SCNT) provides an alternative approach to generating blastomeres and blastocysts. The nuclei of hESCs and adult primary fibroblasts have been transferred to unfertilized human eggs to create blastocyst-stage embryos (French et al. 2008; Stojkovic et al. 2005). While this method has been used to establish animal ESC lines, it has not yet been demonstrated in humans. If the technical and ethical concerns surrounding therapeutic cloning can be addressed, nuclear transfer may allow the establishment of patient-specific hESC lines.

Parthenogenesis, stimulation of an unfertilized oocyte to develop via application of chemical or electrical stimuli, can also generate blastocysts for hESC line derivation. Human leukocyte antigen (HLA) homozygous hESC lines have been generated from HLA heterozygous donors via parthenogenetic embryos (Revazova et al. 2007, 2008). An hESC line originally reported to be derived from an SCNT embryo was found to have arisen from a parthenogenetic embryo (Kim et al. 2007).

Fusion of a somatic cell with an hESC can reprogram the somatic cell to a pluripotent state (Cowan et al. 2005; Tada et al. 2001; Yu et al. 2006). The resulting cells are tetraploid, however, and unless they can be returned to a normal diploid status they are unlikely to be useful for therapeutic approaches or as a developmental model system.

While the gold standard method for deriving hESC lines remains the harvest and culture of the blastocyst ICM, the advances described in this section have improved efficiency of hESC derivation.

9.3 Deriving Induced Pluripotent Stem Cells

Somatic cells can be reprogrammed to a pluripotent state by inducing pluripotency transcriptional programs. These reprogrammed iPSCs can complement hESCs as a source of cells and tissues for *in vitro* and *in vivo* applications. Human iPSCs have several advantages over hESCs, including posing fewer ethical concerns, ease in generating HLA-matched patient-specific lines, and a more straightforward establishment of models of genetic disease. Because of these differences, iPSCs may replace hESCs in development and disease models and as sources of cells for *in vitro* toxicology, and may more easily lead to cell-based therapies than hESCs. However, the effects of reprogramming on the developmental potential of iPSCs have not been rigorously assessed. The technologies to achieve iPSC generation are still in their infancy, although they have been advancing at a rapid pace, and it is not clear what the gold standard reprogramming methods will be in the future.

9.3.1 Genetic Reprogramming Factors

The first human iPSC lines were established by Shinya Yamanaka and James Thomson in 2007 in independent studies. Takahashi et al. induced pluripotency in adult human fibroblasts by expressing *OCT4*, *SOX2*, Kruppel-like factor 4 (*KLF4*), and *c-MYC*, while Yu et al. reprogrammed IMR90 fetal fibroblasts and adult dermal fibroblasts by expression of *OCT4*, *SOX2*, *NANOG*, and *LIN28* (Takahashi et al. 2007; Yu et al. 2007). Takahashi et al. based their choice of reprogramming factors on genes that were able to induce pluripotency in murine somatic cells (Takahashi and Yamanaka 2006). Yu et al. identified genes preferentially expressed in hESCs, and systematically induced the expression of combinations of these genes to isolate their four factors (Yu et al. 2007).

Table 9.1 describes the roles of genetic and chemical factors in reprogramming somatic cells to the pluripotent state. Each of the initial iPSC studies induced expression of *OCT4*, a homeodomain transcription factor in the POU (Pit-Oct-Unc) family, and *SOX2*, a transcription factor that contains a high mobility group (HMG) box. *OCT4* is expressed in ICM cells and hESCs, and is required for maintenance of a pluripotent state (Chambers and Tomlinson 2009). *SOX2* is expressed in hESCs as well as extra-embryonic ectoderm, trophoblast progenitors, and neural progenitor cells (Chambers and Tomlinson 2009). *NANOG*, a transcription factor expressed in the ICM, is a third key transcriptional regulator of pluripotency along with *OCT4* and *SOX2* (Chambers and Tomlinson 2009). *NANOG* expression increases reprogramming efficiency (Yu et al. 2007). *NANOG* is not a necessary reprogramming factor, however, since *NANOG* expression is induced during reprogramming upon transduction with *OCT4* and *SOX2* (Takahashi et al. 2007; Yu et al. 2007). *c-MYC*, an oncogenic transcription factor, improves reprogramming efficiency via an unknown mechanism, but is not required for reprogramming to occur (Nakagawa et al. 2008; Park et al. 2008b). *KLF4* is a zinc finger-containing transcription factor highly expressed in epithelial tissues, primarily in mitotically inactive and terminally differentiated cells (Nandan and Yang 2009). *KLF4* regulates expression of transcription factors directly implicated in pluripotency, including *OCT4*, *SOX2*, *NANOG*, and *c-MYC*, in murine ESCs (Kim et al. 2008). Other Kruppel-like factors have been shown to substitute for *KLF4* murine iPSC generation from somatic cells (Nakagawa et al. 2008). *LIN28* decreases the expression of *let-7* microRNA (miRNA), which has been shown to regulate cell proliferation and development in *Caenorhabditis elegans,* and reduces tumorigenicity in humans (Bussing et al. 2008). The role of *LIN28* in cell reprogramming suggests the importance of miRNAs in inducing and maintaining pluripotency.

The extent of expression of reprogramming factors appears to be important in obtaining high yields of iPSCs. An analysis of the stoichiometry of the reprogramming factor expression found an optimum at equivalent copy numbers of *SOX2*, *KLF4*, and *c-MYC*, but three times as much *OCT4* (Papapetrou et al. 2009). Other genetic factors have also been utilized to improve reprogramming. Senescence induced by the expression of genetic reprogramming factors may limit the efficiency of iPSC generation. Short

TABLE 9.1 Genetic Factors Involved in Somatic Cell Reprogramming to the Pluripotent State

Gene	Role	References
OCT4	Transcription factor of the POU family; required for maintenance of pluripotency	Takahashi et al. (2007); Yu et al. (2007)
SOX2	Transcription factor that regulates self-renewal of pluripotent stem cells	Takahashi et al. (2007); Yu et al. (2007)
NANOG	Transcription factor that regulates self-renewal of pluripotent stem cells	Takahashi et al. (2007)
c-MYC	Transcription factor that regulates cell proliferation and survival; oncogene	Takahashi et al. (2007)
KLF4	Transcription factor that regulates cell proliferation and survival	Yu et al. (2007)
LIN28	RNA binding protein that blocks let-7 microRNA processing	Yu et al. (2007)
SV40LT	Large T antigen involved in cell immortalization	Park et al. (2008b); Yu et al. (2009)
hTERT	Human telomerase reverse transcriptase; maintains telomere length and involved in cell immortalization	Park et al. (2008b)
UTF1	Downstream factor of OCT4/SOX2; chromatin-associated transcriptional repressor	Zhao et al. (2008)
p16 (INK4a) knockdown	Loss of p16 expression represses senescence	Banito et al. (2009); Li et al. (2009a)
p53 knockdown	Loss of p53 expression may promote immortalization, repress senescence, and/or inhibit apoptosis	Banito et al. (2009); Hong et al. (2009); Zhao et al. (2008)
ARF knockdown	INK4/ARF locus knockdown suppresses senescence	Li et al. (2009a)

hairpin (sh) or short interfering (si) RNA knockdown of the senescence effectors *p16* and *p53* improved the yield of iPSCs from IMR90 and human dermal fibroblasts (Banito et al. 2009; Hong et al. 2009; Zhao et al. 2008). The expression of *UTF1* also increased the efficiency of iPSC generation from human fibroblasts, possibly by acting as a transcriptional repressor downstream of *OCT4* and *SOX2* (Zhao et al. 2008). The expression of a short hairpin RNA that targets *p53* enhanced the efficiency of obtaining iPSC lines from CD34+ cord blood cells transduced with *OCT4*, *SOX2*, *KLF4*, and *c-MYC* (Takenaka et al. 2010). However, *p53* expression appears to prevent reprogramming of DNA-damaged cells to a pluripotent state (Marion et al. 2009), so care should be exercised in using *p53* suppression as a mechanism of enhancing reprogramming efficiency. shRNA knockdown of *p16* (*INK4a*) and *ARF* also improves reprogramming efficiency of human fibroblasts to iPSCs (Li et al. 2009a). The expression of the large T antigen represses the pathways activated by INK4/ARF, and also increases reprogramming efficiencies (Li et al. 2009a).

The diverse sets of genetic reprogramming factors that have been successful in reprogramming somatic cells to the pluripotent state suggest that multiple paths to pluripotency exist. Reprogramming technologies are rapidly advancing, and efficient standardized protocols should exist within the next several years.

9.3.2 Delivery of Genetic Reprogramming Factors

Initial iPSC lines were derived using multiple retroviral reprogramming vectors to drive the expression of reprogramming genes via constitutive promoters (Takahashi et al. 2007; Yu et al. 2007). Reprogramming factor expression has also been driven using inducible promoters; in fact, cells differentiated from these iPSC lines can be returned to the pluripotent state at a high efficiency by reinducing expression of the reprogramming genes (Maherali et al. 2008). However, four reprogramming factors (*OCT4*, *SOX2*, *KLF4*, and *c-MYC*) can be delivered to the somatic cell in a single polycistronic viral vector that uses 2A self-cleaving peptide sequences to allow separation of the factors following translation (Carey et al. 2009). These polycistronic vectors can be removed by transient expression of *Cre* to remove *loxP*-flanked sequences (Chang et al. 2009).

Transposons provide an alternative to viral vectors for delivering reprogramming genes. While the efficiency of transposon-mediated delivery is not as high as lentiviral transduction, transposons can carry large payloads and may offer advantages in the removal of reprogramming vectors. For example, the piggyBac transposon, which can deliver ~14.3 kb, has been used to reprogram human fibroblasts by introducing four reprogramming factors linked with 2A peptides (Kaji et al. 2009). The transposed reprogramming genes can be removed from the genome via Cre-loxP excision (Kaji et al. 2009) or by transfer to a recipient plasmid (Lacoste et al. 2009).

An integration-free reprogramming method that utilizes adenoviral vectors has been developed to generate iPSCs from human fibroblasts (Zhou and Freed 2009). Yu et al. used oriP/EBNA1 nonintegrating plasmid vectors to generate iPSC lines that lacked vector and transgene sequences after plasmid selection was discontinued (Yu et al. 2009). Reprogramming efficiency with four factors was very low, but expression of *OCT4*, *SOX2*, *NANOG*, *LIN28*, *c-MYC*, *KLF4*, and *SV40LT* generated stable iPSC lines.

9.3.3 Chemical Reprogramming Factors

Table 9.2 lists chemical factors that have complemented or replaced genetic factors in reprogramming somatic cells to the pluripotent state. Direct protein transduction, rather than the expression of genetic reprogramming factors, can induce pluripotency in somatic cells. Fusion of *OCT4*, *SOX2*, *KLF4*, and *c-MYC* to a cell-penetrating 9R peptide has been used to generate iPSC lines from human newborn fibroblasts (Kim et al. 2009b).

Small molecules have the potential to replace the expression of at least some transgenes during somatic cell reprogramming. Valproic acid, a histone acetylase inhibitor, enhances the efficiency of reprogramming primary human fibroblasts by the expression of just two genes, *OCT4* and *SOX2* (Huangfu et al.

TABLE 9.2 Chemical Factors Involved in Somatic Cell Reprogramming to the Pluripotent State

Compound	Role	References
OCT4, SOX2, KLF4, and c-MYC fused to 9R peptide	Direct protein transduction delivers reprogramming transcription factors to the nucleus	Kim et al. (2009b)
Valproic acid	Inhibits histone acetylase activity	Huangfu et al. (2008)
CHIR99021	Inhibits GSK-3; leads to activation of Wnt signaling	Li et al. (2009b)
Parnate	Inhibits lysine-specific demethylase 1	Li et al. (2009b)
SB431542	Inhibits TGFβ1 receptor ALK5	Lin et al. (2009)
PD0325901	Inhibits MEK	Lin et al. (2009)
Thiazovivin	Improves survival upon trypsinization; mechanism unknown	Lin et al. (2009)
Vitamin C	Antioxidant	Esteban et al. (2010)
Oxygen (hypoxia)		Yoshida et al. (2009)

2008). Primary human keratinocytes transduced with *OCT4* and *KLF4* in the presence of a glycogen synthase kinase 3 (GSK-3) inhibitor and an inhibitor of lysine-specific demethylase 1 generated iPSC lines (Li et al. 2009b); *SOX2* expression was not required. In addition to replacing reprogramming genes, chemical factors can increase the efficiency of reprogramming via transgene expression. When added 7 days after retroviral infection, the inhibitors of mitogen-activated protein kinase kinase (MAPKK, MEK) and the transforming growth factor β (TGFβ) receptor ALK5 dramatically enhanced reprogramming efficiency and kinetics from primary human keratinocytes (Lin et al. 2009). In addition, thiazovivin, a small molecule identified in screen to improve the survival of hESCs upon trypsinization, provided an additional improvement in the yield of iPSCs (Lin et al. 2009).

Antioxidants, including vitamin C, can enhance the reprogramming efficiency of a variety of human somatic cell types by reducing senescence and promoting the transition of preiPSC colonies to the pluripotent state (Esteban et al. 2010). Hypoxia (5% O_2) has also been shown to improve the efficiency of generating iPSCs from human dermal fibroblasts (Yoshida et al. 2009).

9.3.4 Sources of Somatic Cells for Generating iPSC Lines

Initial human iPSC generation efforts reprogrammed fibroblasts, usually of dermal origin. These cells can be easily obtained and cultured. Human iPSC lines have also been generated from other easily obtained cell types, including epidermal keratinocytes (Aasen et al. 2008) and CD34+ peripheral blood cells (Loh et al. 2009). These and other studies have identified the advantages of using particular types of donor cells to ease the reprogramming process or to improve efficiency.

Distinct fibroblast subpopulations are reprogrammed with different yields. An enhanced efficiency of reprogramming of *SSEA3+* dermal fibroblasts as compared to *SSEA3-* dermal fibroblasts was demonstrated, perhaps as a result of higher endogenous expression of *NANOG* in the *SSEA3*-expressing population (Byrne et al. 2009). Human keratinocytes have been reported to be reprogrammed at a higher rate and frequency than human fibroblasts (Aasen et al. 2008; Maherali et al. 2008). The expression of *OCT4* alone is sufficient to reprogram human fetal neural stem cells, which express endogenous *SOX2*, to a pluripotent state (Kim et al. 2009a). Human adipose-derived stem cells (hADSCs) were reprogrammed at a higher efficiency than adult human fibroblasts by expressing *OCT4*, *SOX2*, *KLF4*, and *c-MYC* (Sugii et al. 2010). hADSCs reprogramming does not require the expression of *c-MYC* (Aoki et al. 2010). The dermal papilla cells, which express high levels of endogenous *Sox2* and *c-MYC*, have been reprogrammed to iPSCs via the expression of *OCT4* and *KLF4*. iPSCs have been derived from dental stem cells, including stem cells from deciduous teeth, apical papilla, and dental pulp, at a higher rate than from fibroblasts (Yan et al. 2010). Cells from placental chorionic, amniotic membrane, and umbilical cord were reprogrammed via *OCT4*, *SOX2*, *KLF4*, and *c-MYC* expression (Cai et al. 2010). Newborn human extraembryonic amnion and yolk sac cells can also give rise to iPSC lines (Nagata et al. 2009).

Pluripotency was induced in human cord blood cells via the expression of *OCT4, SOX2, NANOG*, and *LIN28* (Haase et al. 2009), and by the expression of only *OCT4* and *SOX2* (Giorgetti et al. 2009). iPSCs have also been generated from the human aortic vascular smooth muscle cells (Lee et al. 2010). Taken together, these studies indicate that pluripotency can be induced in a wide variety of somatic cell types and suggest that stem and progenitor cells that express subsets of pluripotency regulators are can be reprogrammed more efficiently and with fewer factors than terminally differentiated cells.

9.4 Characterizing Human Pluripotent Stem Cells

A challenge in the derivation of hESC and iPSC lines is verification of the infinite self-renewal and pluripotency that characterize these cells. The initial phenotype typically utilized to identify new lines is cell morphology. These cells pack into tight colonies comprised of small cells with a high nucleus:cytoplasm ratio (Figure 9.1). Additional tests are required to verify self-renewal potential and pluripotency.

9.4.1 Marker Expression and Epigenetic Characterization

Human pluripotent cells express a variety of cell-surface markers and transcription factors characteristic of the pluripotent state. While no single marker is sufficient, combinations of these markers provide a strong evidence of pluripotency. Transcription factors, such as *OCT4, SOX2,* and *NANOG*, directly

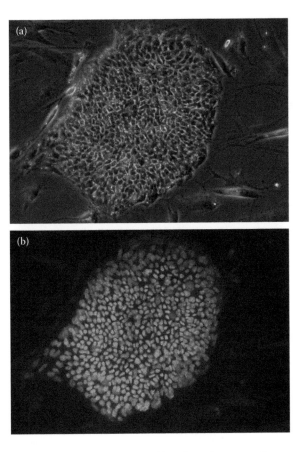

FIGURE 9.1 **(See color insert.)** Phase contrast image (a) and OCT4 immunofluorescence image (b) of an H1 hESC colony cocultured with MEF feeders.

activate the expression of pluripotency genes and/or repress differentiation programs. Common cell-surface markers used to characterize pluripotent cells include SSEA-3, SSEA-4, TRA-1-60, TRA-1-81, GCTM2, GCTM43, THY1, CD9, alkaline phosphatase, and HLA class 1 antigens (Adewumi et al. 2007; Takahashi et al. 2007; Yu et al. 2007). The International Stem Cell Initiative (ISCI) suggested six genes, *NANOG*, *TDFG*, *OCT4*, *GABRB3*, *GDF3*, and *DNMT3B*, as markers to define undifferentiated hESCs based on global gene expression characterization of 59 hESC lines and differentiated hESCs (Adewumi et al. 2007). Another study of putative iPSC colonies reprogrammed from human fibroblasts found that the expression of *TRA-1-60*, *DNMT3B*, and *REX1* distinguished fully reprogrammed colonies from partially reprogrammed colonies that expressed alkaline phosphatase, *SSEA-4*, *GDF3*, *hTERT*, and *NANOG* (Chan et al. 2009).

The ISCI study also discovered the monoallelic expression of the paternally expressed genes *IPW*, *SNRPN*, *KCNQ1OT1*, and *PEG3* in all hESC lines studied (Adewumi et al. 2007). Both hESCs and iPSCs possess highly unmethylated cytosine guanine (CpG) dinucleotides in promoters of core pluripotency regulators, such as *OCT4* and *NANOG* (Takahashi et al. 2007), as well as characteristic histone H3 methylation patterns at the promoters of genes involved in pluripotency (Pan et al. 2007; Takahashi et al. 2007; Yu et al. 2007). At the whole-genome level, hESCs and iPSCs appear to possess greater CpG methylation than differentiated cells such as fibroblasts (Deng et al. 2009).

9.4.2 Characterizing Pluripotency

Since establishing pluripotency via blastocyst injection of an hESC or iPSC is not ethical or practical, pluripotency is typically demonstrated as the capability to differentiate to cells in each of the three germ layers. When injected into an immune-compromised mouse, hESCs and iPSCs will form teratomas containing differentiated cells in the ectoderm, mesoderm, and endoderm lineages (Takahashi et al. 2007; Thomson et al. 1998; Yu et al. 2007). Pluripotency can also be assessed *in vitro* by embryoid body (EB) formation, in which undifferentiated cells are cultured in suspension in the serum-containing medium. The EBs are plated and differentiation to distinct lineages is assessed by marker expression. Alternatively, directed differentiation protocols to generate cell types in the three germ layers may be employed. Teratoma formation is considered a more rigorous demonstration of pluripotency than *in vitro* differentiation methods since some cell types can differentiate to cells in the three germ layers *in vitro* but are not capable of forming teratomas in mice (De Coppi et al. 2007; Lensch et al. 2007).

9.4.3 Characterizing Self-Renewal Potential

Lack of senescence is one hallmark of pluripotent stem cells, resulting in the ability to expand these cells through many, often greater than 100, population doublings (Carpenter et al. 2004). Maintenance of telomere length and high telomerase activity provide additional measures of the self-renewal capabilities of pluripotent stem cells (Rosler et al. 2004; Thomson et al. 1998). Both hESCs and iPSCs also exhibit characteristic patterns of gene expression and DNA methylation near telomeres (Marion et al. 2009).

9.4.4 Similarities and Differences between hESCs and iPSCs

While human iPSCs overcome some of the ethical and technical concerns surrounding translation of hESCs to clinical treatments, iPSCs have not been as well characterized as hESCs and further study of both systems is needed to identify the potential of each cell type. iPSCs and hESCs possess pluripotency and high self-renewal capacity, and in many ways appear very similar in the differentiation potential, but subtle differences are beginning to emerge. While iPSCs have very similar global gene expression patterns to hESCs (Takahashi et al. 2007; Yu et al. 2007), higher levels of neural stem cell-specific genes, genes involved in early embryonic development, and genes directly regulated by reprogramming factors were identified in iPSCs reprogrammed with nonintegrating vectors (Marchetto et al. 2009). Significant expression of reprogramming transgenes has been identified in iPSC-derived cells (Lee et al. 2010), which may have unforeseen consequences on expansion and differentiation. Line-to-line variability exists for both hESC and iPSC differentiation to neural cell types, but iPSCs yielded neural cells with a reduced efficiency as compared to hESCs (Hu et al. 2010). Hemangioblast, endothelial cell, and hematopoietic cell differentiation from iPSCs have also been reported to occur at a lower efficiency than from hESCs, with the resulting differentiated cells undergoing apoptosis at an elevated rate and possessing a more limited expansion capacity (Feng et al. 2010). The reasons for these differences in differentiation efficiency remain unknown.

9.5 Expansion of Human Pluripotent Stem Cells

Substantial progress has been made in developing defined, xeno-free media for human pluripotent stem cell culture. However, implementation of defined extracellular matrices for iPSC and hESC culture lags media development. While lab-scale culture is now relatively straightforward, there are substantial challenges in larger-scale culture for generating industrial- or clinical-quality pluripotent stem cells and tissues derived from these cells.

9.5.1 Coculture with Feeder Cells

The original hESC lines were established and expanded in a medium conditioned by murine embryonic fibroblast feeder cells (Thomson et al. 1998). Coculture with MEFs and a variety of other fibroblast feeder cells (Stacey et al. 2006) remains a common method for expanding undifferentiated hESCs and iPSCs because of the ease and comparatively low cost of this method. However, substantial heterogeneity exists between feeder cell types, and even preparations of a particular feeder cell type (Eiselleova et al. 2008). Also, the derivation and culture of pluripotent stem cells in animal-derived media leads to incorporation of animal components, including the sialic acid *N*-glycolylneuraminic acid (Neu5Gc) (Martin et al. 2005), compromising their utility in clinical applications. These animal components may be reduced in concentration and lost completely by extended culture in humanized medium (Heiskanen et al. 2007).

A number of allogeneic human fibroblast feeder options exist (Stacey et al. 2006), including autologous fibroblasts differentiated from hESCs and iPSCs (Takahashi et al. 2009; Yoo et al. 2005). Human feeder cells, especially allogeneic feeders, still have the potential to modify pluripotent stem cells and potentially elicit an immune response when pluripotent stem cell-derived cells are used *in vivo*. Such feeders can be used in conjunction with human serum to derive and expand new lines without exposure to animal components (Ellerstrom et al. 2006; Skottman et al. 2006).

9.5.2 Defined Media for Pluripotent Stem Cells

The discovery of proteins and chemicals that maintain pluripotency has led to the development of defined, feeder-free culture systems for expansion of hESCs and iPSCs. Basic fibroblast growth factor (bFGF or FGF2) is commonly added to feeder-conditioned or feeder-free media (Amit et al. 2004; Levenstein et al. 2006; Xu et al. 2005b). TGFβ superfamily ligands have also been implicated in maintaining pluripotent stem cells in an undifferentiated state by activation of the SMAD2/3 pathway, and induction of expression of transcription factors regulating pluripotency and cell proliferation pathways (Beattie et al. 2005; James et al. 2005; Prowse et al. 2007; Saha et al. 2008; Vallier et al. 2009). Likewise, inhibition of the SMAD1/5/8 pathway which is induced by bone morphogenetic protein (BMP) signaling assists in maintaining pluripotent stem cells in an undifferentiated state (Wang et al. 2005; Xu et al. 2005a). Insulin-like growth factors also play an important role in self-renewal, perhaps in conjunction with bFGF and TGFβ superfamily signaling (Wang et al. 2007). Lipid molecules also regulate hESC and iPSC self-renewal. Addition of sphingosine-1-phosphate (S1P) to culture medium enhances the expansion of undifferentiated cells (Pebay et al. 2005), perhaps by selectively repressing apoptosis of pluripotent cells while inducing apoptosis of differentiated cells (Inniss and Moore 2006; Salli et al. 2008; Wong et al. 2007). Also, lipids associated with protein carriers in a culture medium, such as albumin, appear to be required for an effective expansion of pluripotent stem cells (Garcia-Gonzalo and Belmonte 2008).

A fully defined medium, mTeSR, was developed to permit self-renewal of hESCs and has been shown to be effective in expanding iPSCs (Ludwig et al. 2006; Yu et al. 2007). A combination of protein and small molecules, including bFGF, TGF-β1, LiCl, GABA, and pipecolic acid, helps maintain cells in a proliferative, self-renewing state when cultured on Matrigel. Alternatively, mTeSR can be used with a defined extracellular matrix consisting of collagen IV, vitronectin, laminin, and fibronectin (Ludwig et al. 2006). STEMPRO, another defined medium for expanding pluripotent stem cells, contains insulin-like growth factor 1 (IGF1), bFGF, Activin A, heregulin-1beta (HRG1β), and is also typically used in conjunction with a growth factor-reduced Matrigel extracellular matrix (Wang et al. 2007).

9.6 Large-Scale Expansion of Human Pluripotent Stem Cells

Translation of scientific advances in pluripotent stem cells to generation of cells and tissues for screening or clinical application will require the development of large-scale culture systems to expand and differentiate these cells. Some of the challenges facing hESC and iPSC expansion have been faced in other mammalian cell systems, while others are relatively unique. Pluripotent stem cells grow in aggregates whose size increases as the cells divide. As the aggregates grow, transport limitations of nutrients and growth factors may be imposed, resulting in death or differentiation. Delivery of self-renewal and differentiation factors must be considered. Since media and extracellular matrices are very expensive, achieving high cell density is important in designing a cost-effective culture system. Perfusion systems may be used to recycle medium and reduce temporal variations in growth factor and metabolite concentrations. Chemical and mechanical methods can be used to dissociate aggregates during passaging, although chemical methods are easier to scale.

Monitoring differentiation status is a critical component of scaling pluripotent stem cell culture. A spontaneous differentiation is undesirable and could be detected via loss of pluripotency markers, while in directed differentiation assessing yield of the desired cell type is important. Human ESCs and iPSCs have the potential to form teratomas *in vivo* (Takahashi et al. 2007; Thomson et al. 1998; Yu et al. 2007), so methods to detect and eliminate any undifferentiated cells must be employed for cultures that will be used in therapeutics. Pluripotent stem cells are prone to acquiring chromosomal abnormalities that confer advantages in growth and maintenance of the undifferentiated state (Draper et al. 2004; Maitra et al. 2005). Monitoring genomic integrity is a critical component of quality control during scaleup.

Compliance with good manufacturing practice (GMP) standards will likely be an important aspect of human pluripotent stem cell expansion for therapeutic applications. Many aspects of GMP cell culture, such as equipment and media sterilization and environmental containment, have been addressed in expanding the culture of other mammalian cells. In addition, the development of animal component-free culture systems, including defined media and matrices, would facilitate GMP cell culture but is not a strict requirement. The derivation and expansion history of the lines may also be an important consideration. Xeno-free derivation of clinical grade hESCs and iPSCs has been demonstrated (Rodriguez-Piza et al. 2010; Ross et al. 2010; Unger et al. 2008). Automated processes for liquid handling and cell passaging may also be components of a scalable, GMP process (Terstegge et al. 2007).

9.6.1 Stirred Culture Vessels

Stirred culture vessels, including stirred-tank bioreactors and spinner flasks, are widely used in the expansion of mammalian cells for protein production. Not surprisingly, stirred culture systems have been adapted to pluripotent stem cell expansion and differentiation. The expansion of undifferentiated

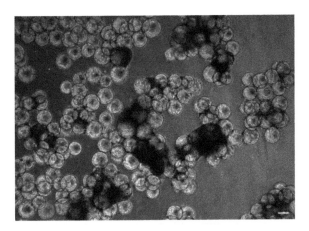

FIGURE 9.2 Phase contrast image of H9 hESC line cultured on Cytodex 3 microcarriers using protocol described previously. Scale bar = 100 µm. (Adapted with permission from Nie, Y. et al. 2009. *Biotechnol Prog* 25:20–31.)

cells typically utilizes microcarriers to provide a suitable substrate for cell attachment. Microcarriers are spherical particles, composed of a variety of materials, including glass, plastic, and cellulose, and typically have a diameter of 100–250 µm. Microcarriers have been used to expand fibroblast feeders and to condition medium for hESC culture (Phillips et al. 2008). Human ESCs have also been expanded in MEF-conditioned and defined media on dextran and cellulose microcarriers seeded with MEFs or coated with extracellular matrix proteins (Figure 9.2) (Fernandes et al. 2009; Nie et al. 2009; Oh et al. 2009). Higher cell expansion rates and densities have been observed in microcarrier culture than in a comparable static 2D culture system (Fernandes et al. 2009; Oh et al. 2009). Pluripotent stem cells can be cryopreserved on microcarriers, and expanded via passaging from microcarriers to microcarriers, illustrating the potential for utilizing stirred culture vessels during all phases of cell stabilization and expansion (Nie et al. 2009).

Stirred culture reactors may also be employed in pluripotent stem cell differentiation. Microcarrier-based culture systems provide a method for generating definitive endoderm from hESCs (Lock and Tzanakakis 2009). Stirred vessels appear well suited for the expansion and maturation of EBs, which do not require a substrate such as a microcarrier. Spinner flasks reduce the heterogeneity of hESC-derived EBs as compared to EBs formed in static culture, and permit differentiation to hematopoietic progenitors (Cameron et al. 2006). Likewise, the rate of cell growth in EBs in stirred culture vessels was reported to be greater than the rate of growth in EBs in static culture (Yirme et al. 2008). One type of stirred bioreactor, the glass ball impeller, led to more cardiac mycocyte differentiation than static culture or several other bioreactors tested (Yirme et al. 2008). Hypoxia in spinner flask culture of hESC-derived EBs has also been shown to increase the expression of mesodermal and cardiac lineage-specific markers, indicating that control of operating conditions is crucial for obtaining the desired lineages (Niebruegge et al. 2009).

9.6.2 Rotary Cell Culture Systems

Rotary cell culture (RCC) systems have been designed to simulate microgravity conditions by maintaining cells in a rotating 3D chamber that keeps the cells in free fall. RCC systems provide a scalable suspension culture vessel that lacks the damaging shear stresses that are often present in stirred vessels. However, RCC systems are more difficult to scale to large sizes than stirred vessels. RCC vessels are suitable for use with microcarriers, permitting the culture of adherent cells. Culture of hESC-derived EBs was compared in two types of RCC systems, the slow-turning lateral vessel (STLV) and

TABLE 9.3 Bioreactors for Expansion of Pluripotent Stem Cells and Their Progeny

Type	Maximum Reported Concentration	Advantages	Disadvantages	References
Stirred culture vessels	5×10^6 cells/mL	Scalable to large volumes, widely utilized in the biotechnology industry, well-mixed environment	High shear, aggregation	Cameron et al. (2006); Fernandes et al. (2009); Lock et al. (2009); Nie et al. (2009); Niebruegge et al. (2009); Oh et al. (2009); Phillips et al. (2008); Yirme et al. (2008)
Rotary cell culture systems	36×10^6 cells/mL	Low shear, efficient gas and nutrient transport	Aggregation, less scalable than stirred vessels	Come et al. (2008); Gerecht-Nir et al. (2004)
Microfluidic devices	3×10^4 cells/cm^2	Precise control of cellular microenvironment	Small scale	Cimetta et al. (2009); Figallo et al. (2007); Korin et al. (2009)

the high aspect rotating vessel (HARV) (Gerecht-Nir et al. 2004). While the HARV led to substantial cell aggregation and necrosis within the EBs, the STLV reactor provided a greater yield of EBs than the traditional static culture (Gerecht-Nir et al. 2004). The STLV has been improved by incorporating perfusion and a dialysis membrane for continuous replenishment of nutrients and removal of wastes (Come et al. 2008). These modifications provide a more temporally uniform environment and reduce the necessary volume of culture medium required to expand the cells. Incorporation of perfusion and dialysis into the STLV improved EB uniformity and enhanced differentiation toward neural lineages (Come et al. 2008).

9.6.3 Microfluidic Culture Systems

Microfluidic devices provide more control over environmental conditions than RCC or stirred vessel reactors, but are not capable of expansion at the scales of RCC and stirred vessels. Thus, microfluidic devices are best suited for optimizing culture conditions and providing precise spatial and temporal control over the cellular microenvironment when necessary for efficient differentiation. A proof-of-concept study demonstrated that single hESC colonies cultured in microfluidic reactors in the presence and absence of flow exhibited similar self-renewal and differentiation potential to standard hESC culture methods (Villa-Diaz et al. 2009). Microfluidic channels can be arrayed in a device for screening chemical factors and flow conditions that promote the desired cellular fate (Figallo et al. 2007). This system was used to culture undifferentiated hESCs and to determine the synergistic effects of cell density and flow on cell response (Cimetta et al. 2009). Another microfluidic system was designed to provide spatial variations in signals in a differentiating EB, and was able to maintain half of an EB in a self-renewing state while inducing differentiation in the other half (Fung et al. 2009). Table 9.3 summarizes the reactor types that have been used to culture human pluripotent stem cells.

9.7 Summary

As the nascent field of human pluripotent stem cells matures, many of the initial technical and ethical limitations surrounding these cells are shrinking. Methods to expand hESCs and iPSCs are becoming more defined and robust. The efficiency of generating new pluripotent stem cell lines is increasing and technologies to derive clinical grade lines are now available. Many technical challenges still remain, notably in efficient expansion and differentiation of pluripotent stem cells at scales suitable for biotechnological applications, real-time monitoring of the status of cells in culture, and assessing the safety of pluripotent stem cells and their progeny in clinical applications. However, it is apparent that pluripotent stem cells will provide a valuable *in vitro* model system to study human development, generate human cells for screening and toxicology, and may serve as a cell source for regenerative therapies in the future.

References

Aasen, T., Raya, A., Barrero, M. J. et al. 2008. Efficient and rapid generation of induced pluripotent stem cells from human keratinocytes. *Nat Biotechnol* 26:1276–84.

Adewumi, O., Aflatoonian, B., Ahrlund-Richter, L. et al. 2007. Characterization of human embryonic stem cell lines by the International Stem Cell Initiative. *Nat Biotechnol* 25:803–16.

Agarwal, S., Loh, Y. H., McLoughlin, E. M. et al. 2010. Telomere elongation in induced pluripotent stem cells from dyskeratosis congenita patients. *Nature* 464:292–6.

Amit, M., Shariki, C., Margulets, V., and Itskovitz-Eldor, J. 2004. Feeder layer- and serum-free culture of human embryonic stem cells. *Biol Reprod* 70:837–45.

Aoki, T., Ohnishi, H., Oda, Y. et al. 2010. Generation of induced pluripotent stem cells from human adipose-derived stem cells without c-MYC. *Tissue Eng Part A* 16:2197–206.

Banito, A., Rashid, S. T., Acosta, J. C. et al. 2009. Senescence impairs successful reprogramming to pluripotent stem cells. *Genes Dev* 23:2134–9.

Beattie, G. M., Lopez, A. D., Bucay, N. et al. 2005. Activin A maintains pluripotency of human embryonic stem cells in the absence of feeder layers. *Stem Cells* 23:489–95.

Bussing, I., Slack, F. J., and Grosshans, H. 2008. let-7 microRNAs in development, stem cells and cancer. *Trends Mol Med* 14:400–9.

Byrne, J. A., Nguyen, H. N., and Reijo Pera, R. A. 2009. Enhanced generation of induced pluripotent stem cells from a subpopulation of human fibroblasts. *PLoS One* 4:e7118.

Cai, J., Li, W., Su, H. et al. 2010. Generation of human induced pluripotent stem cells from umbilical cord matrix and amniotic membrane mesenchymal cells. *J Biol Chem* 285:11227–34.

Cameron, C. M., Hu, W. S., and Kaufman, D. S. 2006. Improved development of human embryonic stem cell-derived embryoid bodies by stirred vessel cultivation. *Biotechnol Bioeng* 94:938–48.

Carey, B. W., Markoulaki, S., Hanna, J. et al. 2009. Reprogramming of murine and human somatic cells using a single polycistronic vector. *Proc Natl Acad Sci USA* 106:157–62.

Carpenter, M. K., Rosler, E. S., Fisk, G. J. et al. 2004. Properties of four human embryonic stem cell lines maintained in a feeder-free culture system. *Dev Dyn* 229:243–58.

Cezar, G. G. 2007. Can human embryonic stem cells contribute to the discovery of safer and more effective drugs? *Curr Opin Chem Biol* 11:405–9.

Chambers, I. and Tomlinson, S. R. 2009. The transcriptional foundation of pluripotency. *Development* 136:2311–22.

Chan, E.M., Ratanasirintrawoot, S., Park, I.H. et al. 2009. Live cell imaging distinguishes bona fide human iPS cells from partially reprogrammed cells. *Nat Biotechnol* 27:1033–37.

Chang, C. W., Lai, Y. S., Pawlik, K. M. et al. 2009. Polycistronic lentiviral vector for "Hit and Run" reprogramming of adult skin fibroblasts to induced pluripotent stem cells. *Stem Cells* 27:1042–9.

Chen, A. E., Egli, D., Niakan, K. et al. 2009. Optimal timing of inner cell mass isolation increases the efficiency of human embryonic stem cell derivation and allows generation of sibling cell lines. *Cell Stem Cell* 4:103–6.

Chung, Y., Klimanskaya, I., Becker, S. et al. 2008. Human embryonic stem cell lines generated without embryo destruction. *Cell Stem Cell* 2:113–7.

Cimetta, E., Figallo, E., Cannizzaro, C., Elvassore, N., and Vunjak-Novakovic, G. 2009. Micro-bioreactor arrays for controlling cellular environments: Design principles for human embryonic stem cell applications. *Methods* 47:81–9.

Come, J., Nissan, X., Aubry, L. et al. 2008. Improvement of culture conditions of human embryoid bodies using a controlled perfused and dialyzed bioreactor system. *Tissue Eng Part C Methods* 14:289–98.

Cowan, C. A., Atienza, J., Melton, D. A., and Eggan, K. 2005. Nuclear reprogramming of somatic cells after fusion with human embryonic stem cells. *Science* 309:1369–73.

De Coppi, P., Bartsch, G. Jr., Siddiqui, M. M. et al. 2007. Isolation of amniotic stem cell lines with potential for therapy. *Nat Biotechnol* 25:100–6.

Deng, J., Shoemaker, R., Xie, B. et al. 2009. Targeted bisulfite sequencing reveals changes in DNA methylation associated with nuclear reprogramming. *Nat Biotechnol* 27:353–60.

Dimos, J. T., Rodolfa, K. T., Niakan, K. K. et al. 2008. Induced pluripotent stem cells generated from patients with ALS can be differentiated into motor neurons. *Science* 321:1218–21.

Draper, J. S., Smith, K., Gokhale, P. et al. 2004. Recurrent gain of chromosomes 17q and 12 in cultured human embryonic stem cells. *Nat Biotechnol* 22:53–4.

Ebert, A. D., Yu, J., Rose, F. F. Jr. et al. 2009. Induced pluripotent stem cells from a spinal muscular atrophy patient. *Nature* 457:277–80.

Eiselleova, L., Peterkova, I., Neradil, J. et al. 2008. Comparative study of mouse and human feeder cells for human embryonic stem cells. *Int J Dev Biol* 52:353–63.

Ellerstrom, C., Strehl, R., Moya, K. et al. 2006. Derivation of a xeno-free human embryonic stem cell line. *Stem Cells* 24:2170–6.

Esteban, M. A., Wang, T., Qin, B. et al. 2010. Vitamin C enhances the generation of mouse and human induced pluripotent stem cells. *Cell Stem Cell* 6:71–9.

Feng, Q., Lu, S. J., Klimanskaya, I. et al. 2010. Hemangioblastic derivatives from human induced pluripotent stem cells exhibit limited expansion and early senescence. *Stem Cells* 28:704–12.

Fernandes, A. M., Marinho, P. A., Sartore, R. C. et al. 2009. Successful scale-up of human embryonic stem cell production in a stirred microcarrier culture system. *Braz J Med Biol Res* 42:515–22.

Figallo, E., Cannizzaro, C., Gerecht, S. et al. 2007. Micro-bioreactor array for controlling cellular microenvironments. *Lab Chip* 7:710–9.

French, A. J., Adams, C. A., Anderson, L. S. et al. 2008. Development of human cloned blastocysts following somatic cell nuclear transfer with adult fibroblasts. *Stem Cells* 26:485–93.

Freund, C. and Mummery, C. L. 2009. Prospects for pluripotent stem cell-derived cardiomyocytes in cardiac cell therapy and as disease models. *J Cell Biochem* 107:592–9.

Fulka, H., St John, J. C., Fulka, J., and Hozak, P. 2008. Chromatin in early mammalian embryos: Achieving the pluripotent state. *Differentiation* 76:3–14.

Fung, W. T., Beyzavi, A., Abgrall, P., Nguyen, N. T., and Li, H. Y. 2009. Microfluidic platform for controlling the differentiation of embryoid bodies. *Lab Chip* 9:2591–5.

Garcia-Gonzalo, F. R. and Belmonte, J. C. 2008. Albumin-associated lipids regulate human embryonic stem cell self-renewal. *PLoS ONE* 3: e1384.

Geens, M., Mateizel, I., Sermon, K. et al. 2009. Human embryonic stem cell lines derived from single blastomeres of two 4-cell stage embryos. *Hum Reprod* 24:2709–17.

Gerecht-Nir, S., Cohen, S., and Itskovitz-Eldor, J. 2004. Bioreactor cultivation enhances the efficiency of human embryoid body (hEB) formation and differentiation. *Biotechnol Bioeng* 86:493–502.

Giorgetti, A., Montserrat, N., Aasen, T. et al. 2009. Generation of induced pluripotent stem cells from human cord blood using OCT4 and SOX2. *Cell Stem Cell* 5:353–7.

Haase, A., Olmer, R., Schwanke, K. et al. 2009. Generation of induced pluripotent stem cells from human cord blood. *Cell Stem Cell* 5:434–41.

Heiskanen, A., Satomaa, T., Tiitinen, S. et al. 2007. N-glycolylneuraminic acid xenoantigen contamination of human embryonic and mesenchymal stem cells is substantially reversible. *Stem Cells* 25:197–202.

Hong, H., Takahashi, K., Ichisaka, T. et al. 2009. Suppression of induced pluripotent stem cell generation by the p53-p21 pathway. *Nature* 460:1132–5.

Hu, B. Y., Weick, J. P., Yu, J. et al. 2010. Neural differentiation of human induced pluripotent stem cells follows developmental principles but with variable potency. *Proc Natl Acad Sci USA* 107:4335–40.

Huangfu, D., Osafune, K., Maehr, R. et al. 2008. Induction of pluripotent stem cells from primary human fibroblasts with only Oct4 and Sox2. *Nat Biotechnol* 26:1269–75.

Inniss, K. and Moore, H. 2006. Mediation of apoptosis and proliferation of human embryonic stem cells by sphingosine-1-phosphate. *Stem Cells Dev* 15:789–96.

James, D., Levine, A. J., Besser, D., and Hemmati-Brivanlou, A. 2005. TGFbeta/activin/nodal signaling is necessary for the maintenance of pluripotency in human embryonic stem cells. *Development* 132:1273–82.

Kaji, K., Norrby, K., Paca, A. et al. 2009. Virus-free induction of pluripotency and subsequent excision of reprogramming factors. *Nature* 458:771–5.

Kim, J., Chu, J., Shen, X., Wang, J., and Orkin, S. H. 2008. An extended transcriptional network for pluripotency of embryonic stem cells. *Cell* 132:1049–61.

Kim, J. B., Greber, B., Arauzo-Bravo, M. J. et al. 2009a. Direct reprogramming of human neural stem cells by OCT4. *Nature* 461:649–3.

Kim, D., Kim, C. H., Moon, J. I. et al. 2009b. Generation of human induced pluripotent stem cells by direct delivery of reprogramming proteins. *Cell Stem Cell* 4:472–6.

Kim, K., Ng, K., Rugg-Gunn, P. J. et al. 2007. Recombination signatures distinguish embryonic stem cells derived by parthenogenesis and somatic cell nuclear transfer. *Cell Stem Cell* 1:346–52.

Klimanskaya, I., Chung, Y., Becker, S., Lu, S. J., and Lanza, R. 2006. Human embryonic stem cell lines derived from single blastomeres. *Nature* 444:481–5.

Korin, N., Bransky, A., Dinnar, U., and Levenberg, S. 2009. Periodic "flow-stop" perfusion microchannel bioreactors for mammalian and human embryonic stem cell long-term culture. *Biomed Microdevices* 11:87–94.

Lacoste, A., Berenshteyn, F., and Brivanlou, A. H. 2009. An efficient and reversible transposable system for gene delivery and lineage-specific differentiation in human embryonic stem cells. *Cell Stem Cell* 5:332–42.

Lee, T. H., Song, S. H., Kim, K. L. et al. 2010. Functional recapitulation of smooth muscle cells via induced pluripotent stem cells from human aortic smooth muscle cells. *Circ Res* 106:120–8.

Lensch, M. W., Schlaeger, T. M., Zon, L. I., and Daley, G. Q. 2007. Teratoma formation assays with human embryonic stem cells: A rationale for one type of human-animal chimera. *Cell Stem Cell* 1:253–8.

Lerou, P. H., Yabuuchi, A., Huo, H. et al. 2008. Human embryonic stem cell derivation from poor-quality embryos. *Nat Biotechnol* 26:212–4.

Levenstein, M. E., Ludwig, T. E., Xu, R. H. et al. 2006. Basic fibroblast growth factor support of human embryonic stem cell self-renewal. *Stem Cells* 24:568–74.

Li, H., Collado, M., Villasante, A. et al. 2009a. The Ink4/Arf locus is a barrier for iPS cell reprogramming. *Nature* 460:1136–9.

Li, W., Zhou, H., Abujarour, R. et al. 2009b. Generation of human-induced pluripotent stem cells in the absence of exogenous Sox2. *Stem Cells* 27:2992–3000.

Lin, T., Ambasudhan, R., Yuan, X. et al. 2009. A chemical platform for improved induction of human iPSCs. *Nat Methods* 6:805–8.

Lock, L. T. and Tzanakakis, E. S. 2009. Expansion and differentiation of human embryonic stem cells to endoderm progeny in a microcarrier stirred-suspension culture. *Tissue Eng Part A* 15:2051–63.

Loh, Y. H., Agarwal, S., Park, I. H. et al. 2009. Generation of induced pluripotent stem cells from human blood. *Blood* 113:5476–9.

Ludwig, T. E., Levenstein, M. E., Jones, J. M. et al. 2006. Derivation of human embryonic stem cells in defined conditions. *Nat Biotechnol* 24:185–7.

Maehr, R., Chen, S., Snitow, M. et al. 2009. Generation of pluripotent stem cells from patients with type 1 diabetes. *Proc Natl Acad Sci USA* 106:15768–73.

Maherali, N., Ahfeldt, T., Rigamonti, A. et al. 2008. A high-efficiency system for the generation and study of human induced pluripotent stem cells. *Cell Stem Cell* 3:340–5.

Maitra, A., Arking, D. E., Shivapurkar, N. et al. 2005. Genomic alterations in cultured human embryonic stem cells. *Nat Genet* 37:1099–103.

Marchetto, M. C., Yeo, G. W., Kainohana, O. et al. 2009. Transcriptional signature and memory retention of human-induced pluripotent stem cells. *PLoS One* 4: e7076.

Marion, R. M., Strati, K., Li, H. et al. 2009. A p53-mediated DNA damage response limits reprogramming to ensure iPS cell genomic integrity. *Nature* 460:1149–53.

Martin, M. J., Muotri, A., Gage, F., and Varki, A. 2005. Human embryonic stem cells express an immunogenic nonhuman sialic acid. *Nat Med* 11:228–32.

Nagata, S., Toyoda, M., Yamaguchi, S. et al. 2009. Efficient reprogramming of human and mouse primary extra-embryonic cells to pluripotent stem cells. *Genes Cells* 14:1395–404.

Nakagawa, M., Koyanagi, M., Tanabe, K. et al. 2008. Generation of induced pluripotent stem cells without MYC from mouse and human fibroblasts. *Nat Biotechnol* 26:101–6.

Nandan, M. O. and Yang, V. W. 2009. The role of Kruppel-like factors in the reprogramming of somatic cells to induced pluripotent stem cells. *Histol Histopathol* 24:1343–55.

Nie, Y., Bergendahl, V., Hei, D. J., Jones, J. M., and Palecek, S. P. 2009. Scalable culture and cryopreservation of human embryonic stem cells on microcarriers. *Biotechnol Prog* 25:20–31.

Niebruegge, S., Bauwens, C. L., Peerani, R. et al. 2009. Generation of human embryonic stem cell-derived mesoderm and cardiac cells using size-specified aggregates in an oxygen-controlled bioreactor. *Biotechnol Bioeng* 102:493–507.

Oh, S. K., Chen, A. K., Mok, Y. et al. 2009. Long-term microcarrier suspension cultures of human embryonic stem cells. *Stem Cell Res* 2:219–30.

Pan, G., Tian, S., Nie, J. et al. 2007. Whole-genome analysis of histone H3 lysine 4 and lysine 27 methylation in human embryonic stem cells. *Cell Stem Cell* 1:299–312.

Papapetrou, E. P., Tomishima, M. J., Chambers, S. M. et al. 2009. Stoichiometric and temporal requirements of Oct4, Sox2, Klf4, and c-MYC expression for efficient human iPSC induction and differentiation. *Proc Natl Acad Sci USA* 106:12759–64.

Park, I. H., Arora, N., Huo, H. et al. 2008a. Disease-specific induced pluripotent stem cells. *Cell* 134:877–86.

Park, I. H., Zhao, R., West, J. A. et al. 2008b. Reprogramming of human somatic cells to pluripotency with defined factors. *Nature* 451:141–6.

Pebay, A., Wong, R. C., Pitson, S. M. et al. 2005. Essential roles of sphingosine-1-phosphate and platelet-derived growth factor in the maintenance of human embryonic stem cells. *Stem Cells* 23:1541–8.

Phillips, B. W., Horne, R., Lay, T. S. et al. 2008. Attachment and growth of human embryonic stem cells on microcarriers. *J Biotechnol* 138:24–32.

Prowse, A. B., McQuade, L. R., Bryant, K. J., Marcal, H., and Gray, P. P. 2007. Identification of potential pluripotency determinants for human embryonic stem cells following proteomic analysis of human and mouse fibroblast conditioned media. *J Proteome Res* 6:3796–807.

Raya, A., Rodriguez-Piza, I., Guenechea, G. et al. 2009. Disease-corrected haematopoietic progenitors from Fanconi anaemia induced pluripotent stem cells. *Nature* 460:53–9.

Revazova, E. S., Turovets, N. A., Kochetkova, O. D. et al. 2007. Patient-specific stem cell lines derived from human parthenogenetic blastocysts. *Cloning Stem Cells* 9:432–49.

Revazova, E. S., Turovets, N. A., Kochetkova, O. D. et al. 2008. HLA homozygous stem cell lines derived from human parthenogenetic blastocysts. *Cloning Stem Cells* 10:11–24.

Rodriguez-Piza, I., Richaud-Patin, Y., Vassena, R. et al. 2010. Reprogramming of human fibroblasts to induced pluripotent stem cells under xeno-free conditions. *Stem Cells* 28:36–44.

Rosler, E. S., Fisk, G. J., Ares, X. et al. 2004. Long-term culture of human embryonic stem cells in feeder-free conditions. *Dev Dyn* 229:259–74.

Ross, P. J., Suhr, S., Rodriguez, R. M. et al. 2010. Human induced pluripotent stem cells produced under xeno-free conditions. *Stem Cells Dev* 19:1221–9.

Saha, S., Ji, L., de Pablo, J. J., and Palecek, S. P. 2008. TGFbeta/Activin/Nodal pathway in inhibition of human embryonic stem cell differentiation by mechanical strain. *Biophys J* 94:4123–33.

Salli, U., Fox, T. E., Carkaci-Salli, N. et al. 2008. Propagation of undifferentiated human embryonic stem cells with nano-liposomal ceramide. *Stem Cells Dev* 18:55–65.

Skottman, H., Dilber, M. S., and Hovatta, O. 2006. The derivation of clinical-grade human embryonic stem cell lines. *FEBS Lett* 580:2875–8.

Snykers, S., Henkens, T., De Rop, E. et al. 2009. Role of epigenetics in liver-specific gene transcription, hepatocyte differentiation and stem cell reprogrammation. *J Hepatol* 51:187–211.

Stacey, G. N., Cobo, F., Nieto, A. et al. 2006. The development of "feeder" cells for the preparation of clinical grade hES cell lines: Challenges and solutions. *J Biotechnol* 125:583–8.

Stojkovic, M., Stojkovic, P., Leary, C. et al. 2005. Derivation of a human blastocyst after heterologous nuclear transfer to donated oocytes. *Reprod Biomed Online* 11:226–31.

Strelchenko, N., Verlinsky, O., Kukharenko, V., and Verlinsky, Y. 2004. Morula-derived human embryonic stem cells. *Reprod Biomed Online* 9:623–9.

Sugii, S., Kida, Y., Kawamura, T. et al. 2010. Human and mouse adipose-derived cells support feeder-independent induction of pluripotent stem cells. *Proc Natl Acad Sci USA* 107:3558–63.

Tada, M., Takahama, Y., Abe, K., Nakatsuji, N., and Tada, T. 2001. Nuclear reprogramming of somatic cells by *in vitro* hybridization with ES cells. *Curr Biol* 11:1553–8.

Takahashi, K., Narita, M., Yokura, M., Ichisaka, T., and Yamanaka, S. 2009. Human induced pluripotent stem cells on autologous feeders. *PLoS One* 4: e8067.

Takahashi, K., Tanabe, K., Ohnuki, M. et al. 2007. Induction of pluripotent stem cells from adult human fibroblasts by defined factors. *Cell* 131:861–72.

Takahashi, K. and Yamanaka, S. 2006. Induction of pluripotent stem cells from mouse embryonic and adult fibroblast cultures by defined factors. *Cell* 126:663–76.

Takenaka, C., Nishishita, N., Takada, N., Jakt, L. M., and Kawamata, S. 2010. Effective generation of iPS cells from CD34+ cord blood cells by inhibition of p53. *Exp Hematol* 38:154–62.

Terstegge, S., Laufenberg, I., Pochert, J. et al. 2007. Automated maintenance of embryonic stem cell cultures. *Biotechnol Bioeng* 96:195–201.

Thomson, J. A., Itskovitz-Eldor, J., Shapiro, S. S. et al. 1998. Embryonic stem cell lines derived from human blastocysts. *Science* 282:1145–7.

Unger, C., Skottman, H., Blomberg, P., Dilber, M. S., and Hovatta, O. 2008. Good manufacturing practice and clinical-grade human embryonic stem cell lines. *Hum Mol Genet* 17: R48–53.

Vallier, L., Mendjan, S., Brown, S. et al. 2009. Activin/Nodal signalling maintains pluripotency by controlling Nanog expression. *Development* 136:1339–49.

Villa-Diaz, L. G., Torisawa, Y. S., Uchida, T. et al. 2009. Microfluidic culture of single human embryonic stem cell colonies. *Lab Chip* 9:1749–55.

Wang, G., Zhang, H., Zhao, Y. et al. 2005. Noggin and bFGF cooperate to maintain the pluripotency of human embryonic stem cells in the absence of feeder layers. *Biochem Biophys Res Commun* 330:934–42.

Wang, L., Schulz, T. C., Sherrer, E. S. et al. 2007. Self-renewal of human embryonic stem cells requires insulin-like growth factor-1 receptor and ERBB2 receptor signaling. *Blood* 110:4111–9.

Winkler, J., Sotiriadou, I., Chen, S., Hescheler, J., and Sachinidis, A. 2009. The potential of embryonic stem cells combined with -omics technologies as model systems for toxicology. *Curr Med Chem* 16:4814–27.

Wong, R. C., Tellis, I., Jamshidi, P., Pera, M., and Pebay, A. 2007. Anti-apoptotic effect of sphingosine-1-phosphate and platelet-derived growth factor in human embryonic stem cells. *Stem Cells Dev* 16:989–1001.

Xu, C., Rosler, E., Jiang, J. et al. 2005b. Basic fibroblast growth factor supports undifferentiated human embryonic stem cell growth without conditioned medium. *Stem Cells* 23:315–23.

Xu, R. H., Peck, R. M., Li, D. S. et al. 2005a. Basic FGF and suppression of BMP signaling sustain undifferentiated proliferation of human ES cells. *Nat Methods* 2:185–90.

Yan, X., Qin, H., Qu, C. et al. 2010. iPS cells reprogrammed from mesenchymal-like stem/progenitor cells of dental tissue origin. *Stem Cells Dev* 19:469–80.

Ye, Z., Zhan, H., Mali, P. et al. 2009. Human-induced pluripotent stem cells from blood cells of healthy donors and patients with acquired blood disorders. *Blood* 114:5473–80.

Yirme, G., Amit, M., Laevsky, I., Osenberg, S., and Itskovitz-Eldor, J. 2008. Establishing a dynamic process for the formation, propagation, and differentiation of human embryoid bodies. *Stem Cells Dev* 17:1227–41.

Yoo, S. J., Yoon, B. S., Kim, J. M. et al. 2005. Efficient culture system for human embryonic stem cells using autologous human embryonic stem cell-derived feeder cells. *Exp Mol Med* 37:399–407.

Yoshida, Y., Takahashi, K., Okita, K., Ichisaka, T., and Yamanaka, S. 2009. Hypoxia enhances the generation of induced pluripotent stem cells. *Cell Stem Cell* 5:237–41.

Yu, J., Hu, K., Smuga-Otto, K. et al. 2009. Human induced pluripotent stem cells free of vector and transgene sequences. *Science* 324:797–801.

Yu, J., Vodyanik, M. A., He, P., Slukvin, II and Thomson, J. A. 2006. Human embryonic stem cells reprogram myeloid precursors following cell-cell fusion. *Stem Cells* 24:168–76.

Yu, J., Vodyanik, M. A., Smuga-Otto, K. et al. 2007. Induced pluripotent stem cell lines derived from human somatic cells. *Science* 318:1917–20.

Zhang, X., Stojkovic, P., Przyborski, S. et al. 2006. Derivation of human embryonic stem cells from developing and arrested embryos. *Stem Cells* 24:2669–76.

Zhao, Y., Yin, X., Qin, H. et al. 2008. Two supporting factors greatly improve the efficiency of human iPSC generation. *Cell Stem Cell* 3:475–9.

Zhou, W. and Freed, C. R. 2009. Adenoviral gene delivery can reprogram human fibroblasts to induced pluripotent stem cells. *Stem Cells* 27:2667–74.

10

Bioreactors for Stem Cell Expansion and Differentiation

Carlos A. V.
Rodrigues
Instituto Superior Técnico

Tiago G. Fernandes
Instituto Superior Técnico

Maria Margarida
Diogo
Instituto Superior Técnico

Cláudia Lobato
da Silva
Instituto Superior Técnico

Joaquim M. S.
Cabral
Instituto Superior Técnico

10.1 Introduction

In the past decades, progress in the biomedical science field has led to an overwhelming increase of products from biological origins for treatment of several diseases (Nagle et al., 2003). The first generation of such therapeutics included recombinant proteins, antibodies, and molecular vaccines. These new products created the need to develop highly controlled and reproducible bioprocesses that would comply with stringent regulatory demands (e.g., FDA and EMEA). Bioengineers have therefore worked toward this goal and have gathered valuable expertise. More recently, however, cell-based therapies have generated great interest in the scientific and medical communities. The increasing number of companies that are engaged in the development of new cell-based therapies for the treatment of several diseases illustrates this recent trend (Parson, 2008). Nevertheless, the number of cell therapy products that have reached the market is still very small, and the vast majority are still under preclinical development. Thus, the success of these products is dependent on the development of novel technologies that would allow the systematic production of cells in a robust and cost-effective manner (Kirouac and Zandstra, 2008).

The starting materials of such manufacturing processes are most likely stem cells, or their differentiated progeny. Stem cells are undifferentiated cells that have unlimited self-renewal capacity and the ability to differentiate into mature cells (Passier and Mummery, 2003). Consequently, these properties make them very attractive for cell therapy approaches. Unfortunately, however, the typical number of cells needed to treat an average adult patient (~70 kg) greatly surpasses the number of cells available from donors (Laflamme and Murry, 2005; Sohn et al., 2003). Therefore, the need to develop fully controlled large-scale *ex vivo* bioreactor systems arises not only from the limited number of cells that can be obtained from the available donor, but also from the need to comply with strict regulatory rules (Cabral, 2001). Additionally, the fact that the desired products are the cells themselves bring forward further challenges related to the required good manufacturing practices (GMP) and product safety. Donor-to-donor variability, microbiological contamination, and potential tumorigenicity of the transplanted cells, among others, are examples of such issues (Ahrlund-Richter et al., 2009).

In addition, the development of stable *in vitro* systems for the expansion and differentiation of stem cells can also contribute valuable tools for the study of the mechanisms controlling such events (Vazin and Schaffer, 2010). For example, scaling out a given process by establishing high-throughput screening platforms can greatly benefit the process development (Amanullah et al., 2010). Furthermore, empirical and mechanistic modeling, along with other rational approaches for process optimization, constitutes an additional means for achieving a clear understanding of the factors affecting a given system. Successful *ex vivo* models will therefore enable the study of the dynamics and mechanisms of cell differentiation and organ development (Abranches et al., 2009). Moreover, meaningful pharmacological studies can also be performed using such systems (Lee et al., 2008).

Thus, the *in vitro* propagation of undifferentiated stem cell populations remains largely undeveloped and is considered a major technical challenge because of the complex kinetics of the heterogeneous starting culture population, the transient nature of the subpopulations of interest, the lack of invariant measures, and the complex interactions between culture parameters. The main goal of this chapter is to present the fundamental concepts for bioprocess and bioreactor development toward the *in vitro* expansion and maintenance of stem cells, while maintaining their functional characteristics, including the ability to differentiate into appropriate receptive tissues. Additionally, recent developments in this area are also described and new approaches are discussed in the following sections.

10.2 Bioprocess Development and Selection

To select an adequate bioprocess for clinical- or pharmacological-grade production of cellular products, a few important considerations should be taken into account. In principle, when compared to molecular therapeutics, cell-based therapies have the potential to provide superior clinical outcomes because of the broad biological activity of cells (growth factor release, contribution to tissue regeneration, release of morphogens, etc.) (Majka et al., 2001). However, this will only be true if cells are produced in a reproducible way, a requirement for consistent clinical outcomes. Therefore, it is possible to foresee some major hurdles related to scientific, technical, regulatory, and commercial aspects (Martin et al., 2009). In fact, the development of bioreactors for stem cell expansion and differentiation is clearly dependent on the ability to mimic the physiological, biochemical, and mechanical cues of the *in vivo* microenvironment (a scientific and technical challenge), in addition to the need for complying with strict regulatory guidelines, while keeping a robust, competitive, and cost-efficient process. These objectives are mainly challenged by the intrinsic variability of cells obtained from different batches or donors (Koller et al., 1996), as well as by the response of cells to variations in the culture environment (Discher et al., 2009).

10.2.1 Stem Cell Isolation and Initial Characterization

The starting point for the design of a new process is the discovery phase, in which product characterization is fundamental (Kirouac and Zandstra, 2008). The isolation of stem cells from donor sources and their functional characterization are key points at this stage. In fact, different types of stem cells can be used for the production of cellular products for clinical applications. These cells can be isolated from embryonic or adult tissues and, more recently, cellular reprogramming can also be used to generate pluripotent cells (Takahashi and Yamanaka, 2006).

Embryonic stem cells (ESCs) are capable of unlimited expansion, and thus have the potential to generate all the cell types derived from the three embryonic germ layers, a property known as pluripotency (Smith, 2001). However, their use in clinical settings is limited by their innate tumorigenicity (e.g., teratoma formation), difficulty to fully control their differentiation, yielding nonhomogeneous cell populations, and ethical considerations due to the destruction of the embryo. Adult stem cells, including those from neonatal source (e.g., umbilical cord blood) on the other hand, do not draw ethical concerns and can be directly obtained from available donors. Nonetheless, cell quality may vary depending on the donor characteristics (e.g., age, sex, genetic background, etc.). Additionally, these cells are multipotent, which means that differentiation is restricted to the original lineage of the cell source, and their proliferative capacity *in vitro* is more restricted compared to ESCs. Nevertheless, adult stem cells (e.g., hematopoietic stem cells (HSCs)) have been used in clinical practice since the 1950s (Thomas et al., 1957). Indeed, bone marrow transplantation has become a commonly used clinical procedure for a number of malignant and nonmalignant diseases since its first successful application in 1968 when an infant with an immune deficiency received a bone marrow transplant from his sister (Bortin, 1970). Another example of adult stem cells already established for cell therapy settings are mesenchymal stem cells (MSCs).

Bone marrow was originally considered the source of choice to obtain adult stem cells for transplantation, but other tissues have drawn recent attention. These include mobilized peripheral blood (PB), adipose tissue, placenta, and umbilical cord blood. The isolation of hematopoietic stem cells (HSCs) from these sources is typically accomplished by the use of magnetic cell sorting based on the expression of surface antigens (CD34+, Thy1+, and CD38-) (Wognum et al., 2003). On the other hand, human MSCs do not express the hematopoietic markers CD45, CD34, CD14, or CD11, but express cell surface markers such as CD105, CD73, CD44, CD90, CD71, and Stro-1 (Chamberlain et al., 2007). MSCs have been isolated based on the expression of specific phenotypes such as Stro1+ (Chamberlain et al., 2007; Gonçalves et al., 2006), but more commonly based on cell adherence to tissue culture plastic (Pittenger, 2008). The use of cell surface antigen expression is not only useful for cell isolation from donor tissues, but also as a quality control measurement during cell culture *ex vivo*.

Human ESCs were first isolated and derived using feeder layers and serum-containing medium (Thomson et al., 1998). Since then, the maintenance of pluripotency in culture has been routinely assessed using the expression of key pluripotency markers, such as the transcription factors Oct4, Nanog, Sox2, or Rex-1, and the cell surface markers SSEA3, SSEA4, TRA-1-60, and TRA-1-81 (Carpenter et al., 2003). Also, several protocols that do not require embryo destruction have been developed (McDevitt and Palecek, 2008). These protocols include fusion of ESCs with somatic cells (Cowan et al., 2005), use of parthenogenesis to generate blastocysts (Revazova et al., 2007), somatic cell nuclear transfer (Byrne et al., 2007), and reprogramming of adult cells to generate induced pluripotent stem cells (iPSCs) (Takahashi et al., 2007). In this context, the generation of iPSCs is of great interest not only due to the ethical concerns surrounding human ESCs, but also because it has become possible to originate patient-specific pluripotent cell lines (Ebert et al., 2009; Soldner et al., 2009), and generating lines suitable for clinical and pharmacological applications. Patient-specific cells will become important models for studying human disease, for testing responses to potential drugs, and might also be used to develop patient-specific cell therapy, establishing a rational basis for personalized medicine in the future (Nishikawa et al., 2008).

10.2.2 Bioprocess Development and Optimization Based on Microscale High-Throughput Profiling

The ability to analyze multiple conditions in a fast and parallel fashion can also enhance our knowledge of a given system, and potentially contribute to accelerate bioprocess development. Microscale high-throughput profiling approaches can therefore contribute to speed up the transition from biological observation to optimized, clinical-scale bioreactor systems. Thus, the development of *in vitro* high-throughput screening methods for evaluating the effects of new growth factors and cytokines, as well as other culture conditions in cell models might assist in the rapid and cost-effective development of novel bioprocesses, and also increase the knowledge on conditions that selectively control cell fate (Fernandes et al., 2009b). Therefore, the use of these methodologies will ultimately aid in the generation of cells in a more reproducible and cost-effective manner.

The first examples of the application of microengineered systems in stem cell research focused mainly on the discovery of combinations of signaling environments that direct stem cell fate (Flaim et al., 2005; Soen et al., 2006). In fact, signals emanating from the stem cell microenvironment, or niche, are crucial in regulating stem cell functions. Nevertheless, advances in microfabrication and microfluidics have also driven the generation of multiple platforms that allow bioprocess optimization (Gómez-Sjöberg et al., 2007). Microbioreactor arrays, containing independent microbioreactors perfused with culture medium, have been fabricated using soft lithography (Hung et al., 2005; Kim et al., 2006). These systems supported the cultivation of cells, either attached to substrates or encapsulated in hydrogels, at variable levels of hydrodynamic shear, and automated image analysis detected the expression of cell differentiation markers. Various conditions and configurations were validated for different cell types, including mouse myoblasts, primary rat cardiac myocytes, and human ESCs (Figallo et al., 2007). Along with cell growth monitoring, design parameters, mass transport phenomena, and shear stress issues can also be examined in these devices via numerical simulations (Korin et al., 2009).

These high-throughput cell culture platforms are thus efficient in the analysis of multiple parameters and parameter interactions that might prove important for bioprocess optimization. However, one major limitation of such systems is the ability to quantify specific cellular responses in an accurate and straightforward manner. Methods to alleviate such limitations have been developed and include, for example, immunofluorescence-based assays for high-throughput analysis of target proteins on three-dimensional (3D) cellular microarray platforms (Fernandes et al., 2008), or cellular microarrays with integrated multifunctional sensing elements that allow immunodetection of secreted proteins (Jones et al., 2008).

Therefore, the development of stable, high-throughput, and high-content screening systems for the study of stem cell fate can also contribute with valuable information related with mechanisms controlling cell proliferation, differentiation, or death (Fernandes et al., 2010; Peerani et al., 2009). This can greatly benefit process development, as the underlying aspects of the stem cell biology are becoming further understood. Additionally, microscale strategies can also be directly employed for parameter measurement and optimization, and ultimately lead to the development of an integrated process for clinical-scale production of stem cells.

10.2.3 Monitoring and Control of Bioreactor Systems

Monitoring and control of bioreactor systems to minimize process and product variability while maximizing productivity are valuable tools in bioprocess development. If physicochemical culture parameters (pH, pO_2, etc.) are monitored and controlled, the required standardization due to quality control and regulatory demands can potentially be improved (Lim et al., 2007). Therefore, fully controlled bioreactors have the potential to increase the robustness and stability of the cellular products obtained in the manufacturing process. Also, models predicting population dynamics by incorporating data such as growth rates, death, differentiation, transition between quiescence and active cycling, concentrations of cytokines, metabolite uptake and production rates, dissolved oxygen, and pH, provide the means to improve process performance (da Silva et al., 2003). In addition, such *in silico* mathematical modeling of stem cell functions is capable of predicting many different cellular events, relating internal parameters and microenvironmental variables to measurable cell fate outcomes (Kirouac et al., 2009). Therefore, models incorporating cell-level kinetics, physicochemical culture parameters, and microenvironmental variables are valuable tools for process development, while providing insights into biological questions important to understand stem cell dynamics, and to explain heterogeneity in culture outputs.

Nevertheless, most models deal with a limited set of inputs and were derived from a few offline experimental measurements in cultures with poorly controlled settings (e.g., T-flasks and tissue culture plates). The implementation of online monitoring and the use of automated control systems are essential not only for the large-scale culture of stem cells, but also as a means to unveil, under controlled and reproducible conditions, the effects of multiple parameters on the growth kinetics of specific cell populations, oxygen consumption, nutrient depletion, metabolic by-product accumulation, and cytokine production and consumption, which are not yet well understood. The kinetic analysis, along with the hydrodynamic and mass transfer characterization of the bioreactor, can then be incorporated in a predictive model that might be useful for establishing optimal operational conditions.

It is now clear that further efforts are needed to overcome the existing difficulties in establishing routine processes for the production of quality-controlled stem cell products under GMP conditions. These difficulties are mainly related with the fundamental understanding of the cellular and molecular aspects underlying the stem cell biology, variability of the starting cellular material, and an important number of technical issues, such as availability of sensing techniques for quantifying important culture parameters (e.g., cell numbers, differentiation stage, and metabolism), bioprocess monitoring and control, and means for predicting the culture outcome based on measured parameters (e.g., bioprocess modeling). Indeed, to better control and standardize key product properties such as cell identity, quality, purity, and potency, all these key points should be taken into consideration (Placzek et al., 2009).

Thus, understanding the scientific aspects of a given stem cell system will help to identify critical features, such as factors involved in stem cell expansion and differentiation, which can then be used for process control and assurance of product quality. With this fundamental knowledge, the next logical step will involve the set up of several process parameters defined by specific biochemical, metabolic, and environmental characteristics of the biological system under study (e.g., supplementation with growth factors, metabolite concentration, dissolved oxygen, etc.). Finally, it is clear that monitoring and control capabilities, which are available in advanced bioreactor systems, provide added means to develop bioprocesses that render cellular products in compliance with GMP practices.

10.3 Bioreactor Configurations

The culture of stem cells is traditionally and usually performed on flat two-dimensional surfaces such as tissue culture flasks (T-flasks), well plates, or gas-permeable blood bags consisting of a single unstirred compartment where nutrients diffuse to cells. Gas exchange (e.g., oxygen and carbon dioxide) occurs at the medium/gas interface. These systems are widely used for research purposes because of their simplicity, ease of handling, and relatively low cost.

Despite their widespread usage, these static systems have serious limitations. First, they lead to concentration gradients (pH, dissolved oxygen, nutrients, metabolites, etc.) in the culture medium. The on-line monitoring of culture parameters such as pH or pO_2 is possible today in these culture systems (Deshpande and Heinzle, 2004; Kensy et al., 2005), but the tight control of these variables is much more difficult. Scale-up is also difficult since these culture platforms present reduced surface area/volume ratios, which limits the number of cells supported per surface area. Multiple plates or flasks (or flasks with multiple trays) are required to obtain high numbers of cells, requiring repeated handling to feed cultures or obtain data on culture performance, making this solution laborious and prone to contamination.

Automation and robotics (Terstegge et al., 2007) could minimize the impact of the last issue, but the static nature of the culture would remain. The use of 3D culture systems that more closely resemble the *in vivo* environment provides increased surface area for cell adhesion and growth, thus leading to higher cellular concentrations but the mass transfer limitations would also increase.

Mass transfer limitations and other problems can be minimized with the use of bioreactors that can accommodate dynamic culture conditions (Figure 10.1 and Table 10.1). These advanced bioreactors are required when large numbers of cells are needed, accessory cells are used, or high cell densities are desired. The next sections describe bioreactor configurations that have been used for the culture of stem cells.

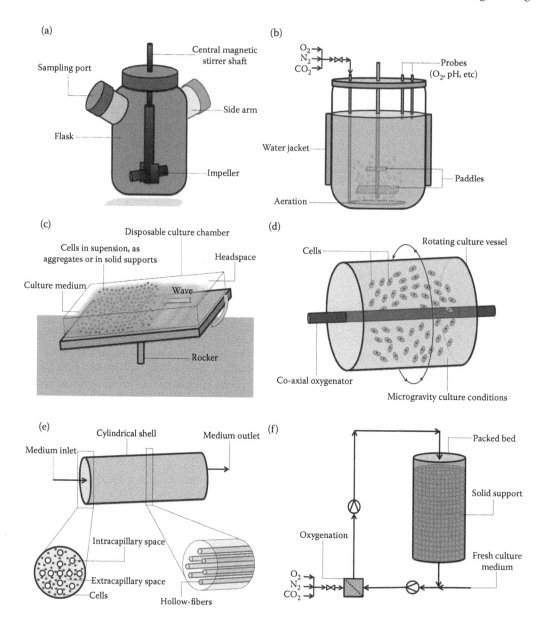

FIGURE 10.1 Schematic representation of several bioreactor configurations used for stem cell culture. (a) Spinner flask; (b) stirred suspension bioreactor; (c) wave bioreactor; and (d) rotating wall vessel. The slow turning lateral vessel (STLV) configuration is shown; (e) hollow-fiber bioreactor; and (f) packed-bed bioreactor.

10.3.1 Roller Bottles

A simple way of culturing anchorage-dependent animal cells, and stem cells in particular, under dynamic conditions is provided by roller bottles systems (Kunitake et al., 1997; Mitaka, 2002). Cell culture in roller bottles consists of placing multiple cylindrical bottles (250–2000 mL) into an apparatus that will rotate the bottles at rotational speeds of 5–60 rph (ECACC/Sigma-Aldrich, 2010). The entire internal surface of the bottle is used for cell growth. With this system, the cells are more efficiently oxygenated due to alternative exposure to the medium and the gas phase. Apparatus are available to

TABLE 10.1 Summary of the Main Advantages and Disadvantages of Different Bioreactor Systems Used for Stem Cell Culture

Bioreactor Configuration	Advantages	Disadvantages
Roller bottles	Simple operation and usage Versatile system Low-cost solution	Monitoring and control is possible, but not straightforward Although minimized, concentration gradients still persist
Stirred suspension bioreactor	Simple design and operation Homogeneous conditions due to agitation Sampling, monitoring, and control are relatively simple Suspension culture and also adherent culture (when microcarriers are used)	Shear stress due to agitation can be harmful to cells Bead bridging and/or cell agglomeration
Wave bioreactor	Disposable system, reduces contamination issues and the need for sterilization of the bioreactor Suitable for GMP operations Easily scalable	Sampling, monitoring, and control are not as simple as with other systems High cost
Rotating wall vessel	Low-shear stress environment Efficient gas transfer	Scalability Complex system
Parallel plates bioreactor	Homogeneous culture conditions High productivity Inhibition of waste accumulation	Continuous removal of secreted factors Unknown effect of hydrodynamic shear stress Medium-intensive culture system
Hollow-fiber bioreactor	Low-shear stress environment Better mimic of the cell microenvironment	Spatial concentration gradients Difficulty in monitoring and control Scale-up difficulties
Fixed- and fluidized-bed bioreactor	Cell–cell or cell–matrix interactions are possible Better mimic of the *in vivo* intricate structure 3D scaffolding for cell culture and attachment	Relatively low volumes Spatial concentration gradients (in the fixed bed configuration) Possible shear stress effects (in the fluidized bed configuration) Difficulties in scaling-up

accommodate four to hundreds of bottles. To prevent oxygen limitation a medium/air volume ratio between 1:5 and 1:10 is often used. It is thus possible to conclude that roller bottles may constitute an easy dynamic culture alternative to well plates or T-flasks but they require more incubator space, are more expensive, and still do not provide control over culture parameters like dissolved oxygen or pH. Moreover, using racks with multiple bottles may lead to heterogeneity among batches.

10.3.2 Stirred Suspension Bioreactors

Stirred suspension bioreactors (SBs) are widely used and characterized for the culture of both microbial and animal cells. This type of bioreactor provides a homogeneous environment and is easy to operate, allowing sampling, data collection, and control of medium conditions. Indeed, SBs allow online monitoring and control of culture variables such as temperature, pH or dissolved oxygen, important for the expansion or controlled differentiation of stem cells, and their simple design allows relatively easy scale-up. Since large-scale mammalian cell culture systems have been developed based on SBs for a number of applications, including production of recombinant proteins and monoclonal antibodies, the requirement of large amounts of stem cells and/or their progeny for clinical use provides an incentive to translate these common mammalian cell culture technologies to commercial stem cell production settings. In conventional stirred suspension bioreactors, concentrations of 10^6–10^7 mammalian cells/mL are common (Kehoe et al., 2009). For clinically relevant applications, 1×10^9–1×10^{10} stem cells and stem cell-derived cells may be required (Kehoe et al., 2009) which means that working volumes of a few hundred milliliters to a few liters would be necessary. However, for proof-of-principle studies or optimization of culture conditions, smaller scale vessels are commonly used. A simple laboratory version of a SB, the spinner flask, consists of a glass or plastic vessel with a central magnetic stirrer shaft and side arms for the addition or removal of cells and medium, being O_2 transfer achieved by surface aeration from the headspace.

In stirred bioreactors, cells can be cultured as single cells, as aggregates, on microcarriers or in scaffolds. Mammalian cells are often grown as aggregates since a better recapitulation of the *in vivo* microenvironment is believed to occur in 3D aggregates rather than in traditional 2D monolayer cultures. These differences may be related to differences in cell–cell and cell–matrix interactions, cell shape and spatial gradients, leading to differences in gene and protein expression (Mohamet et al., 2010). Thus, some attempts to grow stem cells in suspension cultures were based on their tendency to form aggregates (Dang et al., 2004; Reynolds and Weiss, 1992). Oxygen and nutrients are delivered to the cells inside these aggregates mainly by diffusion, which may lead to the formation of necrotic cores inside the spheres if their size becomes too large. The use of stirred bioreactors may allow the control of aggregate size with stirring speed, minimizing this effect (Sen et al., 2001).

The culture of stem cells as aggregates, however, may lead to significant cell agglomeration throughout time, which represents an important drawback of this culture mode (Dang et al., 2004). This excessive agglomeration of aggregates may be circumvented by cell encapsulation. For instance, alginate microbeads have been successfully used to encapsulate mouse ESCs (mESCs) (Magyar et al., 2001) and human-embryonic stem cells (hESCs) (Siti-Ismail et al., 2008). Scaffolds with particular mechanical and biochemical properties have been used in suspension bioreactor cultures to enhance the differentiation of stem/progenitor cells into different tissues and the formation of 3D structures (Goldstein et al., 2001; Lee et al., 2009; Nieponice et al., 2008). By customizing the scaffold, for instance, by immobilization of growth factors or functional groups for the controlled degradation of the matrix, microenvironments can be created within the bioreactors promoting the organization of cells in 3D structures similar to those of native tissues, and constructs less susceptible to immunorejection by the host can be created (Kehoe et al., 2009).

The first attempt of anchorage-dependent animal cell culture on small spheres (microcarriers) kept in suspension by stirring was performed by van Wezel, using the beaded ion exchange medium DEAE Sephadex™ A-50 (van Wezel, 1967). Over the years, the technique has been improved with respect to the surface charge density, dimensions, density, porosity, biomaterial used, and so on. Typically, microcarriers are small particles, 100–300 μm in size, and were initially used for mammalian cell recombinant protein production in stirred vessels (GE-Healthcare, 2005). Stem cells may also be cultured with similar systems (Fernandes et al., 2007). In microcarrier culture, cells grow as monolayers on the surface of small beads or inside the pores of macroporous structures suspended in a culture medium (Kehoe et al., 2009). Microcarrier cultures are characterized by high surface-to-volume ratio, accommodating higher cell densities compared with those in static cultures and, moreover, the area available for cell adhesion and growth can be easily adjusted by changing the amount of microcarriers.

Microcarriers can be classified as nonporous, microporous, or macroporous (GE-Healthcare, 2005). In cultures with nonporous microcarriers, cells are directly exposed to the bulk medium, which facilitates nutrient supply and by-product elimination because of the shorter diffusion paths. Microporous microcarriers have small pores (diameters smaller than 1 μm) in which the cells cannot enter, allowing cells to create a microenvironment inside the beads. Macroporous microcarriers have pores typically with 10–50 μm diameter that allows cells to enter, so that cells can grow in three dimensions at high densities. This internal structure protects the cells from shear forces generated by the stirrer and air/O_2 sparging, and potentially allows culture of both anchorage-dependent and suspension cells. Although microcarrier technology represents one of the most effective techniques to culture anchorage-dependent cells, their application may also have some drawbacks. In the first place, at the end of the culture, attached cells have to be detached from the microcarriers (if the microcarrier matrix cannot be degraded, for instance, by enzymatic means), which may cause cell damage. With higher cell densities, cell counting and harvesting also becomes more difficult than in suspension culture due to the formation of agglomerates of microcarriers, especially later in culture. Finally, cell harvest may be even more difficult in the case of macroporous microcarriers, where cells may also potentially grow inside the pores.

10.3.3 Wave Bioreactor

This bioreactor configuration consists of a presterilized, flexible, and single-use bag that is partially filled with media and inoculated with cells. The remainder of the bag is inflated with air. The air may be continuously passaged through the headspace during the cultivation. Mixing and mass transfer are achieved by rocking the chamber back and forth. This rocking motion generates waves at the liquid–air interface, greatly enhancing oxygen transfer. The wave motion also promotes bulk mixing, and suspension of cells and particles. This concept of using rocking for agitation is used extensively for the agitation of liquids in laboratory assay plates and gels (Singh, 1999). CO_2, temperature, weight, and pH control are possible as well as dissolved oxygen monitoring and data acquisition. Other advantages of this configuration include easy scale-up, up to 500 L, no need for cleaning or sterilization (since the culture bags are disposable), simplicity of use, and versatility. Indeed, the wave bioreactor can be used for suspension or microcarrier culture in batch, fed-batch, or perfusion mode. Moreover, these reactors are suitable for operation under GMP compliancy. The most important drawback of this bioreactor is, however, a relatively high cost of operation, since the culture chambers are expensive and can only be used once.

10.3.4 Rotating Wall Vessel

The expertise of the U.S. National Aeronautics and Space Administration (NASA) has occasionally been applied to biological problems when an engineering solution is required. Probably the best-known example of this collaboration is the "artificial heart" (DeBakey, 1997; Kawahito et al., 1997). Similarly, NASA's biotechnology group has worked on the problem of optimizing mechanical culture conditions in suspension by minimizing shear and turbulence (Goodwin et al., 1993; Hammond and Hammond, 2001). This work led to a new design for suspension culture vessels, where cells remain suspended in near free-fall, simulating microgravity conditions. The bioreactor consists of a horizontally rotated 3D culture vessel oxygenated by a flat silicone rubber gas transfer membrane (high aspect ratio vessel (HARV)) or a coaxial oxygenator in the center (slow turning lateral vessel (STLV)).

These vessels have characteristic features that determine their utility (Hammond and Hammond, 2001). First, these bioreactors do not present, at most operating conditions, the large shear stresses associated with turbulent flow. Shear stress may be further minimized if the inner cylinder and outer cylinders of the rotating wall vessel rotate at the same angular velocity (rpm). Second, the culture medium is gently mixed by rotation, avoiding the necessity for impellers, which may damage cells by both local turbulence at their surface and the high flow rates created between the vessel walls and the impellers. Third, unlike roller bottle culture, there is no headspace. The air in the headspace creates turbulence and bubble formation in the culture medium, which are sources of extra shear and turbulence. Fourth, anchorage-dependent cell types can be grown on microcarriers, just as in other suspension devices. Finally, this bioreactor design offers new approaches for studying the mechanisms of stem cell proliferation, differentiation, and signal transduction of cultured cells under microgravity conditions.

10.3.5 Perfusion Bioreactors

In bioreactors working with perfusion, mass transfer is enhanced by continual exchange of media, as fresh or recycled medium is introduced and exhausted medium removed (King and Miller, 2007). With this strategy, harmful metabolites are removed and growth factors and nutrients are constantly supplied, while cells are retained within the reactor, via retention devices. Different bioreactor configurations were developed based on this concept.

10.3.5.1 Parallel Plates Bioreactors

Parallel plates bioreactors constitute an almost shear-free alternative to stirred bioreactors. A parallel plate bioreactor described for the culture of stem cells consists essentially of two compartments: an upper compartment filled with gas, separated from a bottom compartment by a gas-permeable liquid-impermeable membrane. The bottom compartment is filled with culture medium and contains a tissue culture plastic surface for support of adherent cells (Goltry et al., 2009). Fresh medium flows continuously through the bottom compartment.

Bioreactor geometry was demonstrated to affect cell growth and differentiation in these bioreactors (Peng and Palsson, 1996). Four geometries (slab, gondola, diamond, and radial shapes) for parallel-plate bioreactors were analyzed and the radial-flow type bioreactor was demonstrated to provide the most uniform environment in which cells could grow and differentiate *ex vivo* due to the absence of walls that are parallel to the flow paths creating slow flowing regions. With this geometry, the medium enters in the liquid compartment through a port at the center, flows radially outward and is removed to a waste container.

Parallel plates bioreactors have the advantage of simple automation, providing continuous and automated feeding of the culture. Nevertheless, these systems present difficulties in the collection of representative samples from the system, except through total harvest (Nielsen, 1999).

10.3.5.2 Hollow-Fiber Bioreactors

Hollow-fiber bioreactors can achieve a low shear stress environment with enhanced mass transport properties (Placzek et al., 2009). Modeled after the mammalian circulatory system, hollow-fiber cell culture may offer a more *in vivo*-like manner to grow cells. A hollow-fiber bioreactor is a two-compartment system consisting of intracapillary and extracapillary spaces (Godara et al., 2008). Intracapillary flow is distributed by headers to a hollow-fiber bundle. Hollow fibers are tubular membranes, ~200 μm in diameter with pore sizes ranging from 10 kD to 0.3 μm (FiberCell-Systems, 2010; SpectrumLabs, 2010). The hollow-fiber bundle is encased in a cylindrical shell with ports for flow of medium around hollow fibers.

Cells are grown inside the fibers, with perfusion of medium on the outside or, alternatively, cells are inoculated into the extracellular space with intracapillary perfusion (Godara et al., 2008). When perfused with culture media, the hollow fibers allow oxygen and nutrients to be supplied to the cells while metabolic waste products are eliminated. The process increases the accumulation of the cell-secreted growth factors required for optimal growth (FiberCell-Systems, 2010). The inclusion of membrane technology may increase the surface area per volume available for cell growth over 350 times than of a normal T-flask (Placzek et al., 2009) and the hollow-fiber membrane may be modified with ligands for attachment of anchorage-dependent cell types (Godara et al., 2008).

Hollow-fiber bioreactors present difficulties in scale-up since the spaces between the fiber modules are not kept constant (Sardonini and Wu, 1993). This inconsistency results in fluctuations in oxygen transfer and added complexity when trying to monitor culture parameters (Safinia et al., 2005). Furthermore, these bioreactors face a decrease in mass transfer through the membranes due to cells growing in their periphery (Placzek et al., 2009).

10.3.5.3 Fixed (Packed) and Fluidized-Bed Bioreactors

Fixed-bed bioreactors consist of an immobilized scaffold arranged in a column, the bioreactor bed, where cells are seeded. With the purpose of enhancing mass transfer of substrates and oxygen, culture medium flow permanently through the column, supplied from a reservoir and often in a circulation loop (Portner et al., 2005). The column may consist of particles, either packed (fixed bed) or floating (fluidized bed). Alternatively, mammalian cells can be immobilized on 3D scaffolds as most of these have large interconnected pores, resulting in a quite uniform cell distribution during seeding. A different strategy may consist of encapsulating the cells, for instance in alginate microbeads, which can be packed in a column to create the bioreactor bed Osmokrović et al. (2006).

10.4 Bioreactor Systems for Stem Cell Culture

Several bioreactor configurations have been developed for stem cell expansion and differentiation (Table 10.2). These include systems for both adult and embryonic stem cell culture, and significant examples are illustrated in the following sections.

TABLE 10.2 Bioreactor Systems for Stem Cell Culture

Bioreactor Configuration	Cell Type	Culture Strategy	References
Stirred suspension bioreactor	mESC	Cell aggregates (including EBs)	Zandstra et al. (2003), Fok and Zandstra (2005), Cormier et al., (2006), zur Nieden et al. (2007), Kehoe et al., (2008)
	m and hESC	Cell encapsulation	Dang et al. (2004)
	mESC	Microcarrier culture	Abranches et al. (2007), Fernandes et al. (2007)
	hESC	Cell aggregates	Krawetz et al. (2009), Singh et al. (2010), Olmer et al., (2010)
	hESC	Microcarrier culture	Fernandes et al. (2009a), Lock and Tzanakakis (2009), Oh et al. (2009), Kehoe et al. (2009)
	NSC	Cell aggregates (neurospheres)	Kallos et al., (1999), Sen et al., (2001), Gilbertson et al., (2006), Baghbaderani et al. (2008)
	HSPC	Suspension culture	Sardonini and Wu (1993), Zandstra et al. (1994), Collins et al., (1998a), Collins et al. (1998b)
	MSC	Suspension culture	Baksh et al. (2003)
	MSC	Microcarrier culture	Frauenschuh et al. (2007), Yang et al. (2007), Schop et al. (2008), Sart et al. (2009), (Eibes et al. (2010)
Rotating wall vessel	m and hESC	Cell aggregates (including EBs)	Gerecht-Nir et al. (2004), Hwang et al. (2009)
	NSC	Cell encapsulation	Low et al. (2001), Lin et al. (2004)
	HSPC	Suspension Culture	Liu et al. (2006)
	MSC	Suspension culture	Chen et al. (2006)
	MSC	Osteogenic and chondrogenic differentiation	Duke et al. (1996), Granet et al. (1998), Song et al. (2006)
Micro-bioreactors	hESC	Perfused micro-bioreactors	Figallo et al. (2007), Cimetta et al. (2009), Korin et al. (2009)
	HSPC	Microliter-bioreactors	Luni et al. (2010), Oh et al. (2005)
Parallel plates bioreactor	HSPC	Flat-bed single-step perfusion	Koller et al. (1993b), Palsson et al. (1993), Jaroscak et al. (2003)
	HSPC	Flat-bed multi-step perfusion	Koller et al. (1993a)
	MSC	Flat-bed single-step perfusion	Dennis et al. (2007)
Hollow-fiber bioreactor	HSPC	Extra-capillary cell culture with intra-capillary perfusion	Sardonini and Wu (1993)
Fixed- and fluidized-bed bioreactor	HSPC	Packed bed configuration	Wang et al. (1995), Highfill et al. (1996), Mantalaris et al. (1998), Meissner et al. (1999), Jelinek et al. (2002)
	HSPC	Fluidized bed configuration	Meissner et al. (1999)

10.4.1 Hematopoietic Stem and Progenitor Cells

In its natural environment, hematopoiesis resides in a well-defined microenvironment characterized by local geometry (structure and vasculature), by stromal cells (accessory cells of mixed origin), and by an extracellular matrix composed of collagen-like molecules and proteoglycans, produced by stromal cells (Nielsen, 1999). Thus, it is likely that hematopoietic stem and progenitor cells (HSPCs) are influenced by accessory cells and the microenvironment they create in several ways.

Since the first *in vitro* reconstruction of the *in vivo* hematopoietic microenvironment to culture HSPCs by Dexter et al. (1973), which was later adapted for human cells (Gartner and Kaplan, 1980), hematopoietic cell cultures have been typically performed in multiwell plates, or tissue culture flasks made of polystyrene suitable for cell culture. Gas permeable culture bags are also currently used (Haylock et al., 1992; Lemoli et al., 1992).

The potential of stirred suspension cultures to support hematopoiesis from a starting population of human bone marrow cells has been investigated for a long time. Since HSPCs are relatively shear-sensitive cells, and agitation is thought to affect surface marker expression (McDowell and Papoutsakis, 1998), low agitation rates (30–60 rpm) are necessary in this kind of culture systems in order to avoid cell damage (Collins et al., 1998a; Sardonini and Wu, 1993; Zandstra et al., 1994).

The short-term maintenance of both colony-forming cell (CFC) numbers and their precursors, detected as long-term culture initiating cells (LTC-IC), was initially demonstrated to be possible in stirred suspension (Zandstra et al., 1994). After 4 weeks, the number of LTC-ICs and CFCs present in stirred cultures initiated with 1 million cells increased an average of 7- and 22-fold, respectively. Later on, the same group studied the parameters that possibly limit the cytokine-mediated expansion of primitive hematopoietic cells in stirred suspension cultures (Zandstra et al., 1997). More primitive cells (LTC-ICs) were shown to deplete cytokines from the medium much more rapidly than their more mature progeny using a mechanism that is strongly dependent on the concentration of cytokines to which the cells are exposed.

Cultures of umbilical cord blood (UCB) mononuclear cells (MNCs), peripheral blood (PB) MNCs, and PB CD34+ cells were also carried out in spinner flasks and in T-flasks both in serum-containing and serum-free media (Collins et al., 1997). Glucose and lactate metabolic rates were determined and correlated with the percentage of CFC present in the culture for a broad range of culture conditions. The proliferation and differentiation characteristics of these populations in spinner flask cultures were also examined by the same authors (Collins et al., 1998a). Culture proliferation in spinner flasks was dependent on both agitator design and agitation rate, as well as on the establishment of critical inoculum densities. The expansion of UCB and PB MNCs in a stirred-tank bioreactor system with pH and dissolved oxygen control was also described, as well as oxygen uptake and lactate production in these cultures (Collins et al., 1998b). Expansion of total cells and CFCs was greatly enhanced by the use of a cell-dilution feeding protocol (as compared to a cell-retention feeding protocol). The different metabolic profile of CFCs and more mature cells may allow the prediction of the content of several cell types in culture by monitoring the uptake or production of oxygen, lactate and other metabolites.

A number of perfusion reactors have also been developed for HSPC culture. The greatest success has been achieved with two flat-bed reactor systems: a multipass perfusion system, developed at Northwestern University (Koller et al., 1993a), and one with single-pass perfusion, developed at the University of Michigan (Koller et al., 1993b; Palsson et al., 1993). Both systems support 10- to 20-fold total cell expansion and ~10-fold progenitor expansion, whereas the expansion of primitive cells has only been reported for the second system.

The Northwestern multipass reactor is designed to reduce the formation of gradients. The system was extended for use with or without stroma by the introduction of multiple microgrooves at the chamber bottom (perpendicular to the direction of flow) that retains and protects the cells while allowing for rapid medium exchange with low shear stress (Horner et al., 1998; Sandstrom et al., 1995, 1996).

The Michigan system was the basis for the Aastrom Biosciences' cell production system, employed in the treatment of over 260 patients in phase I/II clinical trials (Goltry et al., 2009). The cell production system consists in a disposable cassette where cells are injected on top of a layer of stromal cells grown on a tissue culture plastic surface. Continuous perfusion of nutrients to the cassette occur while a chamber located above supplies oxygen that diffuses to the cassette through a gas-permeable, liquid-impermeable membrane (Palsson et al., 1993). The device has been integrated into a GMP fully automated, closed system with presterilized, disposable cassettes and automated perfusion and sterile cell recovery for clinical-scale culture (Goltry et al., 2009; Mandalam et al., 1999). The system has been used for expansion of bone marrow aspirates (Palsson et al., 1993) and UCB cells (Jaroscak et al., 2003; Koller et al., 1993a).

A different clinically relevant single-use, closed-system bioprocess capable of generating high numbers of UCB-derived HSPCs was developed (Madlambayan et al., 2006). The system incorporates inline subpopulation selection and medium dilution/exchange capabilities. In addition to expanded numbers of CFCs and LTC-ICs, the bioprocess generated more long-term nonobese diabetic/severe combined immunodeficient repopulating cells (LT-SRC) than present at input.

Other cell culture systems were evaluated for the scale-up of marrow cultures, like airlift reactors and hollow-fiber bioreactors (Sardonini and Wu, 1993). Cell culture in the airlift bioreactor led to MNC expansion, but less extensive than in the static culture used as control. The experiment in the hollow-fiber system demonstrated no observable expansion of hematopoietic cells when compared to control static cultures. Packed bed reactors were also designed to provide 3D scaffolding for cell attachment and culture (Highfill et al., 1996; Jelinek et al., 2002; Mantalaris et al., 1998; Meissner et al., 1999; Wang et al., 1995). In these systems, an initial attachment-dependent stromal cell culture is started on the bed particles, where upon HSPCs can be cocultivated (Cabrita et al., 2003). By using a fixed-bed reactor, where HSPC were co-cultured with stromal cells immobilized in porous glass carriers, the populations of CFU-GEMM (colony-forming units-granulocyte-erythrocyte-macrophage-megakaryocyte), CFU-GM (colony-forming units-granulocyte-macrophage), and BFU-E (burst-forming units-erythrocyte) were expanded up to 4.2-fold, 7-fold, and 1.8-fold, respectively (Meissner et al., 1999). A fluidized bed bioreactor system was also tested (Meissner et al., 1999), but in this case the carrier movement inhibited adhesion of HSPCs to stromal cells. Recently, Andrade-Zaldívar and co-workers have reported the use of roller bottles for the expansion of human HSPCs from the umbilical cord blood (Andrade-Zaldivar et al., 2011), with total CFU expansions up to 17-fold.

Two more recent studies have described HSPC expansion in stirred or rotating wall vessels (Li et al., 2006; Liu et al., 2006). At the end of 200 h of culture, over 400-fold increase in total cell number was observed (Liu et al., 2006) as well as a ~30-fold increase in CD34$^+$ cells, and ~20-fold in CFU-GM. Genetic changes caused by different culture microenvironments were studied by comparing gene expression profiling of CD34$^+$ HSPCs in static and stirred cultures (Li et al., 2006). Genes involved in antioxidation, DNA repair, apoptosis, and chemotactic activity were found to be differently expressed. This kind of data may provide new insights for culture optimization strategies in the future. Rotating wall vessels and spinner flasks were also used to perform simultaneous serum-free harvest and expansion of HSPCs and MSCs derived from the UCB, with the support of glass-coated styrene copolymer microcarriers (Kedong et al., 2010). Finally, a microliter-bioreactor array for HSPC culture was recently presented (Luni et al., 2010) which may also constitute a powerful tool for high-throughput optimization of culture conditions, providing important data that can be translated for larger scale processes.

10.4.2 Mesenchymal Stem Cells

Apart from acting as accessory cells for *ex vivo* HSPC culture, MSCs may be used for clinical applications, for instance, in graft-*versus*-host disease, renal failure, Crohn's disease, or myocardial ischemia (Caplan and Bruder, 2001; Fang et al., 2006; Ringden et al., 2006). However, the low frequency of MSCs in the bone marrow (1:10^4 in young and decaying with age) makes expansion a prerequisite for MSC therapies (Ringden et al., 2006). The time-consuming and labor-intensive nature of conventional tissue-flask culture has limited target doses in clinical trials to about 10^8 cells per patient (Lazarus et al., 2005; Ringden et al., 2006), but to achieve higher therapeutic efficacy more cells will probably be required.

MSCs were cultured as individual cells in stirred suspension, in a cytokine-dependent manner, maintaining their ability to form a functional differentiated bone (Baksh et al., 2003). The authors further demonstrated that MSCs grown under these conditions maintain the ability to differentiate along multiple mesenchymal lineages (Baksh et al., 2005) and, using high-content screening approaches, soluble growth factor combinations that influence MSC growth in serum-free conditions were identified.

Being anchorage-dependent cells, MSCs are also easily cultured on microcarriers, in stirred suspension. Cytodex 1 microcarriers were used for the attachment and growth of porcine MSCs (Frauenschuh et al., 2007), which retained their osteogenic and chondrogenic developmental potential over a cultivation time of 28 days. In another study, the growth and metabolism of goat MSCs in microcarrier spinner flask cultures was studied and the feeding regime was optimized (Schop et al., 2008). During cultivation, nutrient (glucose and glutamine) and metabolite (lactate and ammonia) concentrations in the medium were monitored allowing the determination of a correlation between nutrient consumption, metabolite production and cell growth. Rat BM (Yang et al., 2007) and ear-derived MSCs (Sart et al., 2009) were also successfully cultured on gelatin macroporous microcarriers. Human placenta-derived MSCs were expanded in stirred bioreactors using microcarriers, in serum-containing medium, achieving higher fold expansions and comparable antigenic phenotypes than in static culture (Yu et al., 2009). More recently, a low-serum system was also described for the culture of human MSCs on macroporous microcarriers (Eibes et al., 2010). An almost 10-fold increase in cell number was observed and cells retained their differentiation potential into adipogenic and osteogenic lineages, as well as their clonogenic ability. Human MSCs were also cultured successfully in Cytodex 1 microcarriers, in combination with low serum concentrations, achieving 4.8 population doublings when 50% medium refreshment followed by addition of 30% medium containing microcarriers every 3 days were performed (Schop et al., 2010).

MSCs isolated from bone marrow MNCs were also expanded in other bioreactor systems. The parallel-plate perfusion device described earlier for HSPC culture (Koller et al., 1993b), was used to significantly expand colony-forming efficiency-fibroblast (CFU-F) and progenitor cells with an osteogenic potential from bone marrow MNCs (Dennis et al., 2007). Tubular perfusion systems allowed culture of MSCs in 3D scaffolds and supported early osteoblastic differentiation (Yeatts and Fisher, 2010). Perfusion systems, where cells grow embedded in 3D polymeric matrices, maintaining multi-lineage differentiation potential after extensive expansion at high cell density, were also described (Xie et al., 2006; Zhao et al., 2005, 2007; Zhao and Ma, 2005). A fixed-bed bioreactor, based on nonporous borosilicate glass spheres, was used to expand the model cell line MSC-TERT, with automated inoculation, cultivation and harvesting of the cells (Weber et al., 2010). In order to perform calculations for scaling up, a model describing the process was also developed based in the collected data.

Bone marrow MNCs were cultured in a rotary bioreactor system (Chen et al., 2006) and after 8 days of culture the numbers of Stro-1+ CD34− CD44+MSCs, CD34+ Stro-1− CD44+HSPCs, and total cells increased by 29-, 8-, and 9-fold, respectively. The bioreactor-expanded MSCs expressed primitive mesenchymal cell markers, maintained a high level of CFU-F per day, and were capable of differentiating into chondrocytes, osteoblasts, and adipocytes upon appropriate inductions.

Bioreactors were also used for promoting MSC differentiation. Spinner flasks increase the efficiency of scaffold cell seeding and survival, in comparison to static culture (Godara et al., 2008) and have been used for cultivation of MSCs with osteogenic differentiation (Hofmann et al., 2007; Kim et al., 2007; Meinel et al., 2004; Mygind et al., 2007). The rotating wall reactor has also been successfully used for osteogenic differentiation (Granet et al., 1998; Qiu et al., 1999; Song et al., 2006; Turhani et al., 2005) and cartilage engineering (Marolt et al., 2006). Chondrogenic differentiation of human ESC-derived MSCs was also successfully performed in perfusion bioreactors (Tigli et al., 2011).

A viable alternative approach for undifferentiated MSC culture was recently described where cells are cultured as 3D aggregates or spheroids (Bartosh et al., 2010; Frith et al., 2010). Since it is believed that this approach may lead to an increase of the MSC therapeutic potential (Bartosh et al., 2010), methods were developed for dynamic 3D *in vitro* MSC culture using spinner flasks and rotating wall vessel bioreactors (Frith et al., 2010). Altered cell size and surface antigen expression, together with enhanced osteogenic and adipogenic differentiation potential, were observed, as well as many differences in gene expression between 3D and monolayer cultured MSCs, including those related to cellular architecture and extracellular matrix.

10.4.3 Neural Stem Cells

Murine neural stem cells (mNSCs) typically grow as suspended spherical aggregates, known as neurospheres, and can be induced to grow as spherical aggregates in a stirred-tank bioreactor (Kallos et al., 2003). The development and optimization of these bioreactor protocols for mNSCs have been performed in detail, and this is one of the best-studied systems for stirred suspension bioreactor culture of stem cells. Bioreactor protocol development started with the optimization of a culture medium for mNSCs (Kallos et al., 2003). The next phase was the optimization of inoculation and culture of mNSCs in stirred bioreactors (Kallos and Behie, 1999). Physiochemical growth parameters, such as pH and osmolarity, as well as inoculation parameters, including initial cell density, were determined in this study. When cells are cultured as neurospheres, diffusion of adequate amounts of nutrients to cells in the center of very large-diameter aggregates can be limited. In extreme cases, cell death can occur in the center of the spheres due to necrosis caused by nutrient/oxygen starvation. The diameter of mNSC aggregates in a bioreactor can be controlled below the limit at which necrosis would be expected to occur through manipulation of the agitation rate (Sen et al., 2001). Agitation rates must be high enough to maintain the aggregates in suspension and to create shear levels capable of controlling aggregate diameter, but not high enough to damage the cells.

The authors subsequently developed protocols for the extended culture of mNSCs by successive passaging the cells over 35 days. An overall multiplication ratio greater than 10^7 was achieved with no evident loss in growth potential or stem cell attributes. These protocols were developed for 125–250 mL spinner flasks and were later translated to large-scale (500 mL) computer-controlled bioprocesses (Gilbertson et al., 2006). This was accomplished by following the mass transfer and shear stress guidelines developed in the small-scale studies, and also hydrodynamic criteria. mNSCs obtained with this system had similar characteristics to those obtained from the optimized small-scale systems. The same group also presented protocols for serum-free generation of clinical quantities of human telencephalon-derived neural precursor cells (hNPCs) in 500 mL computer-controlled suspension bioreactors (Baghbaderani et al., 2008). The bioreactor-derived hNPCs retained the expression of nestin, a neural stem/progenitor cell marker, following expansion and were able to differentiate into glial and neuronal phenotypes under defined conditions.

Microcarrier expansion of mouse ESC-derived NSCs in spinner flasks was recently described (Rodrigues et al., 2011). A serum-free medium was used as well as polystyrene microcarrier beads coated with a recombinant peptide containing the RGD motif (Pronectin F). After optimization of the culture, a 35-fold increase in cell number was achieved after 6 days without multipotency loss.

Apart from these studies in stirred vessels, neural stem cell expansion, and differentiation has also been performed in rotary bioreactors (Lin et al., 2004; Low et al., 2001). In these cases, NSCs encapsulated in 3D collagen gels produced cell-collagen constructs containing, after 6 weeks in rotary culture, over 10-fold more Nestin-positive cells than those found in static cultures (Lin et al., 2004). In fact, the rate of proliferation of NSCs decreases with hydrogel stiffness, and a great enhancement in expression of neuronal markers can be achieved in soft hydrogels, which have an elastic modulus comparable to that of brain tissues (Banerjee et al., 2009).

10.4.4 Mouse Embryonic Stem Cells

The expansion of mESCs as aggregates in stirred suspension bioreactors has been reported (Cormier et al., 2006; Fok and Zandstra, 2005; Kehoe et al., 2008; zur Nieden et al., 2007). In the presence of leukemia inhibitory factor (LIF), mESCs proliferate as aggregates without significant loss of viability and with doubling times comparable to those of mESCs cultured in dishes. More importantly, the cells maintain expression of pluripotency markers even after multiple, successive passages.

Stirred bioreactors have also been used for differentiating mESCs. Mouse embryoid bodies (EBs) can be formed directly from enzymatically dissociated mESCs in rotary cell culture systems (E et al., 2006). A similar approach was performed in 250-mL spinner flasks equipped with a paddle-type impeller (Zandstra et al., 2003) and a scaled-up version of the system with a fully automated 2-L bioreactor was reported later (Schroeder et al., 2005). Suspension cultures of mESCs were also expanded and subsequently differentiated into cardiomyocytes in a single process, without an intermediate dissociation step (Fok and Zandstra, 2005). Differentiation of mouse ESCs into osteoblasts in spinner flasks has been described as well (Alfred et al., 2010).

In a recent study (Fridley et al., 2010), spinner flasks and rotary bioreactors were compared in terms of hematopoietic differentiation efficacy and progenitor cell profile were examined. cDNA microarrays were used to monitor mouse ESC gene expression profile during differentiation under dynamic conditions and it was observed that cells from all three germ layers were generated during bioreactor cultures, with distinct profiles in each bioreactor, in particular for hematopoietic differentiation.

Agglomeration of EBs is one concern in these cultures as this phenomenon makes it harder to control the culture environment. In order to control agglomeration in the process of EB formation, strategies like encapsulation of the cells in agarose beads (Bauwens et al., 2005; Dang et al., 2004) or formation of EBs on tantalum scaffolds suspended in a spinner flask (Liu and Roy, 2005) have been proposed. More recently, mESCs encapsulated in alginate beads were cultured in 50 mL HARV bioreactors while being coaxed toward osteogenic lineages (Hwang et al., 2009).

Alternatively mESCs can be cultured in suspension on microcarriers. mESCs can proliferate on microporous collagen-coated dextran beads (Cytodex 3), glass microcarriers, and macroporous gelatin-based beads (Cultispher S) in spinner flasks (Abranches et al., 2007; Fernandes et al., 2007; Fok and Zandstra, 2005) with an increase in cell number up to 70-fold (in 8 days). The scale-up of the spinner flask microcarrier culture system was successfully accomplished by using a fully controlled stirred tank bioreactor and, in these conditions, the concentration of mES cells cultured on microcarriers increased 85-fold over 11 days (Fernandes-Platzgummer et al., 2011). Although microcarrier cultivation requires dissociation of the cells from the carriers once the cells reach confluency, EB suspension cultures require periodic dissociation of the aggregates after a few days, which is more labor intensive and can cause damage to the cells.

The use of perfusion bioreactors, in which the medium is pumped through the culture vessel, has also been reported (Oh et al., 2005) for the expansion of mouse ESC lines on Petriperm (a *Petri* dish with a gas-permeable base). The cell densities obtained were 64-fold greater compared to *Petri* dish controls which only originate a nine-fold increase compared to the initial inoculum, over 6 days. The mESCs that were expanded retained pluripotency markers, had the ability to form derivatives of the three embryonic germ layers in teratomas, and maintained karyotypic stability. Perfusion cultures with cell retention ensure homogeneity of nutrient supplementation, inhibition of waste accumulation, and improved process reproducibility (Thomson, 2007). However, these cultures are still at a small-scale.

Culture and differentiation of mESCs in a perfused 3D fibrous matrix has also been reported (Li et al., 2003). In this study, perfusion led to a higher growth rate and final cell density in relation to static conditions. A polyethylene terephthalate (PET) matrix was applied for construction of the scaffold, which provided a larger surface area for adhesion, growth, and reduced contact inhibition. A bioprocess for efficient ESC-derived cardiomyocyte production was also developed (Bauwens et al., 2005). This system was capable of monitoring and control oxygen tension and pH in 500-mL vessels with continuous medium perfusion. Oxygen tension was shown to be a culture parameter that can be manipulated to improve cardiomyocyte yield.

An innovative system was described to grow mouse ESCs in manual fed-batch shake flask bioreactors, similar to those used for culturing bacteria (Mohamet et al., 2010). Abrogation of the cell surface protein E-cadherin with a blocking antibody leads to loss of cell–cell contact by mouse ESCs and subsequently to reduced cellular aggregation as well as EB agglomeration. Cells were grown for 16 days, with a cumulative expansion of 2775-fold, retaining expression of pluripotency markers as well as potential to differentiate into the three germ layers and normal karyotype. This system may provide an effective alternative to the "usual" bioreactor systems (aggregates or microcarriers), circumventing some of their associated limitations, but at the expense of high costs due to the use of the blocking antibody for E-cadherin.

Although these studies were performed with mESCs, they demonstrated that bioreactors could be promising for the large-scale expansion and differentiation of human pluripotent stem cells. However, severe differences in terms of culture conditions exist between mouse and human pluripotent stem cells. For instance, LIF does not support, at least only by itself, the expansion of these undifferentiated cells and human cells are often cultured in the presence of feeder layers, creating new challenges for their large-scale culture. In fact, a scalable microcarrier-based cell expansion system, using a feeder-dependent murine ESC line adapted to serum-free medium was developed, creating a more realistic model for human ESCs (Marinho et al., 2010). However, various systems were already developed for human pluripotent stem cells and are described in the next section.

10.4.5 Human Pluripotent Stem Cells

The establishment of *in vitro* hESCs and the recent derivation of human iPSCs, have great potential to create a revolution in the fields of tissue engineering and regenerative medicine. The fulfillment of this potential will certainly require high yields of cells that can only be achieved using large-scale culture systems and this issue is currently one of the biggest challenges in the field.

Microscale devices may serve as a tool for optimization of culture conditions while also providing the precise control over the cell microenvironment. Arrays of micro-bioreactors have been developed to study growth and differentiation of hESCs in a perfusion system (Cimetta et al., 2009; Figallo et al., 2007), as well as a micro-bioreactor with a periodic "flow-stop" perfusion system for coculture of hESCs with human feeder cells (Korin et al., 2009).

The first successful attempt of culturing hESCs in suspension bioreactors consisted in the formation of differentiating EBs of hESCs in STLV and HARV rotating bioreactors (Gerecht-Nir et al., 2004). Although agglomeration of EBs was observed in the HARV, a 70-fold expansion occurred after 28 days in the STLV and hESCs could still originate cells of the three germ layers. This system was later improved with two additional features (Come et al., 2008): perfusion, to provide continuous delivery of medium to the cells and a dialysis chamber, to improve the control of the culture environment, and to use less quantity of expensive molecules, such as growth factors. Faster and more synchronized differentiation was observed in the optimized system, in relation to static cultures. Spinner flasks were also used for human EB culture (Cameron et al., 2006), with superior expansion of EB-derived cells in relation to static conditions as well as a more homogenous morphology and size, with comparable hematopoietic differentiation potential. Indeed, different bioreactor systems for EB culture were compared (Yirme et al., 2008) being the highest fold increase in cell number (6.7 in 10 days) obtained with glass bulb impeller-equipped spinner flasks. Spinner flasks equipped with the same kind of bulb-shaped impellers were also used for the generation of cardiomyocytes, from cultures of human iPSCs, reprogrammed without the oncogene c-Myc (Alfred et al., 2010).

More recently, the successful culture of hESCs as aggregates in stirred suspension bioreactors has been achieved (Krawetz et al., 2009). An inhibitor of Rho kinase (ROCK inhibitor Y-27632, Ri), which increases the survival rate of dissociated single hESCs (Watanabe et al., 2007), was used along with continuous treatment with rapamycin. This system was able to maintain cells with high expression levels of pluripotency markers, a normal karyotype, and the ability to form teratomas *in vivo*. Different strategies have been subsequently developed for culturing human PSCs as aggregates in suspension (Olmer et al., 2010; Singh et al., 2010). One of these studies (Olmer et al., 2010) describes a process for culturing both human ESCs and iPSCs in suspension, with an initial step of dissociation into single cells. An almost fully defined serum-free medium (mTeSR1) supplemented with Ri allowed successful long-term expansion of human pluripotent stem cells. Contradictory observations were made in relation to the study by

Krawetz and colleagues (2009) as the continuous exposure to Ri did not inhibit cell growth, as reported in the first study, and supplementation with rapamycin was not required. Although Olmer and collaborators worked mostly in the small scale, pilot studies were done in agitated Erlenmeyer flasks suggesting the scalability of the process (Olmer et al., 2010). Singh et al. (2010) developed also protocols for scalable suspension aggregate culture of human ESCs, relying on Ri (without rapamycin) together with an optimized heat shock treatment. The cells were successfully cultured in spinner flasks with retention of pluripotency marker expression and ability to form teratomas. The authors alert to some variability in behavior among different cell lines, which can explain the discrepancies found in these studies (Singh et al., 2010).

hESCs have also been cultured on dextran (Fernandes et al., 2009a) and cellulose-based (Oh et al., 2009) microcarriers, coated with denatured collagen and Matrigel, respectively, in spinner flasks. Superior expansion was attained compared to static cultures and the pluripotency of the cells was maintained. These results were obtained with mouse embryonic fibroblasts (MEF)-conditioned medium (Fernandes et al., 2009a), as well as with two different types of defined media (Oh et al., 2009). The use of the system developed by Oh and coworkers was afterwards expanded for cardiomyocyte differentiation using GMP-compliant reagents (Lecina et al., 2010). The same group also found cell line specific effects of agitation on cell growth (Leung et al., 2010), in particular for the hESC line HES-3 or the human iPSC line IMR90 which were shown to have increased differentiation in agitated conditions, even with the addition of different cell protective polymers. Successful integration of hESC expansion and differentiation into definitive endoderm was also achieved in stirred bioreactors (Lock and Tzanakakis, 2009), on Matrigel-coated microcarriers.

Perfusion has been shown to improve hESC culture in organ culture dishes (Fong et al., 2005) in an analogous system to what was used for mESCs (Oh et al., 2005). hESCs cultured on MEFs were perfused with supplemented conditioned media, and a 70% improvement in hESC numbers was obtained, when compared to static culture conditions. Perfusion was used as well in controlled stirred tank bioreactors, with O_2 controlled to 30% air saturation, improving the final cell yield by 12-fold when compared to standard colony culture (Serra et al., 2010).

hESCs can be encapsulated in alginate beads (Siti-Ismail et al., 2008) and other matrices (Dang et al., 2004; Gerecht et al., 2007) with positive results in static conditions and, in principle, these strategies may be used as well in dynamic culture. Encapsulation of human ESCs in poly-L-lysine-coated alginate capsules was already tested in stirred suspension bioreactors and led to the generation of heart cells, yielding higher fractions of Nkx2.5 and GATA4-positive cells in the bioreactor when compared to dish cultures (Jing et al., 2010).

10.5 Future Directions

Even though some success was already achieved in the large-scale culture of stem cells, several other issues have to be addressed before bioreactor systems can be used for commercial applications (Kehoe et al., 2009; Ulloa-Montoya et al., 2005).

In the case of human pluripotent stem cells, mainly because they were only recently made available, the challenges are clear. First, seeding of hPSCs as clumps and not as single cells may condition the performance of hPSC culture in bioreactors, both when cultured as aggregates and on microcarriers (Kehoe et al., 2009). Even though some methods already exist to improve survival upon dispersion of hESC colonies into single cells, prior to seeding into the reactor, more efficient protocols are still required. Second, most of the methods described for culture of hPSCs in microcarriers require the coating of the particles with MEFs or Matrigel. However, given the presence of animal origin components, these approaches are unsuitable for therapeutic application of the cells produced. The use of conditioned medium, also sometimes reported, for large-scale cell culture would require a separate complex bioprocess. Clinical-grade human fibroblasts (Phillips et al., 2008) could provide a good alternative. Nevertheless, the functionalization of the beads with defined molecules to enhance the initial attachment of cells would constitute the optimal animal product-free, feeder-free system for large-scale bioreactor culture of hPSCs. Third, the culture media for the large-scale expansion of hPSCs should be not only animal-free but also cost afford-able, thereby making the bioprocess economically attractive and competitive. The high cost of culture media requires a rational and systematic optimization, namely the minimization of the use of growth factors, for example, by including natural or synthetic small molecules capable of supporting stem cell self-renewal or differentiation, and that can be isolated/synthesized economically. With the advent of high-throughput screening technologies, small-molecule libraries can be analyzed to identify molecular interactions leading to particular stem cell responses (Ding and Schultz, 2004).

Finally, the design of large-scale bioprocesses should follow GMP requirements for the production of clinical-grade stem cell derivatives (Kehoe et al., 2009). Although stem cells used in the process should also comply with GMP, from a safety standpoint, most hESC lines which are available today have been exposed to animal cells or proteins rendering them unsuitable for therapies. Alternative methods for deriving hESCs and the advent of hiPSCs (Takahashi and Yamanaka, 2006; Yu et al., 2007) may constitute relevant solutions for these problems. Methods for the real-time probing of cultured hPSCs for chromosomal aber-rations and for sorting undifferentiated cells in medically relevant quantities have also to be developed.

Some of the issues here described for hPSCs also apply to other stem cell types. For instance, concerns related with culture medium composition and costs, the exposure to animal origin cells or molecules and, importantly, the fulfillment of the GMP requirements. Only then the cells obtained by bioreactor culture may be safely applied in clinical trials and, subsequently, in commercial products. In the case of HSPCs and MSCs, culture in completely serum-free conditions is not frequently described and even with certified serum lots there may be some risk of contamination. Human NSCs also have an inter-esting potential for cell therapies (e.g., Parkinson's disease) and were already successfully expanded in bioreactors (Baghbaderani et al., 2008). However their application in mouse models of disease revealed that further understanding of the properties of hNSCs derived from different regions of the central nervous system is needed for successful application (Mukhida et al., 2008). The performance of different bioreactor-expanded populations of NSCs (e.g., ESC-derived), which are cultured adherently to micro-carriers, is yet to be tested. Additionally, different bioreactor configurations may have impact on the performance of the transplanted cells. Nevertheless, this is a fast growing field and future developments in the establishment of large-scale systems for clinical or pharmacological grade production of cellular products are expected to occur.

References

Abranches E, Bekman E, Henrique D, Cabral JM. 2007. Expansion of mouse embryonic stem cells on microcarriers. *Biotechnol Bioeng* 96(6):1211–21.

Abranches E, Silva M, Pradier L, Schulz H, Hummel O, Henrique D, Bekman E. 2009. Neural differentia-tion of embryonic stem cells in vitro: A road map to neurogenesis in the embryo. *PLoS One* 4:e6286.

Ahrlund-Richter L, De Luca M, Marshak DR, Munsie M, Veiga A, Rao M. 2009. Isolation and production of cells suitable for human therapy: Challenges ahead. *Cell Stem Cell* 4:20–6.

Alfred R, Gareau T, Krawetz R, Rancourt D, Kallos MS. 2010. Serum-free scaled up expansion and dif-ferentiation of murine embryonic stem cells to osteoblasts in suspension bioreactors. *Biotechnol Bioeng* 106(5):829–40.

Amanullah A, Otero JM, Mikola M, Hsu A, Zhang J, Aunins J, Schreyer HB, Hope JA, Russo AP. 2010. Novel micro-bioreactor high throughput technology for cell culture process development: Reproducibility and scalability assessment of fed-batch CHO cultures. *Biotechnol Bioeng* 106:57–67.

Andrade-Zaldivar H, Kalixto-Sanchez MA, de la Rosa AP, De Leon-Rodriguez A. 2011. Expansion of human hematopoietic cells from umbilical cord blood using roller bottles in CO_2 and CO_2-free atmosphere. *Stem Cells Dev* 20(4):593–8.

Baghbaderani BA, Behie LA, Sen A, Mukhida K, Hong M, Mendez I. 2008. Expansion of human neu-ral precursor cells in large-scale bioreactors for the treatment of neurodegenerative disorders. *Biotechnol Prog* 24(4):859–70.

Baksh D, Davies JE, Zandstra PW. 2003. Adult human bone marrow-derived mesenchymal progenitor cells are capable of adhesion-independent survival and expansion. *Exp Hematol* 31(8):723–32.

Baksh D, Davies JE, Zandstra PW. 2005. Soluble factor cross-talk between human bone marrow-derived hematopoietic and mesenchymal cells enhances in vitro CFU-F and CFU-O growth and reveals heterogeneity in the mesenchymal progenitor cell compartment. *Blood* 106(9):3012–9.

Banerjee A, Arha M, Choudhary S, Ashton RS, Bhatia SR, Schaffer DV, Kane RS. 2009. The influence of hydrogel modulus on the proliferation and differentiation of encapsulated neural stem cells. *Biomaterials* 30(27):4695–9.

Bartosh TJ, Ylostalo JH, Mohammadipoor A, Bazhanov N, Coble K, Claypool K, Lee RH, Choi H, Prockop DJ. 2010. Aggregation of human mesenchymal stromal cells (MSCs) into 3D spheroids enhances their antiinflammatory properties. *Proc Natl Acad Sci USA* 107(31):13724–9.

Bauwens C, Yin T, Dang S, Peerani R, Zandstra PW. 2005. Development of a perfusion fed bioreactor for embryonic stem cell-derived cardiomyocyte generation: Oxygen-mediated enhancement of cardiomyocyte output. *Biotechnol Bioeng* 90(4):452–61.

Bortin MM. 1970. A compendium of reported human bone marrow transplants. *Transplantation* 9(6):571–87.

Byrne JA, Pedersen DA, Clepper LL, Nelson M, Sanger WG, Gokhale S, Wolf DP, Mitalipov SM. 2007. Producing primate embryonic stem cells by somatic cell nuclear transfer. *Nature* 450:497–502.

Cabral JMS. 2001. *Ex vivo* expansion of hematopoietic stem cells in bioreactors. *Biotechnol Lett* 23:741–51.

Cabrita GJ, Ferreira BS, da Silva CL, Goncalves R, Almeida-Porada G, Cabral JM. 2003. Hematopoietic stem cells: From the bone to the bioreactor. *Trends Biotechnol* 21(5):233–40.

Cameron CM, Hu WS, Kaufman DS. 2006. Improved development of human embryonic stem cell-derived embryoid bodies by stirred vessel cultivation. *Biotechnol Bioeng* 94(5):938–48.

Caplan AI, Bruder SP. 2001. Mesenchymal stem cells: Building blocks for molecular medicine in the 21st century. *Trends Mol Med* 7(6):259–64.

Carpenter MK, Rosler E, Rao MS. 2003. Characterization and differentiation of human embryonic stem cells. *Cloning Stem Cells* 5:79–88.

Chamberlain G, Fox J, Ashton B, Middleton J. 2007. Concise review: Mesenchymal stem cells: Their phenotype, differentiation capacity, immunological features, and potential for homing. *Stem Cells* 25:2739–49.

Chen X, Xu H, Wan C, McCaigue M, Li G. 2006. Bioreactor expansion of human adult bone marrow-derived mesenchymal stem cells. *Stem Cells* 24(9):2052–9.

Cimetta E, Figallo E, Cannizzaro C, Elvassore N, Vunjak-Novakovic G. 2009. Micro-bioreactor arrays for controlling cellular environments: design principles for human embryonic stem cell applications. *Methods* 47(2):81–9.

Collins PC, Miller WM, Papoutsakis ET. 1998a. Stirred culture of peripheral and cord blood hematopoietic cells offers advantages over traditional static systems for clinically relevant applications. *Biotechnol Bioeng* 59(5):534–43.

Collins PC, Nielsen LK, Patel SD, Papoutsakis ET, Miller WM. 1998b. Characterization of hematopoietic cell expansion, oxygen uptake, and glycolysis in a controlled, stirred-tank bioreactor system. *Biotechnol Prog* 14(3):466–72.

Collins PC, Nielsen LK, Wong CK, Papoutsakis ET, Miller WM. 1997. Real-time method for determining the colony-forming cell content of human hematopoietic cell cultures. *Biotechnol Bioeng* 55(4):693–700.

Come J, Nissan X, Aubry L, Tournois J, Girard M, Perrier AL, Peschanski M, Cailleret M. 2008. Improvement of culture conditions of human embryoid bodies using a controlled perfused and dialyzed bioreactor system. *Tissue Eng Part C Methods* 14(4):289–98.

Cormier JT, zur Nieden NI, Rancourt DE, Kallos MS. 2006. Expansion of undifferentiated murine embryonic stem cells as aggregates in suspension culture bioreactors. *Tissue Eng* 12(11):3233–45.

Cowan CA, Atienza J, Melton DA, Eggan K. 2005. Nuclear reprogramming of somatic cells after fusion with human embryonic stem cells. *Science* 309:1369–73.

da Silva CL, Gonçalves R, Lemos F, Lemos MA, Zanjani ED, Almeida-Porada G, Cabral JMS. 2003. Modelling of *ex vivo* expansion/maintenance of hematopoietic stem cells. *Bioprocess Biosyst Eng* 25:365–9.

Dang SM, Gerecht-Nir S, Chen J, Itskovitz-Eldor J, Zandstra PW. 2004. Controlled, scalable embryonic stem cell differentiation culture. *Stem Cells* 22(3):275–82.

DeBakey ME. 1997. Development of a ventricular assist device. *Artif Organs* 21(11):1149–53.

Dennis JE, Esterly K, Awadallah A, Parrish CR, Poynter GM, Goltry KL. 2007. Clinical-scale expansion of a mixed population of bone-marrow-derived stem and progenitor cells for potential use in bone-tissue regeneration. *Stem Cells* 25(10):2575–82.

Deshpande RR, Heinzle E. 2004. On-line oxygen uptake rate and culture viability measurement of animal cell culture using microplates with integrated oxygen sensors. *Biotechnol Lett* 26(9):763–7.

Dexter TM, Allen TD, Lajtha LG, Schofield R, Lord BI. 1973. Stimulation of differentiation and proliferation of haemopoietic cells *in vitro*. *J Cell Physiol* 82(3):461–73.

Ding S, Schultz PG. 2004. A role for chemistry in stem cell biology. *Nat Biotechnol* 22(7):833–40.

Discher DE, Mooney DJ, Zandstra PW. 2009. Growth factors, matrices, and forces combine and control stem cells. *Science* 324:1673–7.

Duke J, Daane E, Arizpe J, Montufar-Solis D. 1996. Chondrogenesis in aggregates of embryonic limb cells grown in a rotating wall vessel. *Adv Space Res* 17(6–7):289–93.

E LL, Zhao YS, Guo XM, Wang CY, Jiang H, Li J, Duan CM, Song Y. 2006. Enrichment of cardiomyocytes derived from mouse embryonic stem cells. *J Heart Lung Transplant* 25(6):664–74.

Ebert A, Yu J, Rose FJ, Mattis V, Lorson C, Thomson J, Svendsen C. 2009. Induced pluripotent stem cells from a spinal muscular atrophy patient. *Nature* 457:277–80.

European Collection of Cell Cultures (ECACC)/Sigma-Aldrich. 2010. Fundamental Techniques in Cell Culture—Laboratory Handbook, Health Protection Agency, 2nd edition, pp. 32–33. Available at: http://www.sigmaaldrich.com/life-science/cell-culture/learning-center/ecacc-handbook-2nd-edition.html

Eibes G, dos Santos F, Andrade PZ, Boura JS, Abecasis MM, da Silva CL, Cabral JM. 2010. Maximizing the *ex vivo* expansion of human mesenchymal stem cells using a microcarrier-based stirred culture system. *J Biotechnol.* 146(4):194–7.

Fang B, Song YP, Liao LM, Han Q, Zhao RC. 2006. Treatment of severe therapy-resistant acute graft-versus-host disease with human adipose tissue-derived mesenchymal stem cells. *Bone Marrow Transplant* 38(5):389–90.

Fernandes AM, Fernandes TG, Diogo MM, da Silva CL, Henrique D, Cabral JM. 2007. Mouse embryonic stem cell expansion in a microcarrier-based stirred culture system. *J Biotechnol* 132(2):227–36.

Fernandes AM, Marinho PA, Sartore RC, Paulsen BS, Mariante RM, Castilho LR, Rehen SK. 2009a. Successful scale-up of human embryonic stem cell production in a stirred microcarrier culture system. *Braz J Med Biol Res* 42(6):515–22.

Fernandes TG, Diogo MM, Clark DS, Dordick JS, Cabral JMS. 2009b. High-throughput cellular microarray platforms: applications in drug discovery, toxicology and stem cell research. *Trends Biotechnol* 27:342–9.

Fernandes TG, Kwon SJ, Bale SS, Lee MY, Diogo MM, Clark DS, Cabral JMS, Dordick JS. 2010. Three-dimensional cell culture microarray for high-throughput studies of stem cell fate. *Biotechnol Bioeng* 106:106–18.

Fernandes TG, Kwon SJ, Lee M-Y, Clark DS, Cabral JMS, Dordick JS. 2008. On-chip, cell-based microarray immunofluorescence assay for high-throughput analysis of target proteins. *Anal Chem* 80:6633–9.

Fernandes-Platzgummer AM, Baptista RP, Diogo MM, da Silva CL, Cabral JM. 2011. Scale-up of mouse embryonic stem cell expansion in stirred bioreactors. *Biotechnol Prog* DOI: 10.1002/btpr.658.

FiberCell-Systems. 2010. Available at: http://www.fibercellsystems.com/

Figallo E, Cannizzaro C, Gerecht S, Burdick JA, Langer R, Elvassore N, Vunjak-Novakovic G. 2007. Microbioreactor array for controlling cellular microenvironments. *Lab Chip* 7(6):710–9.

Flaim CJ, Chien S, Bhatia SN. 2005. An extracellular matrix microarray for probing cellular differentiation. *Nat Methods* 2:119–25.

Fok EY, Zandstra PW. 2005. Shear-controlled single-step mouse embryonic stem cell expansion and embryoid body-based differentiation. *Stem Cells* 23(9):1333–42.

Fong WJ, Tan HL, Choo A, Oh SK. 2005. Perfusion cultures of human embryonic stem cells. *Bioprocess Biosyst Eng* 27(6):381–7.

Frauenschuh S, Reichmann E, Ibold Y, Goetz PM, Sittinger M, Ringe J. 2007. A microcarrier-based cultivation system for expansion of primary mesenchymal stem cells. *Biotechnol Prog* 23(1):187–93.

Fridley KM, Fernandez I, Li MT, Kettlewell RB, Roy K. 2010. Unique differentiation profile of mouse embryonic stem cells in rotary and stirred tank bioreactors. *Tissue Eng Part A* 16(11):3285–98.

Frith JE, Thomson B, Genever PG. 2010. Dynamic three-dimensional culture methods enhance mesenchymal stem cell properties and increase therapeutic potential. *Tissue Eng Part C Methods* 16(4):735–49.

Gartner S, Kaplan HS. 1980. Long-term culture of human bone marrow cells. *Proc Natl Acad Sci USA* 77(8):4756–9.

GE-Healthcare. 2005. Microcarrier cell culture—Principles and methods, GE Healthcare, Amersham Biosciences AB, Handbook 18-1140-62, pp. 7–53. Available at: http://www.gelifesciences.com/webapp/wcs/stores/servlet/productById/en/GELifeSciences-pt/18114062

Gerecht S, Burdick JA, Ferreira LS, Townsend SA, Langer R, Vunjak-Novakovic G. 2007. Hyaluronic acid hydrogel for controlled self-renewal and differentiation of human embryonic stem cells. *Proc Natl Acad Sci USA* 104(27):11298–303.

Gerecht-Nir S, Cohen S, Itskovitz-Eldor J. 2004. Bioreactor cultivation enhances the efficiency of human embryoid body (hEB) formation and differentiation. *Biotechnol Bioeng* 86(5):493–502.

Gilbertson JA, Sen A, Behie LA, Kallos MS. 2006. Scaled-up production of mammalian neural precursor cell aggregates in computer-controlled suspension bioreactors. *Biotechnol Bioeng* 94(4):783–92.

Godara P, McFarland CD, Nordon RE. 2008. Design of bioreactors for mesenchymal stem cell tissue engineering. *J Chem Technol Biotechnol* 83:408–420.

Goldstein AS, Juarez TM, Helmke CD, Gustin MC, Mikos AG. 2001. Effect of convection on osteoblastic cell growth and function in biodegradable polymer foam scaffolds. *Biomaterials* 22(11):1279–88.

Goltry K, Hampson B, Venturi N, Bartel R. 2009. Large-scale production of adult stem cells for clinical use. In: Lakshmipathy U, Chesnut JD, Thyagarajan B, editors. *Emerging Technology Platforms for Stem Cells*, John Wiley and Sons, Hoboken, New Jersey, pp. 153–168.

Gómez-Sjöberg R, Leyrat AA, Pirone DM, Chen CS, Quake SR. 2007. Versatile, fully automated, microfluidic cell culture system. *Anal Chem* 79:8557–63.

Gonçalves R, Lobato da Silva C, Cabral JMS, Zanjani ED, Almeida-Porada G. 2006. A Stro-1+ human universal stromal feeder layer to expand/maintain human bone marrow hematopoietic stem/progenitor cells in a serum–free culture system. *Exp Hematology* 34:1353–9.

Goodwin TJ, Prewett TL, Wolf DA, Spaulding GF. 1993. Reduced shear stress: A major component in the ability of mammalian tissues to form three-dimensional assemblies in simulated microgravity. *J Cell Biochem* 51(3):301–11.

Granet C, Laroche N, Vico L, Alexandre C, Lafage-Proust MH. 1998. Rotating-wall vessels, promising bioreactors for osteoblastic cell culture: Comparison with other 3D conditions. *Med Biol Eng Comput* 36(4):513–9.

Hammond TG, Hammond JM. 2001. Optimized suspension culture: The rotating-wall vessel. *Am J Physiol Renal Physiol* 281(1):F12–25.

Haylock D, To L, Dowse T, Juttner C, Simmons P. 1992. *Ex vivo* expansion and maturation of peripheral blood CD34+ cells into the myeloid lineage. *Blood* 80(6):1405–12.

Highfill JG, Haley SD, Kompala DS. 1996. Large-scale production of murine bone marrow cells in an airlift packed bed bioreactor. *Biotechnol Bioeng* 50(5):514–20.

Hofmann S, Hagenmuller H, Koch AM, Muller R, Vunjak-Novakovic G, Kaplan DL, Merkle HP, Meinel L. 2007. Control of *in vitro* tissue-engineered bone-like structures using human mesenchymal stem cells and porous silk scaffolds. *Biomaterials* 28(6):1152–62.

Horner M, Miller WM, Ottino JM, Papoutsakis ET. 1998. Transport in a grooved perfusion flat-bed bioreactor for cell therapy applications. *Biotechnol Prog* 14(5):689–98.

Hung PJ, Lee PJ, Sabounchi P, Lin R, Lee LP. 2005. Continuous perfusion microfluidic cell culture array for high-throughput cell-based assays. *Biotechnol Bioeng* 89:1–8.

Hwang YS, Cho J, Tay F, Heng JY, Ho R, Kazarian SG, Williams DR, Boccaccini AR, Polak JM, Mantalaris A. 2009. The use of murine embryonic stem cells, alginate encapsulation, and rotary microgravity bioreactor in bone tissue engineering. *Biomaterials* 30(4):499–507.

Jaroscak J, Goltry K, Smith A, Waters-Pick B, Martin PL, Driscoll TA, Howrey R. et al. 2003. Augmentation of umbilical cord blood (UCB) transplantation with *ex vivo*-expanded UCB cells: Results of a phase 1 trial using the AastromReplicell System. *Blood* 101(12):5061–7.

Jelinek N, Schmidt S, Hilbert U, Thoma S, Biselli M, Wandrey C. 2002. Novel bioreactors for the *ex vivo* cultivation of hematopoietic cells. *Chem Eng Technol* 25(1):A15–A18.

Jing D, Parikh A, Tzanakakis ES. 2010. Cardiac cell generation from encapsulated embryonic stem cells in static and scalable culture systems. *Cell Transplant* 19(11):1397–412.

Jones CN, Lee JY, Zhu J, Stybayeva G, Ramanculov E, Zern MA, A. R. 2008. Multifunctional protein microarrays for cultivation of cells and immunodetection of secreted cellular products. *Anal Chem* 80:6351–7.

Kallos MS, Behie LA. 1999. Inoculation and growth conditions for high-cell-density expansion of mammalian neural stem cells in suspension bioreactors. *Biotechnol Bioeng* 63(4):473–83.

Kallos MS, Behie LA, Vescovi AL. 1999. Extended serial passaging of mammalian neural stem cells in suspension bioreactors. *Biotechnol Bioeng* 65(5):589–9.

Kallos MS, Sen A, Behie LA. 2003. Large-scale expansion of mammalian neural stem cells: A review. *Med Biol Eng Comput* 41(3):271–82.

Kawahito K, Benkowski R, Ohtsubo S, Noon GP, Nose Y, DeBakey ME. 1997. Improved flow straighteners reduce thrombus in the NASA/DeBakey axial flow ventricular assist device. *Artif Organs* 21(4):339–43.

Kedong S, Xiubo F, Tianqing L, Macedo HM, LiLi J, Meiyun F, Fangxin S, Xuehu M, Zhanfeng C. 2010. Simultaneous expansion and harvest of hematopoietic stem cells and mesenchymal stem cells derived from umbilical cord blood. *J Mater Sci Mater Med* 21(12):3183–93.

Kehoe DE, Jing D, Lock LT, Tzanakakis EM. 2009. Scalable stirred-suspension bioreactor culture of human pluripotent stem cells. *Tissue Eng Part A.* 16(2):405–21.

Kehoe DE, Lock LT, Parikh A, Tzanakakis ES. 2008. Propagation of embryonic stem cells in stirred suspension without serum. *Biotechnol Prog* 24(6):1342–52.

Kensy F, John GT, Hofmann B, Buchs J. 2005. Characterisation of operation conditions and online monitoring of physiological culture parameters in shaken 24-well microtiter plates. *Bioprocess Biosyst Eng* 28(2):75–81.

Kim HJ, Kim UJ, Leisk GG, Bayan C, Georgakoudi I, Kaplan DL. 2007. Bone regeneration on macroporous aqueous-derived silk 3-D scaffolds. *Macromol Biosci* 7(5):643–55.

Kim L, Vahey MD, Lee HY, Voldman J. 2006. Microfluidic arrays for logarithmically perfused embryonic stem cell culture. *Lab Chip* 6:394–406.

King JA, Miller WM. 2007. Bioreactor development for stem cell expansion and controlled differentiation. *Curr Opin Chem Biol* 11(4):394–8.

Kirouac DC, Madlambayan GJ, Yu M, Sykes EA, Ito C, Zandstra PW. 2009. Cell-cell interaction networks regulate blood stem and progenitor cell fate. *Mol Syst Biol* 5:293.

Kirouac DC, Zandstra PW. 2008. The systematic production of cells for cell therapies. *Cell Stem Cell* 3:369–381.

Koller MR, Bender JG, Miller WM, Papoutsakis ET. 1993a. Expansion of primitive human hematopoietic progenitors in a perfusion bioreactor system with IL-3, IL-6, and stem cell factor. *Biotechnology (NY)* 11(3):358–63.

Koller MR, Emerson SG, Palsson BO. 1993b. Large-scale expansion of human stem and progenitor cells from bone marrow mononuclear cells in continuous perfusion cultures. *Blood* 82(2):378–84.

Koller MR, Manchel I, Brott DA, Palsson B. 1996. Donor-to-donor variability in the expansion potential of human bone marrow cells is reduced by accessory cells but not by soluble growth factors. *Exp Hematol* 24:1484–93.

Korin N, Bransky A, Dinnar U, Levenberg S. 2009. Periodic "flow-stop" perfusion microchannel bio-reactors for mammalian and human embryonic stem cell long-term culture. *Biomed Microdev* 11(1):87–94.

Krawetz R, Taiani JT, Liu S, Meng G, Li X, Kallos MS, Rancourt DE. 2010. Large-scale expansion of pluripotent human embryonic stem cells in stirred-suspension bioreactors. *Tissue Eng Part C Methods* 16(4):573–82.

Kunitake R, Suzuki A, Ichihashi H, Matsuda S, Hirai O, Morimoto K. 1997. Fully-automated roller bottle handling system for large scale culture of mammalian cells. *J Biotechnol* 52(3):289–94.

Laflamme MA, Murry CE. 2005. Regenerating the heart. *Nat Biotechnol* 23:845–56.

Lazarus HM, Koc ON, Devine SM, Curtin P, Maziarz RT, Holland HK, Shpall EJ. et al. 2005. Cotrans-plantation of HLA-identical sibling culture-expanded mesenchymal stem cells and hematopoietic stem cells in hematologic malignancy patients. *Biol Blood Marrow Transplant* 11(5):389–98.

Lecina M, Ting S, Choo A, Reuveny S, Oh S. 2010. Scalable platform for human embryonic stem cell differentiation to cardiomyocytes in suspended microcarrier cultures. *Tissue Eng Part C Methods* 16(6):1609–19.

Lee M-Y, Kumar RA, Sukumaran SM, Hogg MG, Clark DS, Dordick JS. 2008. Three-dimensional cellular microarrays for high-throughput toxicology assays. *Proc Natl Acad Sci USA* 105:59–63.

Lee SH, Hao E, Savinov AY, Geron I, Strongin AY, Itkin-Ansari P. 2009. Human beta-cell precursors mature into functional insulin-producing cells in an immunoisolation device: Implications for dia-betes cell therapies. *Transplantation* 87(7):983–91.

Lemoli RM, Tafuri A, Strife A, Andreeff M, Clarkson BD, Gulati SC. 1992. Proliferation of human hema-topoietic progenitors in long-term bone marrow cultures in gas-permeable plastic bags is enhanced by colony-stimulating factors. *Exp Hematol* 20(5):569–75.

Leung HW, Chen A, Choo AB, Reuveny S, Oh SK. 2011. Agitation can induce differentiation of human pluripotent stem cells in microcarrier cultures. *Tissue Eng Part C Methods* 17(2):165–72.

Li Q, Liu Q, Cai H, Tan WS. 2006. A comparative gene-expression analysis of CD34+ hematopoietic stem and progenitor cells grown in static and stirred culture systems. *Cell Mol Biol Lett* 11(4):475–87.

Li Y, Kniss DA, Lasky LC, Yang ST. 2003. Culturing and differentiation of murine embryonic stem cells in a three-dimensional fibrous matrix. *Cytotechnology* 41(1):23–35.

Lim M, Ye H, Panoskaltsis N, Drakakis EM, Yue X, Cass AE, Radomska A, Mantalaris A. 2007. Intelligent bioprocessing for haemotopoietic cell cultures using monitoring and design of experiments. *Biotechnol Adv* 25:353–68.

Lin HJ, O'Shaughnessy TJ, Kelly J, Ma W. 2004. Neural stem cell differentiation in a cell-collagen-bioreac-tor culture system. *Dev Brain Res* 153(2):163–73.

Liu H, Roy K. 2005. Biomimetic three-dimensional cultures significantly increase hematopoietic differen-tiation efficacy of embryonic stem cells. *Tissue Eng* 11(1–2):319–30.

Liu Y, Liu T, Fan X, Ma X, Cui Z. 2006. *Ex vivo* expansion of hematopoietic stem cells derived from umbili-cal cord blood in rotating wall vessel. *J Biotech* 124(3):592–601.

Lock LT, Tzanakakis ES. 2009. Expansion and differentiation of human embryonic stem cells to endoderm progeny in a microcarrier stirred-suspension culture. *Tissue Eng Part A* 15(8):2051–63.

Low HP, Savarese TM, Schwartz WJ. 2001. Neural precursor cells form rudimentary tissue-like structures in a rotating-wall vessel bioreactor. *In Vitro Cell Dev Biol-Animal* 37(3):141–7.

Luni C, Feldman HC, Pozzobon M, De Coppi P, Meinhart CD, Elvassore N. 2010. Microliter-bioreactor array with buoyancy-driven stirring for human hematopoietic stem cell culture. *Biomicrofluidics* 4(3).

Madlambayan GJ, Rogers I, Purpura KA, Ito C, Yu M, Kirouac D, Casper RF, Zandstra PW. 2006. Clinically relevant expansion of hematopoietic stem cells with conserved function in a single-use, closed-sys-tem bioprocess. *Biol Blood Marrow Transplant* 12(10):1020–30.

Magyar JP, Nemir M, Ehler E, Suter N, Perriard JC, Eppenberger HM. 2001. Mass production of embryoid bodies in microbeads. *Ann NY Acad Sci* 944:135–43.

Majka M, Janowska-Wieczorek A, Ratajczak J, Ehrenman K, Pietrzkowski Z, Kowalska MA, Gewirtz AM, Emerson SG, Ratajczak MZ. 2001. Numerous growth factors, cytokines, and chemokines are secreted by human CD34(+) cells, myeloblasts, erythroblasts, and megakaryoblasts and regulate normal hematopoiesis in an autocrine/paracrine manner. *Blood* 97:3075–85.

Mandalam R, Koller M, Smith A. 1999. *Ex vivo* hematopoietic cell expansion for bone marrow transplantation In: Schindhelm K, Nordon R, editors. *Ex Vivo Cell Therapy*. New York: Academic Press. pp. 273–91.

Mantalaris A, Keng P, Bourne P, Chang AY, Wu JH. 1998. Engineering a human bone marrow model: A case study on *ex vivo* erythropoiesis. *Biotechnol Prog* 14(1):126–33.

Marinho PA, Fernandes AM, Cruz JC, Rehen SK, Castilho LR. 2010. Maintenance of pluripotency in mouse embryonic stem cells cultivated in stirred microcarrier cultures. *Biotechnol Prog* 26(2):548–55.

Marolt D, Augst A, Freed LE, Vepari C, Fajardo R, Patel N, Gray M, Farley M, Kaplan D, Vunjak-Novakovic G. 2006. Bone and cartilage tissue constructs grown using human bone marrow stromal cells, silk scaffolds and rotating bioreactors. *Biomaterials* 27(36):6138–49.

Martin I, Smith T, Wendt D. 2009. Bioreactor-based roadmap for the translation of tissue engineering strategies into clinical products. *Trends Biotechnol* 27:495–502.

McDevitt TC, Palecek SP. 2008. Innovation in the culture and derivation of pluripotent human stem cells. *Curr Opin Biotechnol* 19:527–33.

McDowell CL, Papoutsakis ET. 1998. Increased agitation intensity increases CD13 receptor surface content and mRNA levels, and alters the metabolism of HL60 cells cultured in stirred tank bioreactors. *Biotechnol Bioeng* 60(2):239–50.

Meinel L, Karageorgiou V, Fajardo R, Snyder B, Shinde-Patil V, Zichner L, Kaplan D, Langer R, Vunjak-Novakovic G. 2004. Bone tissue engineering using human mesenchymal stem cells: effects of scaffold material and medium flow. *Ann Biomed Eng* 32(1):112–22.

Meissner P, Schroder B, Herfurth C, Biselli M. 1999. Development of a fixed bed bioreactor for the expansion of human hematopoietic progenitor cells. *Cytotechnology* 30(1–3):227–34.

Mitaka T. 2002. Reconstruction of hepatic organoid by hepatic stem cells. *J Hepatobiliary Pancreat Surg* 9(6):697–703.

Mohamet L, Lea ML, Ward CM. 2010. Abrogation of E-cadherin-mediated cellular aggregation allows proliferation of pluripotent mouse embryonic stem cells in shake flask bioreactors. *PLoS One* 5(9):e12921.

Mukhida K, Baghbaderani BA, Hong M, Lewington M, Phillips T, McLeod M, Sen A, Behie LA, Mendez I. 2008. Survival, differentiation, and migration of bioreactor-expanded human neural precursor cells in a model of Parkinson disease in rats. *Neurosurgical Focus* 24(3–4):E8.

Mygind T, Stiehler M, Baatrup A, Li H, Zou X, Flyvbjerg A, Kassem M, Bunger C. 2007. Mesenchymal stem cell ingrowth and differentiation on coralline hydroxyapatite scaffolds. *Biomaterials* 28(6):1036–47.

Nagle T, Berg C, Nassr R, Pang K. 2003. The further evolution of biotech. *Nat Rev Drug Discov* 2:75–79.

Nielsen LK. 1999. Bioreactors for hematopoietic cell culture. *Annu Rev Biomed Eng* 1:129–52.

Nieponice A, Soletti L, Guan J, Deasy BM, Huard J, Wagner WR, Vorp DA. 2008. Development of a tissue-engineered vascular graft combining a biodegradable scaffold, muscle-derived stem cells and a rotational vacuum seeding technique. *Biomaterials* 29(7):825–33.

Nishikawa S, Goldstein R, Nierras C. 2008. The promise of human induced pluripotent stem cells for research and therapy. *Nat Rev Mol Cell Biol* 9:725–9.

Oh SK, Chen AK, Mok Y, Chen X, Lim UM, Chin A, Choo AB, Reuveny S. 2009. Long-term microcarrier suspension cultures of human embryonic stem cells. *Stem Cell Res*. 2(3):219–30.

Oh SK, Fong WJ, Teo Y, Tan HL, Padmanabhan J, Chin AC, Choo AB. 2005. High density cultures of embryonic stem cells. *Biotechnol Bioeng* 91(5):523–33.

Olmer R, Haase A, Merkert S, Cui W, Palecek J, Ran C, Kirschning A. et al. 2010. Long term expansion of undifferentiated human iPS and ES cells in suspension culture using a defined medium. *Stem Cell Res* 5(1):51–64.

Osmokrović A, Obradović B, Bugarski D, Bugarski B, Vunjak-Novaković G. 2006. Development of a packed bed bioreactor for cartilage tissue engineering. *FME Transactions* 34(2):65–70.

Palsson BO, Paek SH, Schwartz RM, Palsson M, Lee GM, Silver S, Emerson SG. 1993. Expansion of human bone marrow progenitor cells in a high cell density continuous perfusion system. *Biotechnology (NY)* 11(3):368–72.

Parson AB. 2008. Stem cell biotech: Seeking a piece of the action. *Cell* 132:511–3.

Passier R, Mummery C. 2003. Origin and use of embryonic and adult stem cells in differentiation and tissue repair. *Cardiovasc Res* 58:324–35.

Peerani R, Onishi K, Mahdavi A, Kumacheva E, Zandstra PW. 2009. Manipulation of signaling thresholds in "Engineered Stem Cell Niches" identifies design criteria for pluripotent stem cell screens. *PLoS One* 7:e6438.

Peng CA, Palsson BO. 1996. Cell growth and differentiation on feeder layers is predicted to be influenced by bioreactor geometry. *Biotechnol Bioeng* 50(5):479–92.

Phillips BW, Lim RY, Tan TT, Rust WL, Crook JM. 2008. Efficient expansion of clinical-grade human fibroblasts on microcarriers: Cells suitable for *ex vivo* expansion of clinical-grade hESCs. *J Biotechnol* 134(1–2):79–87.

Pittenger MF. 2008. Mesenchymal stem cells from adult bone marrow. *Methods Mol Biol* 449:27–44.

Placzek MR, Chung IM, Macedo HM, Ismail S, Mortera Blanco T, Lim M, Cha JM. et al. 2009. Stem cell bioprocessing: Fundamentals and principles. *J R Soc Interface* 6(32):209–32.

Portner R, Nagel-Heyer S, Goepfert C, Adamietz P, Meenen NM. 2005. Bioreactor design for tissue engineering. *J Biosci Bioeng* 100(3):235–45.

Qiu QQ, Ducheyne P, Ayyaswamy PS. 1999. Fabrication, characterization and evaluation of bioceramic hollow microspheres used as microcarriers for 3-D bone tissue formation in rotating bioreactors. *Biomaterials* 20(11):989–1001.

Revazova ES, Turovets NA, Kochetkova OD, Kindarova LB, Kuzmichev LN, Janus JD, Pryzhkova MV. 2007. Patient-specific stem cell lines derived from human parthenogenetic blastocysts. *Cloning Stem Cells* 9:432–49.

Reynolds BA, Weiss S. 1992. Generation of neurons and astrocytes from isolated cells of the adult mammalian central nervous system. *Science* 255(5052):1707–10.

Ringden O, Uzunel M, Rasmusson I, Remberger M, Sundberg B, Lonnies H, Marschall HU. et al. 2006. Mesenchymal stem cells for treatment of therapy-resistant graft-versus-host disease. *Transplantation* 81(10):1390–7.

Rodrigues CAV, Diogo M, M., Lobato da Silva C, Cabral JMS. 2011. Microcarrier expansion of mouse embryonic stem cell-derived neural stem cells in stirred bioreactors. *Biotechnol Appl Biochem* DOI: 10.1002/bab.37.

Safinia L, Panoskaltsis N, Mantalaris A. 2005. Haemotopoietic culture systems. In: Chaudhuri J, Al-Rubeai M, editors. *Bioreactors for Tissue Engineering*. Dordrecht: Springer. pp. 309–334.

Sandstrom CE, Bender JG, Miller WM, Papoutsakis ET. 1996. Development of novel perfusion chamber to retain nonadherent cells and its use for comparison of human "mobilized" peripheral blood mononuclear cell cultures with and without irradiated bone marrow stroma. *Biotechnol Bioeng* 50(5):493–504.

Sandstrom CE, Bender JG, Papoutsakis ET, Miller WM. 1995. Effects of CD34+ cell selection and perfusion on ex vivo expansion of peripheral blood mononuclear cells. *Blood* 86(3):958–70.

Sardonini CA, Wu YJ. 1993. Expansion and differentiation of human hematopoietic cells from static cultures through small-scale bioreactors. *Biotechnol Prog* 9(2):131–7.

Sart S, Schneider YJ, Agathos SN. 2009. Ear mesenchymal stem cells: An efficient adult multipotent cell population fit for rapid and scalable expansion. *J Biotechnol* 139(4):291–9.

Schop D, Janssen FW, Borgart E, de Bruijn JD, van Dijkhuizen-Radersma R. 2008. Expansion of mesenchymal stem cells using a microcarrier-based cultivation system: Growth and metabolism. *J Tissue Eng Regen Med* 2(2-3):126–35.

Schop D, van Dijkhuizen-Radersma R, Borgart E, Janssen FW, Rozemuller H, Prins HJ, de Bruijn JD. 2010. Expansion of human mesenchymal stromal cells on microcarriers: Growth and metabolism. *J Tissue Eng Regen Med* 4(2):131–40.

Schroeder M, Niebruegge S, Werner A, Willbold E, Burg M, Ruediger M, Field LJ, Lehmann J, Zweigerdt R. 2005. Differentiation and lineage selection of mouse embryonic stem cells in a stirred bench scale bioreactor with automated process control. *Biotechnol Bioeng* 92(7):920–33.

Sen A, Kallos MS, Behie LA. 2001. Effects of hydrodynamics on cultures of mammalian neural stem cell aggregates in suspension bioreactors. *Ind Eng Chem Res* 40(23):5350–7.

Serra M, Brito C, Sousa MF, Jensen J, Tostoes R, Clemente J, Strehl R, Hyllner J, Carrondo MJ, Alves PM. 2010. Improving expansion of pluripotent human embryonic stem cells in perfused bioreactors through oxygen control. *J Biotechnol* 148(4):208–15.

Singh H, Mok P, Balakrishnan T, Rahmat SN, Zweigerdt R. 2010. Up-scaling single cell-inoculated suspension culture of human embryonic stem cells. *Stem Cell Res* 4(3):165–79.

Singh V. 1999. Disposable bioreactor for cell culture using wave-induced agitation. *Cytotechnology* 30(1-3):149–58.

Siti-Ismail N, Bishop AE, Polak JM, Mantalaris A. 2008. The benefit of human embryonic stem cell encapsulation for prolonged feeder-free maintenance. *Biomaterials* 29(29):3946–52.

Smith AG. 2001. Embryo-derived stem cells: Of mice and men. *Annu Rev Cell Dev Biol* 17:435–62.

Soen Y, Mori A, Palmer TD, Brown PO. 2006. Exploring the regulation of human neural precursor cell differentiation using arrays of signaling microenvironments. *Mol Syst Biol* 2:37.

Sohn SK, Kim JG, Kim DH, Lee NY, Suh JS, Lee KB. 2003. Impact of transplanted CD34+ cell dose in allogeneic unmanipulated peripheral blood stem cell transplantation. *Bone Marrow Transplant* 31:967–972.

Soldner F, Hockemeyer D, Beard C, Gao Q, Bell G, Cook E, Hargus G. et al. 2009. Parkinson's disease patient-derived induced pluripotent stem cells free of viral reprogramming factors. *Cell* 136:964–77.

Song K, Yang Z, Liu T, Zhi W, Li X, Deng L, Cui Z, Ma X. 2006. Fabrication and detection of tissue-engineered bones with bio-derived scaffolds in a rotating bioreactor. *Biotechnol Appl Biochem* 45(Pt 2):65–74.

SpectrumLabs. 2010. Available at: http://www.spectrumlabs.com/

Takahashi K, Tanabe K, Ohnuki M, Narita M, Ichisaka T, Tomoda K, Yamanaka S. 2007. Induction of pluripotent stem cells from adult human fibroblasts by defined factors. *Cell* 131:861–72.

Takahashi K, Yamanaka S. 2006. Induction of pluripotent stem cells from mouse embryonic and adult fibroblast cultures by defined factors. *Cell* 126(4):663–76.

Terstegge S, Laufenberg I, Pochert J, Schenk S, Itskovitz-Eldor J, Endl E, Brustle O. 2007. Automated maintenance of embryonic stem cell cultures. *Biotechnol Bioeng* 96(1):195–201.

Thomas E, Lochte H, Lu W, Ferrebee J. 1957. Intravenous infusion of bone marrow in patients receiving radiation and chemotherapy. *N Engl J Med* 157:491–6.

Thomson H. 2007. Bioprocessing of embryonic stem cells for drug discovery. *Trends Biotechnol* 25(5):224–30.

Thomson J, Itskovitz-Eldor J, Shapiro S, Waknitz M, Swiergiel J, Marshall V, Jones J. 1998. Embryonic stem cell lines derived from human blastocysts. *Science* 282:1145–7.

Tigli RS, Cannizaro C, Gumusderelioglu M, Kaplan DL. 2011. Chondrogenesis in perfusion bioreactors using porous silk scaffolds and hESC-derived MSCs. *J Biomed Mater Res A* 96(1):21–8.

Turhani D, Watzinger E, Weissenbock M, Cvikl B, Thurnher D, Wittwer G, Yerit K, Ewers R. 2005. Analysis of cell-seeded 3-dimensional bone constructs manufactured *in vitro* with hydroxyapatite granules obtained from red algae. *J Oral Maxillofac Surg* 63(5):673–81.

Ulloa-Montoya F, Verfaillie CM, Hu WS. 2005. Culture systems for pluripotent stem cells. *J Biosci Bioeng* 100(1):12–27.

van Wezel AL. 1967. Growth of cell-strains and primary cells on micro-carriers in homogeneous culture. *Nature* 216(5110):64–5.

Vazin T, Schaffer DV. 2010. Engineering strategies to emulate the stem cell niche. *Trends Biotechnol* 28(3):117–24.

Wang TY, Brennan JK, Wu JH. 1995. Multilineal hematopoiesis in a three-dimensional murine long-term bone marrow culture. *Exp Hematol* 23(1):26–32.

Watanabe K, Ueno M, Kamiya D, Nishiyama A, Matsumura M, Wataya T, Takahashi JB, Nishikawa S, Muguruma K, Sasai Y. 2007. A ROCK inhibitor permits survival of dissociated human embryonic stem cells. *Nat Biotechnol* 25(6):681–6.

Weber C, Freimark D, Portner R, Pino-Grace P, Pohl S, Wallrapp C, Geigle P, Czermak P. 2010. Expansion of human mesenchymal stem cells in a fixed-bed bioreactor system based on non-porous glass carrier–part A: Inoculation, cultivation, and cell harvest procedures. *Int J Artif Organs* 33(8):512–25.

Wognum AW, Eaves AC, Thomas TE. 2003. Identification and isolation of hematopoietic stem cells. *Arch Med Res* 34:461–75.

Xie Y, Hardouin P, Zhu Z, Tang T, Dai K, Lu J. 2006. Three-dimensional flow perfusion culture system for stem cell proliferation inside the critical-size beta-tricalcium phosphate scaffold. *Tissue Eng* 12(12):3535–43.

Yang Y, Rossi FM, Putnins EE. 2007. *Ex vivo* expansion of rat bone marrow mesenchymal stromal cells on microcarrier beads in spin culture. *Biomaterials* 28(20):3110–20.

Yeatts AB, Fisher JP. 2011. Tubular perfusion system for the long-term dynamic culture of human mesenchymal stem cells. *Tissue Eng Part C Methods* 17(3):337–48.

Yirme G, Amit M, Laevsky I, Osenberg S, Itskovitz-Eldor J. 2008. Establishing a dynamic process for the formation, propagation, and differentiation of human embryoid bodies. *Stem Cells Dev* 17(6):1227–41.

Yu J, Vodyanik MA, Smuga-Otto K, Antosiewicz-Bourget J, Frane JL, Tian S, Nie J. et al. 2007. Induced pluripotent stem cell lines derived from human somatic cells. *Science* 318(5858):1917–20.

Yu Y, Li K, Bao C, Liu T, Jin Y, Ren H, Yun W. 2009. *Ex vitro* expansion of human placenta-derived mesenchymal stem cells in stirred bioreactor. *Appl Biochem Biotechnol* 159(1):110–8.

Zandstra PW, Bauwens C, Yin T, Liu Q, Schiller H, Zweigerdt R, Pasumarthi KB, Field LJ. 2003. Scalable production of embryonic stem cell-derived cardiomyocytes. *Tissue Eng* 9(4):767–78.

Zandstra PW, Eaves CJ, Piret JM. 1994. Expansion of hematopoietic progenitor cell populations in stirred suspension bioreactors of normal human bone marrow cells. *Biotechnology (NY)* 12(9):909–14.

Zandstra PW, Petzer AL, Eaves CJ, Piret JM. 1997. Cellular determinants affecting the rate of cytokine in cultures of human hematopoietic cells. *Biotechnol Bioeng* 54(1):58–66.

Zhao F, Chella R, Ma T. 2007. Effects of shear stress on 3-D human mesenchymal stem cell construct development in a perfusion bioreactor system: Experiments and hydrodynamic modeling. *Biotechnol Bioeng* 96(3):584–95.

Zhao F, Ma T. 2005. Perfusion bioreactor system for human mesenchymal stem cell tissue engineering: Dynamic cell seeding and construct development. *Biotechnol Bioeng* 91(4):482–93.

Zhao F, Pathi P, Grayson W, Xing Q, Locke BR, Ma T. 2005. Effects of oxygen transport on 3-d human mesenchymal stem cell metabolic activity in perfusion and static cultures: Experiments and mathematical model. *Biotechnol Prog* 21(4):1269–80.

zur Nieden NI, Cormier JT, Rancourt DE, Kallos MS. 2007. Embryonic stem cells remain highly pluripotent following long term expansion as aggregates in suspension bioreactors. *J Biotechnol* 129(3):421–32.

Index

9 780367 380649